Advances in Computer Vision and Pattern Recognition

More information about this series at http://www.springer.com/series/4205

Amir R. Zamir · Asaad Hakeem
Luc Van Gool · Mubarak Shah
Richard Szeliski
Editors

Large-Scale Visual Geo-Localization

Springer

Editors
Amir R. Zamir
Department of Computer Science
Stanford University
Stanford, CA
USA

Mubarak Shah
Department of Computer Science
University of Central Florida
Orlando, FL
USA

Asaad Hakeem
Machine Learning Division
Decisive Analytics Corporation
Arlington, VA
USA

Richard Szeliski
Facebook
Seattle, WA
USA

Luc Van Gool
Computer Vision Laboratory
ETH Zürich
Zürich
Switzerland

ISSN 2191-6586 ISSN 2191-6594 (electronic)
Advances in Computer Vision and Pattern Recognition
ISBN 978-3-319-79840-0 ISBN 978-3-319-25781-5 (eBook)
DOI 10.1007/978-3-319-25781-5

Printed on acid-free paper

This Springer imprint is published by SpringerNature
The registered company is Springer International Publishing AG Switzerland

Preface

One of the greatest inventions of all times is *camera*, a device for capturing images of the world. Computer vision emerged as the scientific field of understanding the images that cameras produce. Another influential invention of modern technology is *positioning systems*, instruments (namely GPS) for identifying one's location. Even though these two devices are often viewed as belonging to completely different realms, their products (images and locations) have a close relationship. For instance, by just looking at an image, one can sometimes guess its location, e.g., when a rain forest is in view or looking at Eiffel Tower. The opposite is also often true: knowing the location of an image can provide a rich context and significantly assist the process of understanding the visual content. The goal of this book is to explore this bidirectional relationship between images and locations.

In this book, we provide a comprehensive map of the state of the art in large scale visual geo-localization and discuss the emerging trends in this area. Visual geo-localization is defined as the problem of identifying the location of the camera that captured an image and/or the content of the image. Geo-localization finds numerous applications in organizing photo collections, law enforcement, or statistical analysis of commercial market trends.

The book is divided into four main parts: Data-Driven Geo-localization, Semantic Geo-localization, Geometric Matching based Geo-localization, and Real-World Applications. The first part (Data-Driven Geo-localization) discusses recent methods that exploit internet-scale image databases for devising geographically rich features and geo-localizing query images at a city, country, or global scale. The second part (Semantic Geo-localization) presents geo-localization techniques that are built upon high-level and semantic cues. The third part (Geometric Matching based Geo-localization) focuses on the methods that perform localization by geometrically aligning the query image against a 3D model (namely digital elevation model or structure-from-motion reconstructed cities).

In the fourth part of the book (Real-World Applications), methods that use the geo-location of the image, whether extracted automatically or acquired from a GPS-chip, for understanding the image content are presented. Such frameworks are of great importance as cell phones and cameras are now being equipped with built-in localization devices. Therefore, it is of particular interest to develop techniques that accomplish image analysis *assisted by geo-location*. We also discuss several approaches for geo-localization under more practical settings, e.g., when the supervision of an expert in the form of a user-in-the-loop system is available.

The editors sincerely appreciate everyone who assisted in the process of preparing this book. In particular, we thank the authors, who are among the leading and most active researchers in the field of computer vision. Without their contribution in writing and peer reviewing the chapters, this book would not have been possible. We are also thankful to Springer for providing us with excellent support. Lastly, the editors are truly grateful to their families and loved ones for their continual support and encouragement throughout the process of writing this book.

Stanford, CA, USA — Amir R. Zamir
Arlington, VA, USA — Asaad Hakeem
Zürich, Switzerland — Luc Van Gool
Orlando, FL, USA — Mubarak Shah
Seattle, WA, USA — Richard Szeliski
May 2015

Contents

Contributors

Mathieu Aubry LIGM (UMR CNRS 8049), ENPC/Université Paris-Est, Marne-la-Vallée, France

Georges Baatz Google Inc., Zürich, Switzerland

Mayank Bansal Vision Technologies Lab, SRI International, Princeton, NJ, USA; Center for Vision Technologies, SRI International, Princeton, NJ, USA

Serge Belongie Cornell University, Ithaca, NY, USA

Liangliang Cao IBM Watson Research, New York, USA

Mei Chen Intel Corporation, Hillsboro, OR, USA

Yi Chen Object Video, Inc., Reston, VA, USA

Hui Cheng Vision Technologies Lab, SRI International, Princeton, NJ, USA

Matthew Clements UC Berkeley, Berkeley, CA, USA

Dave Conger Object Video, Inc., Reston, VA, USA

David J. Crandall Indiana University, Bloomington, IN, USA

Kostas Daniilidis Department of Computer and Information Science, University of Pennsylvania, Philadelphia, PA, USA

Alexei A. Efros Department of Electrical Engineering and Computer Science, UC Berkeley, Berkeley, CA, USA

Pascal Fua EPFL, Lausanne, Switzerland

Kiran Gunda Object Video, Inc., Reston, VA, USA

Himaanshu Gupta Object Video, Inc., Reston, VA, USA

Asaad Hakeem Decisive Analytics Corporation, Arlington, VA, USA

James Hays Brown University, Providence, RI, USA

Martial Hebert Robotics Institute, Carnegie Mellon University, Pittsburgh, PA, USA

Varsha Hedau Apple, Cupertino, CA, USA

James C. Hoe Carnegie Mellon University, Pittsburgh, PA, USA

Daniel P. Huttenlocher Cornell University, Ithaca, NY, USA

Rahul Kumar Jha University of Michigan, Ann Arbor, USA

Leif Kobbelt RWTH Aachen University, Aachen, Germany

Jana Košecká George Mason University, Fairfax, VA, USA

Kevin Köser GEOMAR, Kiel, Germany

Neeraj Kumar Department of Computer Science and Engineering, University of Washington, Seattle, WA, USA

L'ubor Ladický ETH Zürich, Zürich, Switzerland

Stefan Lee Indiana University, Bloomington, IN, USA

Yong Jae Lee Department of Computer Science, UC Davis, Davis, CA, USA

Bastian Leibe RWTH Aachen University, Aachen, Germany

Li-Jia Li Yahoo! Research, Sunnyvale, USA

Yunpeng Li École Polytechnique Fédérale de Lausanne, Lausanne, Switzerland

Tsung-Yi Lin Cornell University, Ithaca, NY, USA

Eriko Nurvitadhi Intel Corporation, Hillsboro, OR, USA

Hyun Soo Park Carnegie Mellon University, Pittsburgh, PA, USA

Minwoo Park Object Video, Inc., Reston, VA, USA

Marc Pollefeys ETH Zürich, Zürich, Switzerland

Gang Qian Object Video, Inc., Reston, VA, USA

Bryan Russell Adobe Research, Lexington, KY, USA

Torsten Sattler Department of Computer Science, ETH Zürich, Zürich, Switzerland

Olivier Saurer ETH Zürich, Zürich, Switzerland

Harpreet Sawhney Center for Vision Technologies, SRI International, Princeton, NJ, USA

Steven Seitz Department of Computer Science and Engineering, University of Washington, Seattle, WA, USA

Khurram Shafique Object Video, Inc., Reston, VA, USA

Mubarak Shah University of Central Florida, Orlando, FL, USA

David Ayman Shamma Yahoo! Research, San Francisco, USA

Yaser Sheikh Carnegie Mellon University, Pittsburgh, PA, USA

Gautam Singh George Mason University, Fairfax, VA, USA

Sudipta N. Sinha Microsoft Research, Redmond, WA, USA

Josef Sivic Inria, WILLOW Project-team, Département d'Informatique de l'Ecole Normale Supérieure, ENS/INRIA/CNRS UMR, Paris, France

Noah Snavely Cornell University, Ithaca, NY, USA

Richard Szeliski Facebook, Seattle, WA, USA

Bart Thomee Yahoo! Research, San Francisco, USA

Raphael Townshend Stanford University, Stanford, CA, USA

Eric Tzeng UC Berkeley, Berkeley, CA, USA

Luc Van Gool ETH Zurich, Zurich, Switzerland

Nick Vander Valk Vision Technologies Lab, SRI International, Princeton, NJ, USA

Yang Wang University of Manitoba, Winnipeg, Canada

Yu Wang Carnegie Mellon University, Pittsburgh, PA, USA

Avideh Zakhor UC Berkeley, Berkeley, CA, USA

Amir R. Zamir Stanford University, Stanford, CA, USA

Andrew Zhai UC Berkeley, Berkeley, CA, USA

Jiejie Zhu Vision Technologies Lab, SRI International, Princeton, NJ, USA

C. Lawrence Zitnick Facebook AI Research, Palo Alto, CA, USA

Chapter 1
Introduction to Large-Scale Visual Geo-localization

Amir R. Zamir, Asaad Hakeem, Luc Van Gool, Mubarak Shah and Richard Szeliski

Abstract Despite recent advances in computer vision and large-scale indexing techniques, automatic geo-localization of images and videos remains a challenging task. The majority of existing computer vision solutions for geo-localization are limited to highly-visited urban regions for which a significant amount of geo-tagged imagery is available, and therefore, do not scale well to large and ordinary geospatial regions. In this chapter, we provide an overview of the major research themes in visual geo-localization, investigate the challenges, and point to problem areas that will benefit from common synthesis of perspectives from these research themes. In particular, we discuss how the availability of web-scale geo-referenced data affects visual geo-localization, what role semantic information plays in this problem, and how precise localization can be achieved using large-scale textured (RGB) and untextured (non-RGB) 3D models. We also introduce a few real-world applications which became feasible as a result of the capability of estimating an image's geo-location. We conclude this chapter by providing an overview of the emerging trends in visual geo-localization and a summary of the rest of the chapters of the book.

A.R. Zamir (✉)
Stanford University, Stanford, CA, USA
e-mail: zamir@cs.stanford.edu

A. Hakeem
Decisive Analytics Corporation, Arlington, VA, USA
e-mail: asaad.hakeem@dac.us

L. Van Gool
ETH Zurich, Zurich, Switzerland
e-mail: vangool@vision.ee.ethz.ch

M. Shah
University of Central Florida, Orlando, FL, USA
e-mail: shah@crcv.ucf.edu

R. Szeliski
Facebook, Seattle, WA, USA
e-mail: szeliski@fb.com

A.R. Zamir et al. (eds.), *Large-Scale Visual Geo-Localization*,
Advances in Computer Vision and Pattern Recognition,
DOI 10.1007/978-3-319-25781-5_1

1.1 Introduction

Geo-localization is the problem of discovering the location where an image or video was captured. This task arises in a variety of real-world applications. Consumers of imagery may be interested in determining where and when an image/video was taken, who is in the image, what the different landmarks and objects in the depicted scene are, and how they are related to each other. Local government agencies may be interested in using large-scale imagery to automatically obtain and index useful geographic and geological features and their distributions in a region of interest. Similarly, local businesses may utilize content statistics to target their marketing based on 'where', 'what', and 'when' that may automatically be extracted using visual analysis and geo-localization of large-scale imagery. Law enforcement agencies are often interested in finding the location of an incident captured in an image, for forensic purposes. Many of these applications require establishing a relationship between the visual *content* and the location; a task that demands more than simply a GPS-location and requires careful content analysis.

Despite the enormous recent progresses in the field of computer vision, automatic geo-localization is still a difficult problem. It involves identifying, extracting, and indexing geo-informative features from a large variety of datasets, discovering subtle geo-location cues from the imagery, and searching in massive reference visual and nonvisual databases, such as Geographic Information System (GIS). To solve these problems, significant technological advancements are needed in the areas of data-driven discovery of geo-informative features, geometric modeling and reasoning, semantic scene understanding, context-based reasoning, and exploitation of cross-view aerial imagery and cross-modality GIS data. In addition, complementary viewpoints and techniques from these diverse areas will provide additional insight into the problem domain and may spur new research directions, which will be likely to not remain limited to the geo-localization topic.

The central aim of this monograph is to facilitate the exchange of ideas on how to develop visual geo-localization and geo-spatial capabilities. The following are some of the scientific questions and challenges we hope to address:

- What are the general principles that help in characterization and geo-localization of visual data? What features, ranging from low-level and data-driven to high-level and semantic, are geo-spatially informative?
- How can verifiable mathematical models of visual analysis and geo-localization be developed based on these principles? What are the proper search techniques for matching the extracted informative features?
- How can these principles be used to enhance the performance of generic mid- to high-level vision tasks, such as object recognition, scene segmentation/understanding, geometric modeling, alignment, and matching, or empower new applications?

To address these challenges, a number of peer-reviewed chapters from leading researchers in the fields of computer vision, multimedia, and machine learning are compiled. These chapters provide an understanding of the state of the art, open problems, and future directions related to visual geo-spatial analysis and geo-localization of large-scale imagery as a scientific discipline.

The rest of the chapter is organized as follows: In Sect. 1.2, we introduce the central themes and topics in the book. In Sect. 1.3, a brief overview of the emerging trends in the field of geo-localization is provided. In Sect. 1.4, we summarize of the organization of the rest of the book, giving a brief introduction to each chapter.

1.2 Central Themes and Topics

Until the early 2000s, the majority of automatic visual geo-localization methods were targeted toward airborne and satellite imagery. That is, the furnished datasets were mainly composed of images with nadir or limb views, and the query data was captured from either satellites or aircraft. Toward this end, many successful methods [1–3] were developed which were primarily based on planar-scene registration techniques and often utilized subsidiary models, such as Digital Elevation Model (DEM) Digital Elevation Map (DEM), and Digital Terrain Model (DTM), along with regular aerial imagery and Digital Orthophoto Quadrangles (DOQ).

However, during the past decade, the production of visual data has undergone a major shift with the sudden surge of consumer imagery, which mainly retains a ground-level viewpoint. The shift was mostly due to the plummeting cost of the photographic devices as well as the increasing convenience of sharing the multimedia material, including pictures. Currently, ground-level images and videos are primarily produced either through systematic efforts by governments and the private sector (e.g., Google Street View [4]) or directly by consumers (crowdsourcing [5–7]). Google Street View, which provides dense spherical views of public roadways, and Panoramio Collection [5], which is composed of crowdsourced images, are two notable examples of such structured and unstructured databases, respectively.

This considerable shift poses new challenges in the context of automatic geo-localization and demands novel techniques capable of coping with this substantial change. For instance, unlike the traditional geo-registration problem where the scene was mainly assumed to be planar, the ground-level images show heavy nonplanarity. In addition, in traditional aerial images, each pixel is associated with a GPS-location while in the ground-level user images, the entire image is often associated with one GPS-tag which is the location of camera. The following are some of the primary challenges that have emerged as a result of this shift in the production of visual data:

- **Large-scale data handling**: The amount of produced data is prohibitively massive and is increasing sharply. This makes devising techniques that are efficient in preprocessing (leveraging the large amount of reference data) and query processing (sifting through the preprocessed data) essential.
- **Necessity of an accurate geo-location**: Many of the procedures that use geo-tags as their input require a precise geo-location, particularly in the urban areas.

In general, extracting a coarse geographical location, e.g., which continent the image was captured at or distinguishing between desert and coast, has limited applications, and performing the geo-localization with an accuracy comparable to, or better than, handheld GPS devices is desirable.

- **Ambiguity and excessive similarity of visual features**: Unless the data include distinctive objects, such as landmarks, discovering the location merely based on visual information is often challenging due to the significant similarity between man-made structures. This issue becomes critical when city or country scale geo-localization is of interest.
- **Undesirable photography effects**: Unwanted effects, such as suboptimal lighting, frequent occlusions by moving objects, lens distortions, or stitching artifacts, often introduce additional complexities.
- **Lack of unified reference data**: Unlike the conventional geo-registration problem where the reference data (e.g., DOQ) was often unified and enjoyed a homogeneous format, the reference resources for geo-localization of ground-level images are commonly diverse, cross-modality (e.g., GIS which is semantic), and cross-view (e.g., bird's eye imagery which is oblique).

We investigate these challenges in-depth and provide a set of solutions and directions for the future research for each. The chapters of the book are distributed over four major themes and topics described next.

1.2.1 Part I: Data-Driven Geo-localization

Methods for exploiting web-scale datasets for geo-localization and extraction of geographically informative features.

The recent availability of massive repositories of public geo-tagged imagery from around the world has paved the way for geo-localization methods that adopt a data-driven approach. One example of such strictly data-driven methods are those that employ an *image retrieval*-based strategy. That is, match the content of the query image against a dataset of geo-tagged images, identify the reference images with high similarity to the query, and estimate the location of the query based on the geo-tags of the retrieved reference images [8–16]. This approach has been observed to yield promising results as image retrieval and image geo-localization share a great deal of similarity in terms of the problem definition and the challenges faced. The requirement of handling a massive volume of data, or the necessity of dealing with undesirable photography effects, nonplanarity of the scene, and frequent occlusions by irrelevant objects are some of these common challenges.

However, image geo-localization and image retrieval differ in a few fundamental aspects: In image retrieval, the ultimate goal is to find all of the images that match the query with different amounts of similarity. On the contrary, in image geo-localization, the goal is to propose the best location for the query that does not necessarily require finding a large number of matching images. For instance, estimating the location of the query is deemed easier having a few geo-tagged images with relatively similar viewpoints (and probably substantially similar in content) as compared to a large number of not-so-similar images. Additionally, all forms of resemblance, such as semantic similarity (e.g., sharing generic objects), is typically in the interest of image retrieval. In contrast, the primary objective in image geo-localization is to find the images that indeed show the *same* location, and not just a similar one. Such divergences signify that adopting image retrieval and matching techniques off-the-shelf is not an adequate solution for geo-localization, and devising methods specifically intended for the task of localization, even if inspired by image retrieval, is vital. The primary challenges and opportunities in this context are: expanding the applicability area of geo-localization techniques, and identifying geographically informative features. Each of these are elaborated below:

- **Expanding the localization coverage using cross-view imagery and web-scale repositories of user-shared images**: One of the most critical characteristics of any geo-localization system is the size of its covered area. The techniques that are applicable to a larger region are significantly more desirable and practically useful [14, 17, 18]. Databases that are collected in a systematical and controlled manner, such as Google Street View [4], have a limited coverage when it comes to arbitrary regions (e.g., Sahara desert or unpaved roads). This signifies the necessity of leveraging new resources of reference data that do not suffer from this limitation. One remedy to this issue is utilizing aerial imagery (satellite or bird's eye), which is known to be typically available for any arbitrary location. However, this requires performing the matching in a challenging cross-view manner which will be discussed in detail in Chaps. 4 and 5. Adopting other reference modalities with a ubiquitous coverage, such as DEM, GIS, or DTM, are also additional solutions that we will discuss in Chaps. 6, 11–13. Moreover, another possible approach is to utilize the (rapidly expanding) ground-level user-shared image repositories as they typically include images from any point that could be visited by humans. This is discussed in Chap. 3.
- **Devising geographically informative features:** Employing features that are rich and distinctive in capturing geo-spatial information is essential for successful and efficient geo-localization [19, 20]. This is of particular interest as the image features that are devised for generic content matching are not necessarily a good fit for capturing location-dependent information. For instance, the subtle architectural styles that indeed encode a lot of information about the location are typically lost in feature extraction, e.g., SIFT, and quantization, e.g., Bag of Words modeling, steps of existing techniques. Luckily, availability of large-scale geo-tagged data empowers adopting a bottom-up approach to this problem. In Chap. 2, we will discuss how a set of *data-driven* mid-level features that are specifically devised to be geo-spatially discriminative can be extracted from a web-scale dataset of geo-tagged images. In addition, leveraging the massive amount of available data, such features can be simultaneously learned to be invariant with respect to undesirable effects, such as lighting or viewpoint changes.

1.2.2 Part II: Semantic Geo-localization

Identifying the geo-location based on high-level and semantic cues.

In Part I, the methods that adopt a data-driven approach to geo-localization, and consequently, do not make use of cues that are necessarily semantically meaningful were discussed. However, high-level and semantic features of an image carry a significant amount of location-dependent information. Such cues range from human-related characteristics, such as text, architectural style, vehicle types, or urban structures, to natural properties, such as foliage type or weather. Each of these are discriminative geo-location cues: the language of a sign can specify the country/region the picture was captured at; the architectural style of buildings can narrow down the search to certain countries or even cities; observing a particular type of foliage will significantly reduce the search to certain geographical areas. The *aggregation* of such diverse cues, especially when coupled with their geometric arrangement [21], indeed leads to a surprisingly discriminative signature for geo-location [21, 22].

Even though we, as humans, often perform visual geo-localization in such semantic manner and a similar approach for aerial geo-registration has been developed before [23, 24], automatic geo-localization of ground-level imagery based on semantic cues is relatively under-explored [25]. Fortunately, there are reference databases where such features can be extracted from; Geographical Information System (GIS) which contains the location of a large and detailed set of semantic object/structures/land covers, Wikipedia and Wikimapia that provide almost every type of location specific information about events and landmarks, and hashtags of GPS-tagged images in online photo repositories which establish a link between textual tags and geo-locations are some of the nominal examples.

The three main challenges of semantic geo-localization can be summarized as: what features to employ, how to match them, and how to consolidate the diverse semantic cues in a unified system. In Chap. 6, we discuss how a set of semantic features based on scene segments, e.g., road, building facade, etc., along with a histogram representation can significantly narrow down the search area. Also, in Chap. 2, we investigate a set of mid-level features that are found in a data-driven manner but encode beyond low-level pixel information by capturing architectural and temporal information. In addition, in Chap. 6, it will be shown that textual tags and temporal orders of images can considerably improve geo-localization, especially for landmarks.

1.2.3 Part III: Geometric Matching-Based Geo-localization

Geometric alignment of query image to geo-referenced 3D and elevation models.

Data-driven and semantic matching techniques typically provide a rather coarse estimation of the geo-location. Geometric matching-based methods, on the other hand, match the query image against a precise scene model and provide a means for acquiring a more accurate geo-location. Such methods typically involve, first, constructing a geo-referenced 3D scene model (commonly acquired from Structure from Motion) [26–29], and then, performing 2D-to-3D matching between the features extracted from the query image and the reference scene model [30–32]. This approach often comes with additional byproducts of a full 6 DoF camera parameters estimation (i.e., camera rotation as well as the location) and an estimation of the location of the image *content* (besides the location of the camera). Matching against a textured (RGB) 3D point cloud or an untextured (non-RGB) model, such as DEM, constitute the two main classes of approaches in this area; the latter is specifically a good fit for the areas where dense image coverage is not available. The challenges of each are elaborated below:

- **Constructing high fidelity 3D textured (RGB) scene models and performing 2D-to-3D matching against them**: A significant portion of the geometric matching-based geo-localization methods require a 3D point cloud associated with RGB information as their reference input. The sheer size of image datasets from which such 3D models are constructed, and the large variations in imaging and scene parameters pose unique challenges for a wide-area reconstruction. Recent work in Structure from Motion (SfM) has successfully built 3D models from city-scale unstructured collections of images from publicly available photo repositories [26, 27, 33]. This formulation first uses a hybrid discrete-continuous optimization for finding a coarse initial solution and then improves it using bundle adjustment. It naturally incorporates various sources of information about both the cameras and the points, including noisy geo-tags and vanishing point estimates. By using all of the available data at once (rather than incrementally), and by allowing additional forms of constraints, the approach has been shown to be quite robust under realistic settings.
  Once the reconstruction of the reference model is done, the query image features need to be matched against it in a 2D-to-3D manner. The primary challenges of this step are: coping with the massive size of the reference 3D model, effective utilization of the geometric information of the 3D model during matching (besides the RGB information), and performing the matching in a robust way despite the large variation in imaging conditions. In Chaps. 8 and 10, we discuss how an accurate large-scale point cloud can be formed and its size can be significantly reduced without hurting the overall geo-localization performance. Also, in Chap. 9, we will discuss how the accuracy of 2D-to-3D matching can be improved at no or little computational cost, by utilizing the regularities (specifically, co-visibility, and spatial constraints) inherent in point clouds. In Chap. 14, it will be demonstrated that geometric matching-based geo-localization can be extended to aligning even different modalities, e.g., paintings, against image-based 3D models. The techniques developed for this robust extension can be adopted for performing image alignment in presence of severe lighting, viewpoint, and imaging condition changes.
- **Geometric alignment of images to untextured (non-RGB) 3D models**: The aforementioned point clouds are typically formed using SfM techniques which

require a significant number of images for reconstruction. That is why the best performance of such methods are reported for heavily visited areas where a large number of user-shared images are available. However, a significant portion of the Earth does not have such dense image coverage that would yield a 3D point cloud. On the other hand, Digital Elevation Models (DEM), Digital Terrain Model (DTM), and similar non-RGB 3D models that are typically acquired from satellite multispectral analysis and remote sensing are commonly available for any spot on Earth. Therefore, several approaches for utilizing this type of reference data for performing geo-localization in sparsely visited/populated locations have been developed [34–36]. Such methods typically involve extracting a set of relevant features (e.g., mountain folds and creases) from the query and matching them against DEM. The main challenges in this area lie in: what image features would allow performing such cross-modality matching, and how to cope with the large-scale search process. In Chaps. 11, 12, and 13, three methods adopting this approach are presented, which are applicable to mountainous areas (mountain contour matching) as well as desert regions (terrain contour matching).

1.2.4 Part IV: Real-World Applications

Practical tools for geo-localization of web-scale databases and utilizing the extracted geo-location.

Thus far, we focused on fully automated geo-localization frameworks and discussed how this could be accomplished using data-driven features, semantic cues, or geometric matching against 3D models. However, a fundamental follow-up question would be: *What can this extracted geo-location be used for?* This question is of particular importance as the modern cameras and camera phones are GPS enabled and generate photos that are automatically geo-tagged at the time of collection. This signifies the importance of developing methods that can *make use of the geo-location* of images in a principled manner, for instance for understanding the content. In addition, the cameras being GPS enabled means at least a coarse location for the image is known at the time of collection; therefore, there is a growing interest in visual geo-localization techniques that perform the localization *aided by the initial location*, leading to an accuracy far beyond GPS. In the final part of this monograph, we focus on methods for utilizing the geo-location in real-world, performing the visual localization aided by an initial GPS location, and human-in-the-loop localization frameworks that can leverage multimodal cues for a practical geo-localization.

It is of particular interest to leverage the geo-location of an image, whether found visually or using a GPS chip, for understanding its content. The majority of the current applications which utilize the geo-location of images, such as location-based retrieval [5] or large-scale 3D reconstruction [37], are not intended to provide a high-level understanding of the image content. With the exception of a few recent methods which use the geo-tags in connection with the image content (e.g., GIS-assisted object detection [21] or GPS-assisted visual business recognition [38]), the potential impact of the image geo-tags on understating the image content is largely unexplored. This is quite unproductive as knowing the location of an image immediately provides a strong context about the image content: an image taken in Manhattan is expected to show large buildings and maybe yellow cabs; an image taken in a national forest cannot include a skyscraper; knowing a picture was taken in East Asia would give a strong prior about what type of architecture to expect to see.

In Chap. 17, a framework that analyzes the image content in connection with its GPS-tag is presented. The aim of this method is to provide a convenient and automatic way for browsing personal photo collections based on user's textual queries (e.g., "the pictures I took at Radiohead concert"), but without the need for any manual annotation. This is accomplished by using the GPS-tag of the images in the personal photo collection along with their time/date stamp for searching in the web databases (e.g., Wikimapia). This search yields a set of hypotheses for the events that could have possibly happened at that particular time and location (e.g., a concert). These hypotheses are then pruned and verified by matching them against the image content.

As another real-world application, we discuss the situations where performing the visual geo-localization fully from scratch is inefficient and even unnecessary. An example is the cases where some level of human supervision or a prior knowledge about the location of the image is available. In Chap. 15, we will discuss a framework that effectively makes use of the location the GPS chip of the camera provides for performing fast landmark recognition on low-powered mobile device. In Chap. 16, a human-in-the-loop geo-localization framework in which an analyst's feedback is looped back into the system is introduced. The key idea behind this method is to combine a minimal intervention of an analyst with the power of automatically mined web-scaled databases to achieve a localization accuracy that outperforms both fully automated methods and purely manual geo-localization.

As discussed earlier, the end goal of the methods discussed in this part is not an entirely automated geo-localization, but either utilizing the geo-location for content understanding or performing the localization in an assisted manner under practical settings.

1.3 Emerging Trends

Geo-spatial localization is a fast evolving field, mostly due to availability of unprecedented data resources, new applications, and novel techniques. In this section, we briefly overview some of the emerging trends in this area each centered around one of the aforementioned aspects.

1.3.1 New Geo-referenced Data Resources

Geo-localization is constantly affected by the available resources of geo-referenced data [14, 39–42]. It experienced a major shift with the emergence of massive ground-level imagery from users and street view during the past decade. Similar changes in the geo-referenced data are expected to continue and define new opportunities and challenges. One of the new resources that is becoming extensively popular is personal drones. Such inexpensive, GPS-enabled, and widely available UAVs can provide an HD, and sometimes live, aerial imagery of an arbitrary location. This is considered a major change as it takes collection of aerial imagery to a personal level, while it used to be exclusive to large companies and governments until not too long ago. This is similar to what cell phones and user-shared imagery brought in ground-level geo-localization about a decade ago, when collection of large-area datasets was limited to governments and large corporations, such as Google.

Another shift in geo-referenced data resources is taking place as a result of significant improvements in the spatial and temporal resolution of satellite images. As of now, typical GSD (Ground Sampling Distance that is the metric distance represented by one pixel) values have reached the accuracy of <0.35 m, and the temporal resolution (satellite revisit time) has reduced to <1 day [43, 44]. In addition, impressive near-live HD videos from space have been recently offered in the commercial market [45] that provide new grounds for performing temporal modeling along with spatial geo-localization. Such data resources are all new and rapidly changing which require novel approaches to geo-spatial analysis in order to effectively accommodate and utilize them.

1.3.2 Temporal Geo-localization and New Applications

The majority of frameworks for geo-localization are targeted toward extracting the spatial location. With the exception of a few efforts [46–48], the temporal aspect of location is largely ignored, whereas providing information about "*when*" a particular event takes place is as important as "*where*." Appending time information to geo-spatial localization is expected to be of growing interest in the future. Government agencies for law enforcement purposes or local businesses for analyzing their target market are only some of the entities that are interested in the temporal aspect as much as they are interested in the spatial location. Performing the geo-localization in a timely manner and joint modeling of temporal and geo-spatial changes in visual data are some of the potential tasks in this area.

In particular, the temporal aspect of cross-view geo-localization is further brought to interest by the notable improvements in the temporal resolution of satellite imagery (currently as low as <1 day and decreasing) and the near-live HD videos from space, as elaborated in the previous subsection.

1.3.3 Deep Learning Based Geo-localization

Convolutional neural networks have remarkably improved the state of the art in several computer vision problems, namely object detection [49], scene recognition [50], and depth from monocular vision [51]. In essence, a convolutional neural network is a cascade of several layers of convolution operations followed by nonlinearity with fully learned parameters. It essentially provides a means for end-to-end and unified feature learning and classification/regression. With the notable abilities that deep learning-based methods have recently shown in solving basic computer vision problems, it is expected that they will bring substantial improvements to the field of geo-localization as well. Very recently, a few attempts have shown promising results for cross-view image geo-localization [52] and semantic feature learning [53] employing deep Learning-based techniques that demonstrate the potential merits in this approach. Considering the basic characteristics of deep learning, it is expected to be particularly beneficial in cross-view/cross-modality matching, end-to-end geo-spatial feature learning, and temporal geo-localization using recurrent neural networks.

1.4 Organization of the Book

An overview of each of the chapters is provided below:

Part I: Data-Driven Geo-localization: *Methods for exploiting web-scale datasets for geo-localization and extraction of geographically informative features.*

Chapter 2: Discovering Mid-level Visual Connections in Space and Time

This chapter explores what a mid-level visual representation can bring to geo-spatial and longitudinal analyses. Specifically, the authors present a weakly-supervised visual data mining approach that discovers connections between recurring mid-level visual elements in historic (temporal) and geographic (spatial) image collections, and attempts to capture the underlying visual style. In contrast to existing discovery methods that mine for patterns that remain visually consistent throughout the dataset, the goal here is to discover visual elements whose appearance changes due to change in time or location; i.e., exhibit consistent stylistic variations across the label space (date or geo-location). To discover these elements, the authors first identify groups of patches that are style-sensitive. Then, they incrementally build correspondences to find the same element across the entire dataset. Finally, they train style-aware regressors that model each element's range of stylistic differences. The authors demonstrate the method's effectiveness on the related task of fine-grained classification.

Chapter 3: Where the Photos were Taken: Location Prediction by Learning from Flickr Photos

This chapter investigates the characteristics of geographically tagged Internet photos and determines their location based on the visual content. To build reliable geographical estimators, it is important to find distinguishable geographical clusters in the world. These clusters cover general geographical regions not limited to just landmarks. Geographical clusters provide more training samples and hence lead to better recognition accuracy. To solve this estimation problem, the authors built an efficient solver to find the clusters employing a set of latent variables. They demonstrate detailed qualitative results obtained from beach photos taken in different continents.

Chapter 4: Cross-View Image Geo-localization

In this chapter, a cross-view feature translation approach that greatly extends the reach of image geo-localization methods to a vast majority of the Earth's land area that has no ground level reference photos available is presented. The authors present a method that can often localize a query even if it has no corresponding ground-level images in the database. The key idea is to learn a mapping from ground-level appearance to overhead appearance and land cover attributes. This relationship is learned from sparsely available geo-tagged ground-level images and the corresponding aerial and land cover data at those locations. The authors demonstrate their method over a 1135 km^2 region containing a variety of scenes and land cover types. For each query, their algorithm produces a probability density over the region of interest.

Chapter 5: Ultrawide Baseline Facade Matching for Geo-localization

This chapter describes a technique that matches street-level images to a database of airborne images. This problem is hard because of extreme viewpoint and illumination differences. Color/gradient distributions or local descriptors fail to match under such setting forcing us to rely on the structure of self-similarity of patterns on facades. The authors propose to capture this structure with a novel "scale-selective self-similarity" (S^4) descriptor which is computed at each point on the facade at its inherent scale. To achieve this, the authors introduce a new method for scale selection which enables the extraction and segmentation of facades as well. They also introduce a novel geometric method that aligns satellite and bird's-eye-view imagery to extract building facade regions in a stereo graph-cuts framework. Matching of the query facade to the database facade regions is done with a Bayesian classification of the street-view query S^4 descriptors given all labeled descriptors in the bird's-eye view database. The authors also discuss geometric techniques for camera pose-estimation using correspondences between building corners in the query and the matched aerial imagery. They demonstrate the retrieval accuracy on a challenging set of publicly available imagery and compare with standard SIFT-based techniques.

Part II: Semantic Reasoning-based Geo-localization: *Identifying the geo-location based on high-level and semantic cues.*

Chapter 6: Semantically Guided Geo-localization and Modeling in Urban Environments

This chapter presents a technique that enables semantic labeling of both query views and the reference dataset through semantic segmentation that can aid (1) retrieval of views similar and possibly overlapping with the query and (2) guiding the recognition and discovery of commonly occurring scene layouts in the reference dataset.

The authors demonstrate the effectiveness of these semantic representations on examples of localization, semantic concept discovery, and intersection recognition in the images of urban scenes.

Chapter 7: Recognizing Landmarks in Large-Scale Social Image Collections

In this chapter, a technique that can recognizes popular landmarks in large-scale datasets of unconstrained consumer images from Flickr by formulating a classification problem involving nearly 2 million images and 500 categories is described. The dataset and categories are formed automatically from geo-tagged photos from Flickr by looking for peaks in the spatial geo-tag distribution corresponding to frequently photographed landmarks. The authors learn models for these landmarks with a multiclass support vector machine, using classic vector-quantized interest point descriptors as features. They also incorporate the semantic non-visual metadata available on modern photo-sharing sites, showing that *textual tags* and *temporal constraints* lead to significant improvements in classification rate. Finally, they apply recent breakthroughs in deep learning with Convolutional Neural Networks, finding that these models can dramatically outperform the traditional recognition approaches to this problem, and even beat human observers in some cases.

Part III: Geometric Matching Based Geo-localization: *Geometric alignment of query image to geo-referenced 3D and elevation models.*

Chapter 8: Worldwide Pose Estimation Using 3D Point Clouds

This chapter addresses the problem of determining where a photo was taken by estimating a full 6-DOF-plus-intrinsic camera pose with respect to a large geo-registered 3D point cloud, bringing together research on image localization, landmark recognition, and 3D pose estimation. The authors propose a method that scales to datasets with hundreds of thousands of images and tens of millions of 3D points through the use of two new techniques: a co-occurrence prior for RANSAC and bidirectional matching of image features with 3D points. The authors evaluate their method on several large data sets, and show state-of-the-art results on landmark recognition as well as the ability to locate cameras to within meters, requiring only seconds per query.

Chapter 9: Exploiting Spatial and Co-visibility Relations for Image-Based Localization

This chapter describes a technique that can increase the effectiveness of approaches based on prioritized 2D-to-3D matching at little to no additional run-time costs by exploiting both spatial and co-visibility relations between the 3D points in the model. Geometry matching based localization techniques aim to estimate the position and orientation from which a given query image was taken with respect to a 3D model of the scene. Recent advances in Structure-from-Motion, which allow us to reconstruct large scenes in little time, create a need for image-based localization approaches that handle large-scale models consisting of millions of 3D points both efficiently and effectively in order to localize as many query images as possible in as little time as possible. While multiple efficient localization methods based on prioritized feature matching have been proposed recently, they lack the effectiveness of slower approaches. The authors demonstrate that the resulting localization framework incorporates both 2D-to-3D and 3D-to-2D matching and achieves state-of-the-art efficiency and effectiveness.

Chapter 10: 3D Point Cloud Reduction using Mixed-Integer Quadratic Programming

A method to accelerate the matching process and reduce the memory footprint by analyzing the view-statistics of points in a training corpus is provided in this chapter. Given a training image set that is representative of common views of a scene, the authors propose an approach that identifies a compact subset of the 3D point cloud for efficient localization, while achieving comparable localization performance to using the full 3D point cloud. The authors demonstrate that the problem can be precisely formulated as a mixed-integer quadratic program and present a point-wise descriptor calibration process to improve matching. They also show that their algorithm outperforms the state-of-the-art greedy algorithm on standard datasets, on measures of both point-cloud compression and localization accuracy.

Chapter 11: Image Based Large-Scale Geo-localization in Mountainous Regions

This chapter describes a technique that can geo-localize images in mountainous terrain and use digital elevation models to extract representations for fast visual database lookup. The authors propose an automated approach for very large-scale visual localization that can efficiently exploit visual information (contours) and geometric constraints (consistent orientation) at the same time. They demonstrate and validate their approach at the scale of Switzerland (40,000 km^2) using over 1000 landscape query images.

Chapter 12: Adaptive Rendering for Large-Scale Skyline Characterization and Matching

This chapter explores an adaptive rendering approach for large-scale skyline characterization and matching. Given an image, the authors propose a system that automatically extracts the skyline and then matches it to a database of reference skylines extracted from rendered images using digital elevation data (DEM). The sampling density of these rendering locations determines both the accuracy and the speed of skyline matching. The proposed approach successfully combines global planning and local greedy search strategies to select new rendering locations incrementally. The authors report quantitative and qualitative results from synthesized and real experiments, where a $4\times$ computational speedup is achieved.

Chapter 13: User-Aided Geo-localization of Untagged Desert Imagery

This chapter presents a system for user-aided visual localization of desert imagery without the use of any metadata such as GPS readings, camera focal length, or field-of-view. The system makes use only of publicly available datasets—in particular, digital elevation models (DEMs)—to rapidly and accurately locate photographs in nonurban environments such as deserts. The authors propose a system that generates synthetic skyline views from a DEM and extracts stable concavity-based features from these skylines to form a database. To localize queries, a user manually traces the skyline on an input photograph. The skyline is automatically refined based on this estimate, and the same concavity-based features are extracted. They then apply a variety of geometrically constrained matching techniques to efficiently and accurately match the query skyline to a database skyline, thereby localizing the query image. The authors evaluate their system using a test set of 44 ground-truthed images over a 10,000 km^2 region of interest in a desert and show that in many cases, queries can be localized with precision as fine as 100 m^2.

Chapter 14: Visual Geo-localization of Non-photographic Depictions via 2D-3D Alignment

In this chapter, a technique that can geo-localize arbitrary 2D depictions of architectural sites, including drawings, paintings, and historical photographs is described. This goal is achieved by aligning the input depiction with a 3D model of the corresponding site. The task is difficult as the appearance and scene structure in the 2D depictions can be very different from the appearance and geometry of the 3D model, e.g., due to the specific rendering style, drawing error, age, lighting, or change of seasons. In addition, it is a hard search problem: the number of possible alignments of the painting to a set of 3D models from different architectural sites is huge. To address these issues, the authors have developed a compact representation of complex 3D scenes. It is demonstrated that the proposed approach can automatically identify the correct architectural site as well as recover an approximate viewpoint of historical photographs and paintings with respect to the 3D model of the site.

Part IV: Real-World Applications: *Practical Tools for Geo-Localization of Web-Scale Databases and Utilizing the Extracted Geo-location.*

Chapter 15: A Memory Efficient Discriminative Approach for Location-Aided Recognition

In this chapter, a GPS-assisted visual recognition technique for fast recognition of urban landmarks on a GPS-enabled mobile device is presented. Most of the existing similar methods offload their computation to a server by uploading the query image. Over a slow network, this can cause a latency of several seconds. In contrast, the authors present an approach that requires uploading only the approximate GPS location to a server after which a compact, location-specific classifier is downloaded to the device and all subsequent computation is performed onboard. Their approach is supervised and involves training compact random forest classifiers (RDF) on a database of geo-tagged images. The feature vector for the RDF is computed by densely searching the image for the presence of selective discriminative local image patches extracted from the training images. The images are rectified using detected vanishing points and binary descriptors allow for an efficient search for the discriminative patches, a step that is further accelerated using min-hash. The authors have evaluated the performance of their approach on representative urban datasets where it outperforms traditional methods based on bag of visual words representation or direct matching of local feature descriptors, neither of which are feasible when processing must occur on a low-power mobile device.

Chapter 16: A Real-World System for Image/Video Geo-localization
This chapter presents WALDO (Wide Area Localization of Depicted Objects), a system that solves the challenging problem of image geo-localization by combining the insight of analysts with the power of automated analysis for Internet-scale, geo-location-driven data mining. WALDO's goal-driven constrained resource management leverages a full spectrum of data-driven, semantic, and geometric geo-localization experts and user tools.

Chapter 17: Photo Recall: Using the Internet to Label Your Photos

In this chapter, a system for searching your personal photos using an extremely wide range of text queries is proposed. The authors accomplish this by finding the correlations between the information in the photos—the timestamps, GPS locations, and image pixels—and the information mined from the Internet. This includes matching dates to holidays listed on Wikipedia, GPS coordinates to places listed on Wikimapia, places, and dates to find named events using Google, visual categories using classifiers either pre-trained on ImageNet or trained on-the-fly using results from Google Image Search, and object instances using interest point-based matching, again using results from Google Images. The authors tie all of these disparate sources of information together in a unified way, allowing for fast and accurate searches using whatever information the user remembers about a photo. They represent all of this information in a layered graph which prevents duplication of effort and data storage, while simultaneously allowing for fast searches, generating meaningful descriptions of search results, and even suggesting query completions to the user as he/she types, via auto-complete.

References

1. Sheikh Y, Khan S, Shah M (2004) Feature-based georegistration of aerial images. GeoSensor Netw 4
2. Zitova B, Flusser J (2003) Image registration methods: a survey. Image Vis Comput 21(11): 977–1000
3. Kumar R, Sawhney H, Asmuth J, Pope A, Hsu S (1998) Registration of video to georeferenced imagery. In: Proceedings of fourteenth international conference on pattern recognition, vol 2, pp 1393–1400
4. https://www.google.com/maps/streetview/
5. http://www.panoramio.com/
6. https://www.flickr.com/
7. http://picasa.google.com/
8. Grant S, Brown M, Szeliski R (2007) City-scale location recognition. In: IEEE conference on computer vision and pattern recognition, CVPR'07
9. Zamir AR, Shah M (2014) Image geo-localization based on multiple nearest neighbor feature matching using generalized graphs. In: T-PAMI
10. Torii A, Sivic J, Pajdla T, Okutomi M (2013) Visual place recognition with repetitive structures. In: IEEE conference on computer vision and pattern recognition (CVPR)

11. Gronat P et al (2013) Learning and calibrating per-location classifiers for visual place recognition. In: IEEE conference on computer vision and pattern recognition (CVPR)
12. Knopp J, Sivic J, Pajdla T (2010) Avoiding confusing features in place recognition. Computer vision-ECCV. Springer, Heidelberg, pp 748–761.
13. Chen DM et al (2011) City-scale landmark identification on mobile devices. In: IEEE conference on computer vision and pattern recognition (CVPR)
14. Lin T-Y, Belongie S, Hays J (2013) Cross-view image geolocalization. In: IEEE conference on computer vision and pattern recognition (CVPR)
15. Hays J, Efros AA (2008) IM2GPS: estimating geographic information from a single image. In: IEEE conference on computer vision and pattern recognition (CVPR)
16. Vaca-Castano G, Zamir AR, Shah M (2012) City scale geo-spatial trajectory estimation of a moving camera. In: Computer vision and pattern recognition (CVPR)
17. Bansal M et al (2011) Geo-localization of street views with aerial image databases. In: Proceedings of the 19th ACM international conference on multimedia
18. Bansal M, Daniilidis K, Sawhney H (2012) Ultra-wide baseline facade matching for geo-localization computer vision-ECCV 2012. In: Workshops and demonstrations. Springer, Berlin
19. Doersch C et al (2012) What makes Paris look like Paris? ACM Trans Graph (TOG) 31(4):101
20. Lee YJ, Efros AA, Hebert M (2013) Style-aware mid-level representation for discovering visual connections in space and time. In: ICCV
21. Ardeshir S, Zamir AR, Shah M (2014) GIS-assisted object detection and geospatial localization. In: European conference on computer vision (ECCV)
22. Sullivan A. The view from your window contest. http://dish.andrewsullivan.com/vfyw-contest/
23. Wang C, Croitoru A, Stefanidis A, Agouris P (2007) Image-to-X registration using linear features and networks. In: FUZZ-IEEE
24. Wang C, Stefanidis A, Agouris P (2007) Relaxation matching for georegistration of aerial and satellite imagery. In: IEEE ICIP
25. Castaldo F, Zamir A, Angst R, Palmieri F, Savarese S (2015) The IEEE international conference on computer vision (ICCV) workshops, pp 9–17
26. Agarwal S, Snavely N, Simon I, Seitz SM, Szeliski R (2009) Building rome in a day. In: ICCV
27. Crandall D, Owens A, Snavely N, Huttenlocher D (2011) Discrete-continuous optimization for large-scale structure from motion. In: CVPR 2011. Best paper award runner-up
28. Snavely N, Simon I, Goesele M, Szeliski R, Seitz SM (2010) Scene reconstruction and visualization from community photo collections. In: Proceedings of the IEEE
29. Grzeszczuk R, Kosecka J, Hile H, Vedantham R (2009) Creating compact architectural models by geo-registering image collections. In: IEEE international workshop on 3-D digital imaging and modeling, ICCV
30. Li Y et al (2012) Worldwide pose estimation using 3d point clouds. Computer vision-ECCV 2012. Springer, Berlin, pp 15–29
31. Snavely N, Garg R, Seitz SM, Szeliski R (2008) Finding paths through the world's photos, SIGGRAPH
32. Kosecka J, Zhang W (2007) Image based localization. IEEE Trans Rob
33. Li, Y, Snavely N, Huttenlocher DP (2010) Location recognition using prioritized feature matching. Computer vision-ECCV 2010. Springer, Berlin, pp 791–804
34. Baatz G et al (2012) Large scale visual geo-localization of images in mountainous terrain. Computer vision-ECCV 2012. Springer, Berlin, pp 517–530
35. Zakhor A et al (2013) User-driven geolocation of untagged desert imagery using digital elevation models
36. Baboud L et al (2011) Automatic photo-to-terrain alignment for the annotation of mountain pictures. In: IEEE conference on computer vision and pattern recognition (CVPR)
37. Snavely N, Seitz SM, Szeliski R (2006) Photo tourism: exploring photo collections in 3D. ACM Trans Graph (TOG) 25(3):835–846
38. Zamir AR, Dehghan A, Shah M (2013) Visual business recognition: a multimodal approach. In: Proceedings of the 21st ACM international conference on multimedia

39. Jacobs N, Miskell K, Pless R (2011) Webcam geo-localization using aggregate light levels. In: Applications of computer vision (WACV)
40. Jacobs N, Satkin S, Roman N, Speyer R, Pless R (2007) Geolocating static cameras. In: ICCV
41. Lalonde J-F, Narasimhan SG, Efros AA (2010) What do the sun and the sky tell us about the camera. IJCV 88(1):24–51
42. Zamir A, Shah M (2010) Accurate image localization based on google maps street view. Computer vision–ECCV 2010. Springer, Berlin, pp 255–268
43. http://www.satimagingcorp.com/
44. https://www.digitalglobe.com/
45. http://www.skyboximaging.com/
46. Schindler G, Dellaert F (2012) 4D cities: analyzing, visualizing, and interacting with historical urban photo collections. J Multimed 7(2):124–131
47. Matzen K, Snavely N (2014) Scene chronology. Computer vision-ECCV 2014. Springer International Publishing, pp 615–630
48. Torii A et al (2015) 24/7 place recognition by view synthesis. In: CVPR 2015–28th IEEE conference on computer vision and pattern recognition
49. Krizhevsky A, Sutskever I, Hinton GE (2012) Imagenet classification with deep convolutional neural networks. In: Advances in neural information processing systems
50. Zhou B et al (2014) Learning deep features for scene recognition using places database. In: Advances in neural information processing systems
51. Eigen D, Puhrsch C, Fergus R (2014) Depth map prediction from a single image using a multi-scale deep network. In: Advances in neural information processing systems
52. Lin T-Y et al (2015) Learning deep representations for ground-to-aerial geolocalization. In: Proceedings of the IEEE conference on computer vision and pattern recognition
53. Workman S, Souvenir R, Jacobs N (2015) Wide-area image geolocalization with aerial reference imagery. arXiv:1510.03743

Part I
Data-Driven Geo-localization

Methods for exploiting web-scale datasets for geo-localization and extraction of geographically informative features

Unlike the traditional geo-registration problem that was restricted to aerial imagery, the availability of large-scale ground-level images, e.g., street view or user-shared images, revolutionized the way we thought of geo-localization. Compared to only a decade ago, the current data resources are significantly broader in coverage, more diverse, and cheaper to access. These characteristics empower large-area localization and enable extensive data-driven processes on geo-referenced data. However, some of the main remaining challenges to overcome are developing new methods for effective utilization of multi-modal data (e.g., for cross-view matching) and automatic learning of geographically informative features. In this part of the book, we investigate several techniques that address geo-localization with utilizing massive datasets at their core, particularly for geographical feature learning and cross-view geo-localization.

Chapter 2
Discovering Mid-level Visual Connections in Space and Time

Yong Jae Lee, Alexei A. Efros and Martial Hebert

Abstract Finding recurring visual patterns in data underlies much of modern computer vision. The emerging subfield of visual category discovery/visual data mining proposes to cluster visual patterns that capture more complex appearance than low-level blobs, corners, or oriented bars, without requiring any semantic labels. In particular, mid-level visual elements have recently been proposed as a new type of visual primitive, and have been shown to be useful for various recognition tasks. The visual elements are discovered automatically from the data, and thus, have a flexible representation of being either a part, an object, a group of objects, etc. In this chapter, we explore what the mid-level visual representation brings to geo-spatial and longitudinal analyses. Specifically, we present a weakly supervised visual data mining approach that discovers connections between recurring mid-level visual elements in historic (temporal) and geographic (spatial) image collections, and attempts to capture the underlying *visual style*. In contrast to existing discovery methods that mine for patterns that remain visually consistent throughout the dataset, the goal is to discover visual elements whose appearance changes due to change in time or location, i.e., exhibit consistent stylistic variations across the label space (date or geo-location). To discover these elements, we first identify groups of patches that are style-sensitive. We then incrementally build correspondences to find the same element across the entire dataset. Finally, we train style-aware regressors that model each element's range of stylistic differences. We apply our approach to date and geo-location prediction and show substantial improvement over several baselines that do

Y.J. Lee (✉)
Department of Computer Science, UC Davis, Davis, CA, USA
e-mail: yjlee@cs.ucdavis.edu

A.A. Efros
Department of Electrical Engineering and Computer Science,
UC Berkeley, Berkeley, CA, USA
e-mail: efros@eecs.berkeley.edu

M. Hebert
Robotics Institute, Carnegie Mellon University, Pittsburgh, PA, USA
e-mail: hebert@cs.cmu.edu

A.R. Zamir et al. (eds.), *Large-Scale Visual Geo-Localization*,
Advances in Computer Vision and Pattern Recognition,
DOI 10.1007/978-3-319-25781-5_2

not model visual style. We also demonstrate the method's effectiveness on the related task of fine-grained classification.

2.1 Introduction

Long before the age of "data mining," historians, geographers, anthropologists, and paleontologists have been discovering and analyzing patterns in data. One of their main motivations is finding patterns that correlate with spatial (geographical) and/or temporal (historical) information, allowing them to address two crucial questions: *where?* (geo-localization) and *when?* (historical dating). Interestingly, many such patterns, be it the shape of the handle on an Etruscan vase or the pattern of bark of a Norwegian pine, are predominantly *visual*. The recent explosion in the sheer volume of visual information that humanity has been capturing poses both a challenge (it is impossible to go through by hand), and an opportunity (discovering things that would never have been noticed before) for these fields. In this chapter, we take the first step in considering temporally as well as spatially varying visual data and developing a method for *automatically discovering* visual patterns that correlate with time and space.

Of course, finding recurring visual patterns in data underlies much of modern computer vision itself—it is what connects the disparate fragments of our visual world into a coherent narrative. At the low level, this is typically done via simple unsupervised clustering (e.g., k-means in visual words [32]). But clustering visual patterns that are more complex than simple blobs, corners, and oriented bars turns out to be rather difficult because everything becomes more dissimilar in higher dimensions. The emerging subfield of visual category discovery/visual data mining [6–8, 12, 19, 25, 27, 30, 31] proposes ways to address this issue. Most such approaches look for tight clumps in the data, discovering visual patterns that stay globally consistent throughout the dataset. More recent discriminative methods, such as [6, 30], take advantage of weak supervision to divide the dataset into discrete subsets (e.g., kitchen vs. bathroom [30], Paris vs. Not-Paris [6]) to discover specific visual patterns that repeatedly occur in one subset while *not* occurring in others.

But in addition to the globally consistent visual patterns (e.g., the Pepsi logo is exactly the same all over the world) and the specific ones (e.g., toilets are only found in bathrooms), much in our visual world is neither global nor specific, but rather undergoes a *gradual visual change*. This is nowhere more evident than in the visual changes across large extents of space (geography) and time (history). Consider the three cars shown in Fig. 2.1: one antique, one classic, and one from the 1970s. Although these cars are quite different visually, they clearly share some common elements, e.g., a headlight or a wheel. But notice that even these "common" elements differ substantially in their appearance across the three car types, making this a very challenging correspondence problem. Notice further that the way in which they differ is not merely random (i.e., a statistical "noise term"). Rather, these subtle yet consistent differences (curvy vs. boxy hood, the length of the ledge under the

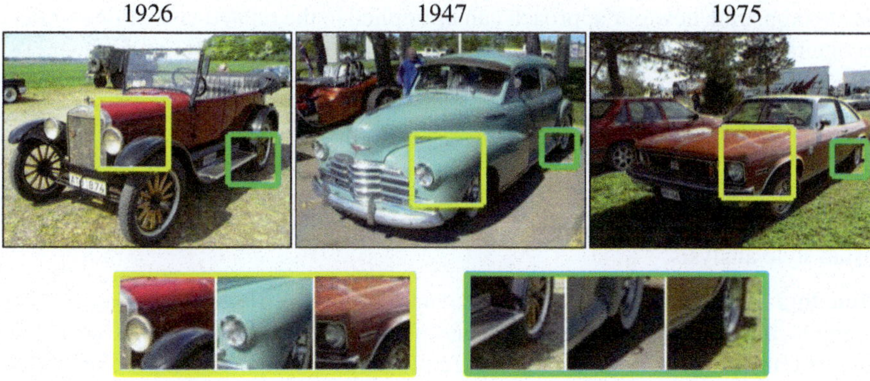

Fig. 2.1 Given images of historic cars, our algorithm is not only able to automatically discover corresponding visual elements (e.g., *yellow* and *green boxes*) despite the large visual variations, but can model these variations to capture the changes in visual style across time

door, etc.) tend to reflect the particular *visual style* that is both specific to an era yet changing gradually over time (Fig. 2.9). If now we were given a photo of a different car and asked to estimate its model year, we would not only need to detect the common visual elements on the new car but also understand what its stylistic differences (e.g., the length of that ledge) tell us about its age.

2.1.1 Overview

We propose a method for discovering connections between similar mid-level visual elements in temporally and spatially varying datasets and modeling their "visual style." Here, we define visual style as appearance variations of the same visual element due to change in time or location. Our central idea is to (1) create reliable *generic* visual element detectors that "fire" across the entire dataset independent of style, and then (2) model their *style-specific* differences using weakly supervised image labels (date, geo-location, etc.). The reason for doing the first step is that each generic detector puts all of its detections into correspondence (lower right in Fig. 2.1), creating a "closed world" focused on one visual theme, where it is much easier to "subtract away" the commonalities and focus on the stylistic differences. Furthermore, without conditioning on the generic detector, it would be very difficult to even detect the stylistically informative features. For instance, the ledge in Fig. 2.1 (green box) is so tiny that it is unlikely to be detectable in isolation, but in combination with the wheel and part of the door (the generic part), it becomes highly discriminable.

We evaluate our method on the task of date and geo-location prediction in three scenarios: two historic car datasets with model year annotations and a Street View imagery dataset annotated with GPS coordinates. We show that our method outperforms several baselines, which do not explicitly model visual style. Moreover, we also demonstrate how our approach can be applied to the related task of fine-grained recognition of birds. This chapter expands upon our previous conference paper [18].

2.2 Related Work

We review related work in modeling geo-location and time, visual data mining, and visual style analysis.

Modeling Space and Time Geo-tagged datasets have been used for geo-localization on the local [16, 28], regional [3, 4], and planetary [13, 14] scales, but we are not aware of any prior work on improving geo-location by explicitly capturing stylistic differences between geo-informative visual elements (but see [6] for anecdotal evidence of such possibility). Longitudinal (i.e., long-term temporal) visual modeling has received relatively little attention. Most previous research has been on the special case of age estimation for faces (see [10] for a survey). Recent work includes modeling the temporal evolution of Web image collections [15] and dating of historical color photographs [23]. We are not aware of any prior work on modeling historical visual style.

Visual data mining Existing visual data mining/object discovery approaches have been used to discover object categories [8, 12, 20, 25, 31], mid-level patches [6, 26, 30], attributes [7, 27], and low-level foreground features [19]. Typically, an appropriate similarity measure is defined between visual patterns (i.e., images, patches, or contours) and those that are most similar are grouped into discovered entities. Of these methods, mid-level discriminative patch mining [6, 30] shares the most algorithmic similarities with our work; we also represent our visual elements with HOG patches [5] and refine the clusters through discriminative cross-validation training. However, unlike [6, 30] and all existing discovery methods, we go beyond simply detecting recurring visual elements, and model the stylistic differences among the common discovered elements.

Visual style analysis The seminal paper on "style-content separation" [34] uses bilinear models to factor out the style and content components in pre-segmented, prealigned visual data (e.g., images of letters in different fonts). While we also use the term "style" to describe the differences between corresponding visual elements, we are solving a rather different problem. Our aim is to automatically *discover* recurring visual elements despite their differences in visual style, and then model those differences. While our "generic detectors" could perhaps be thought of as capturing "content" (independent of style), we do not explicitly factor out the style, but model it *conditioned* on the content.

Fine-grained categorization can also be viewed as a form of style analysis, as subtle differences within the same basic-level category differentiate one subordinate category from another. Existing approaches use human-labeled attributes and keypoint annotations [1, 9, 36, 40] or template matching [38, 39]. Because these methods are focused on classification, they limit themselves to the simpler visual world of manually annotated object bounding boxes, whereas our method operates on full images. Furthermore, discovering one-to-one correspondences is given a primary role in our method, whereas in most fine-grained approaches the correspondences are already provided. While template matching methods [38, 39] also try to discover correspondences, unlike our approach, they do not explicitly model the style-specific differences within each correspondence set. Finally, these approaches have not been applied to problems with continuous labels (regression), where capturing the range of styles is particularly important.

Lastly, relative/comparative attributes [24, 29] model how objects/scenes relate to one another via ordered pairs of labels (A is "furrier" than B). We also share the idea of relating things. However, instead of using strong supervision to define these relationships, we automatically mine for visual patterns that exhibit such behavior.

2.3 Approach

Our goal is to discover and connect mid-level visual elements across temporally and spatially varying image collections and model their style-specific differences. We assume that the image collections are weakly supervised with date or location labels.

There are three main steps to our approach: First, as initialization, we mine for "style-sensitive" image patch clusters, that is, groups of visually similar patches with similar labels (date or location). Then, for each initial cluster, we try to generalize it by training a generic detector that computes correspondences across the entire image collection to find the same visual element independent of style. Finally, for each set of correspondences, we train a style-aware regression model that learns to differentiate the subtle stylistic differences between different instances of the same generic element. In the following sections, we describe each of the steps in turn. We will use an image collection of historic cars as our running example, but note that there is nothing specific to cars in our algorithm, as will be shown in the results section.

2.3.1 *Mining Style-Sensitive Visual Elements*

Most recurring visual patterns in our dataset will be extremely boring (sky, asphalt, etc.). They will also not exhibit any stylistic variation over time (or space), and not be of any use in historical dating (or geo-localization)—after all, asphalt is always just asphalt! Even some parts of the car (e.g., a window) do not really change much over the decades. On the other hand, we would expect the shape of the hood between two 1920s cars to be more similar than between a 1920s and a 1950s car. Therefore, our first task is to mine for visual elements whose appearance somehow correlates with its labels (i.e., date or location). We call visual elements that exhibit this behavior *style-sensitive*.

Since we do not know a priori the correct scale, location, and spatial extent of the style-sensitive elements, we randomly sample patches across various scales and locations from each image in the dataset. Following [6], we represent each patch with a histogram of gradients (HOG) descriptor [5], and find its top N nearest neighbor patches in the database (using normalized correlation) by matching it to each image in a sliding window fashion over multiple scales and locations. To ensure that redundant overlapping patches are not chosen more than once, for each matching image we only take its best matching patch.

Each sampled patch and its N nearest neighbors ideally form a cluster of a recurring visual element; although, in practice, many clusters will be very noisy due to inadequacies of simple HOG matching. To identify the style-sensitive clusters, we can analyze the temporal distribution of labels for each cluster's instances. Intuitively, a cluster that has a tightly grouped ("peaky") label distribution suggests a visual element that prefers a particular time period, and is thus style-sensitive, while a cluster that has a uniform label distribution suggests a pattern that does not change over time. As extra bonus, most noisy clusters will also have a uniform distribution since it is very unlikely to be style-sensitive by random chance. To measure the style sensitivity of cluster c, we histogram its labels and compute its entropy:

$$E(c) = -\sum_{i=1}^{n} H(i) \cdot \log_2 H(i), \tag{2.1}$$

where $H(i)$ denotes the histogram count for bin i and n denotes the number of quantized label bins (we normalize the histogram to sum to 1). We then sort the clusters in ascending order of entropy. Figure 2.2a, b shows examples of the highest and lowest ranked clusters for the car dataset images. Notice how the highest ranked clusters correspond to style-sensitive car elements, while the lowest ranked clusters contain noisy or style-insensitive ones. We take the top M clusters as our discovered style-sensitive visual elements, after rejecting near-duplicate clusters. A cluster is considered to be a near-duplicate of a higher ranked cluster if it has at least five cluster members that spatially overlap by more than 25 % with any of the instances from the higher ranked cluster.

2.3.2 *Establishing Correspondences*

Each of the top M clusters corresponds to a style-sensitive visual element in a local region of the label space. A few of these elements represent very specific visual features that just do not occur in other parts of the data (e.g., car tailfins from 1960s). But most others have similar counterparts in other time periods and our goal is to connect them together, which will allow us to model the change in style of the same visual element over the entire label space. For instance, one of the style-sensitive

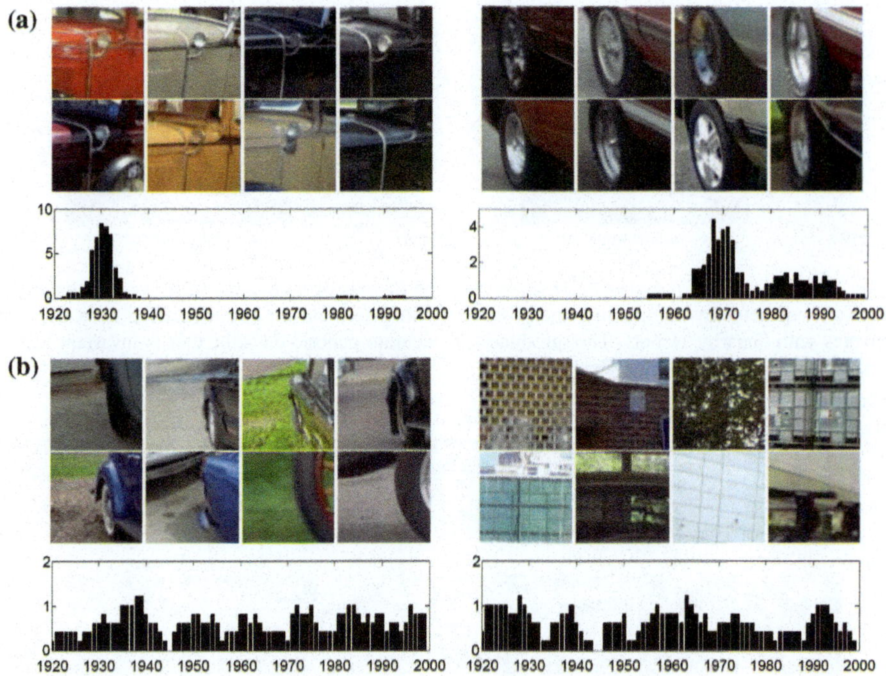

Fig. 2.2 Mining style-sensitive visual elements. Clusters are considered style-sensitive if they have "peaky" (low-entropy) distribution across time (**a**) and style-insensitive if their instances are distributed more uniformly (**b**). Notice how the high-entropy distributions (**b**) represent not only style insensitivity (e.g., nondescript side of car) but also visually noisy clusters. Both are disregarded by our method. **a** Peaky (low-entropy) clusters, **b** uniform (high-entropy) clusters

elements could represent frontal cars from 1920s. We want to find corresponding frontal car patches across all time periods.

The same visual element, however, can look quite different across the label space, especially over larger temporal extents (Fig. 2.1). To obtain accurate correspondences across all style variations, we propose to train a discriminative detector using an iterative procedure that exploits the continuous nature of the label space. In general, we expect the appearance of a visual element to change *gradually* as a function of its label. Our key idea is to initialize the detector using a style-sensitive cluster as the initial positive training data, but then incrementally revise it by augmenting the positive set with detections fired only on images with "nearby" labels (e.g., decades), as shown in Fig. 2.3.

Specifically, we first train a linear support vector machine (SVM) detector with the cluster patches as positives and patches sampled from thousands of random Flickr images as negatives. These negatives will make the detector discriminative against generic patterns occurring in the "natural world" [30], which helps it to fire accurately on unseen images. We then incrementally revise the detector. At each step, we run the current detector on a new subset of the data that covers a slightly broader range

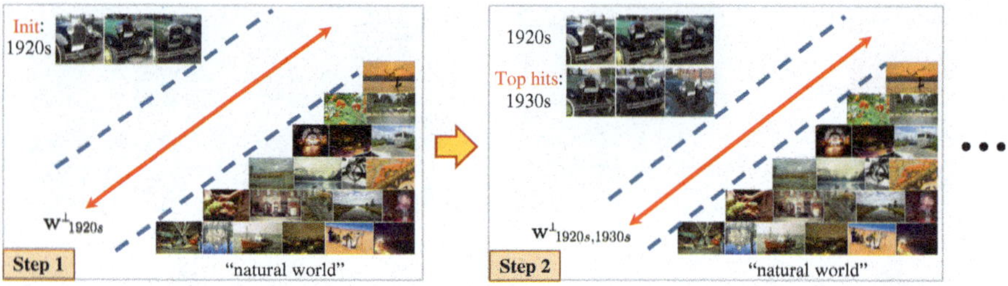

Fig. 2.3 To account for a visual element's variation in style over space or time, we incrementally revise its detector by augmenting the positive training set with the top detections fired only on images with "nearby" labels. This produces an accurate generic detector that is invariant to the visual element's changes in style

Fig. 2.4 Establishing correspondences across time. **a** Correspondences made using the discriminative patch mining approach [6, 30] using a positive set of 1920s frontal cars. Note how the correspondences break down a third of the way through. **b** Starting with the same initial set of 1920s frontal cars, our algorithm gradually expands the positive set over the continuous label space until it is able to connect the same visual element across the entire temporal extent of the dataset. **a** Singh et al., **b** our approach

in label space, and retrain it by augmenting the positive training set with the top detections. We repeat this process until all labels have been accounted for. Making these transitive connections produces a final generic detector that fires accurately across the entire label space, as shown in Fig. 2.4b. Note that automatic discovery of transitive visual correspondences across a dataset is very much in the spirit of the Visual Memex [22] opening up several promising future directions for investigation.

There is an important issue that we must address to ensure that the detector is robust to noise. The initial cluster can contain irrelevant, outlier patches, since some of the top N nearest neighbors of the query patch could be bad matches. To prune out the noisy instances, at each step of the incremental revision of our detector, we apply cross-validation training [6, 30]. Specifically, we create multiple partitions of the training set and iteratively refine the current detector by: (1) training on one partition; (2) testing on another; (3) taking the resulting top detections as the new training instances; and (4) repeating steps 1–3 until convergence, i.e., the top detections do not change. Effectively, at each iteration, the detector learns to boost the common patterns shared across the top detections and down-weights their

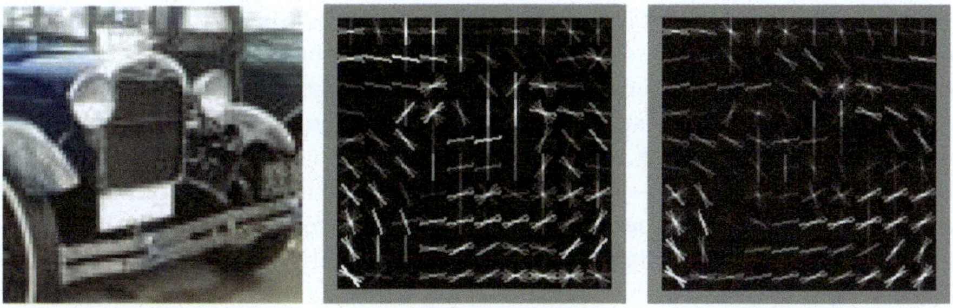

Fig. 2.5 Visualization of the positive HOG weights learned for the discovered frontal car visual element. Compared to the initial model, which was trained on the 1920s images, the final model cares much less about the car's global shape but still prefers to see a wheel in the *lower left corner*

discrepancies without overfitting, which leads to more accurate detections in the next iteration. After several iterations, we obtain a robust detector.

Note that a direct application of [6, 30] will not work for our case of continuous, style-varying data because the variability can be too great. Figure 2.4a shows detections made by a detector trained with [6, 30], using the same initial style-sensitive cluster of 1920s cars as positives. The detector produces accurate matches in nearby decades, but the correspondence breaks down across larger temporal extents because it fails to model the variation in style. Figure 2.5 visualizes the positive components of the weights learned by the initial model trained only with the 1920s frontal car, and those of our final model trained incrementally over all decades. The final model has automatically learned to be invariant to the change in global shape of the car over time. It still prefers to see a wheel in the lower left corner, since its appearance changes much less over time.

Finally, we fire each trained generic detector on all images and take the top detection per image (and with SVM score greater than -1) to obtain the final correspondences.

2.3.3 Training Style-Aware Regression Models

The result of the previous step is a set of generic mid-level detectors, each tuned to a particular visual element and able to produce a set of corresponding instances under many different styles. Now we are finally ready to model that variation in style. Because the correspondences are so good, we can now forget about the larger dataset and focus entirely on each set of corresponding instances in isolation, making our modeling problem much simpler. The final step is to train a style-aware regressor for each element that models its stylistic variation over the label space.

In general, style will not change linearly over the label space (e.g., with cars, it is possible that stylistic elements from one decade could be reintroduced as "vintage" in a later decade). To account for this, we train a standard nonlinear support vector regressor (SVR) [33] with an ϵ-insensitive loss function using ground-truth weakly supervised image labels (e.g., date, geo-location) as the target score. We use Gaussian kernels: $K(x_i, x_j) = \exp(-\gamma^{-1}||x_i - x_j||^2)$, where γ is the mean of the pairwise distances among all instances and x_i is the HOG feature, for instance, i. Under this kernel, instances with similar appearance are most likely to have similar regression outputs. Furthermore, to handle possible misdetections made by the generic detector which could add noise, we weight each instance proportional to its detection score when training the SVR. (We map a detection score s to a weight in [0, 1], via a logistic function $1/(1 + \exp(-2s))$.) Each resulting model captures the stylistic differences of the same visual element found by the generic detector.

2.4 Results

In this section, we (1) evaluate our method's ability to predict date/location compared to several baselines, (2) provide in-depth comparisons to the discriminative patch mining approach of [30], (3) show qualitative examples of discovered correspondences and learned styles, and (4) apply our approach to fine-grained recognition of birds.

Datasets We use three datasets: (1) Car Database (CarDb): 13,473 photos of cars made in 1920–1999 crawled from cardatabase.net; (2) Internet Movie Car Database (IMCDb): 2,400 movie images of cars made in 1920–1999 crawled from imcdb.org; and (3) East Coast Database (EDb): 4,455 Google Street View images along the eastern coasts of Georgia, South Carolina, and North Carolina. Example images are shown in Fig. 2.6. CarDb and IMCDb images are labeled with the model year of the main car in the image, and EDb images are labeled with their GPS coordinates. These are the "style" labels and the only supervisory information we use. For EDb, since our SVRs expect 1D outputs (although a multivariate regression method could also be used), we project the images' 2D GPS coordinates to 1D using principal component analysis; this works because the area of interest is roughly linear, i.e., long and narrow, see Fig. 2.6c. These datasets exhibit a number of challenges including clutter, occlusion, scale, location and viewpoint change, and large appearance variations of the objects. Importantly, unlike standard object recognition datasets, ours have continuous labels. We partition the CarDb and EDb datasets into train/test sets with 70/30 % splits. We evaluate on all datasets, and focus additional analysis on CarDb since it has the largest number of images.

Image-level date/location prediction To evaluate on a label prediction task, we need to combine all of our visual element predictors together. We train an image-level prediction model using as features the outputs of each style-aware regressor on an image. Specifically, we represent an image I with feature $\phi(I)$, which is

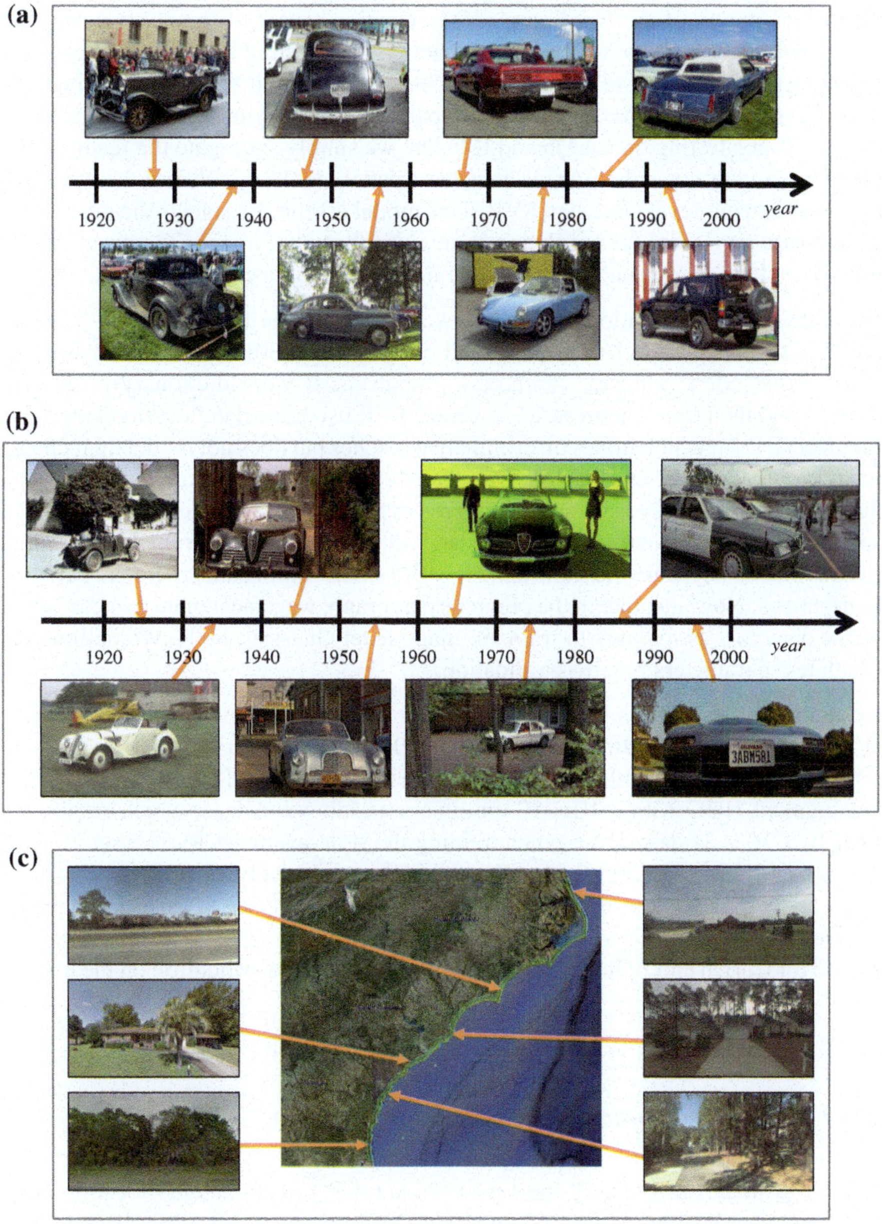

Fig. 2.6 Example images of the **a** Car Database (CarDb), **b** Internet Movie Car Database (IMCDb), and **c** East coast Database (EDb). Each image in CarDb and IMCDb is labeled with the car's model year. Each image in EDb is labeled with its GPS coordinate

the concatenation of the maximum SVM detection scores of the generic detectors (over the image) and the SVR scores of their corresponding style-aware regressors. When testing on EDb, we aggregate the features in spatial bins via a spatial pyramid [17, 21], since we expect there to be spatial consistency of visual patterns across images. When testing on CarDb and IMCDb, we simply aggregate the features over the entire image, since the images have less spatial regularity. We use these features to train an image-level Gaussian SVR. This model essentially selects the most useful style-aware regressors for predicting style given the entire image. To ensure that the image-level model does not overfit, we train it on a separate validation set.

Baselines For date/location prediction, we compare three baselines: bag-of-words (BOW), spatial pyramid (SP) [17], and Singh et al. [30]. For the first two, we detect dense SIFT features, compute a global visual word dictionary on the full dataset, and then train an intersection kernel SVR using the date/location labels. For Singh et al. [30], which mines discriminative patches but does not model their change in style, we adapt the approach to train date/location-specific patch detectors using the initial style-sensitive clusters discovered in Sect. 2.3.1. Specifically, we take each specific cluster's instances as positives and all patches from the remaining training images that do not share the same labels (with a small "don't care" region in between) as negatives. Now, just like in the previous paragraph, we concatenate the max output of the detectors as features to train an image-level Gaussian SVR. We optimize all baselines' parameters by cross-validation.

Implementation details We sample 80×80 pixel patches over an image pyramid at five scales (i.e., min/max patch is 80/320 pixels wide in original image), and represent each patch with a $10 \times 10 \times 31$ HOG descriptor [5]. For EDb patches, we augment HOG with a 10×10 tiny image in lab colorspace, which results in a final $10 \times 10 \times 34$ descriptor, when training the style-aware SVRs. We set $N = 50$, $n = 80$, and $M = 80,\ 315$ for CarDb and EDb, respectively. For our generic SVM detectors, we fix $C_{svm} = 0.1$, and cover 1/8 of the label space at each training step; CarDb: 10 years, EDb: 66 miles. For our SVRs, we fix $\epsilon = 0.1$ and set $C_{svr} = 100$ and 10 for CarDb and EDb, respectively, tuned using cross-validation on the training set.

2.4.1 Date and Location Prediction Accuracy

We first evaluate our method's ability to predict the correct date/geo-location of the images in CarDb/EDb. Figure 2.7 and Table 2.1 (top rows) show the absolute error rates for all methods. This metric is computed by taking the absolute difference between the ground-truth and predicted labels.

Our approach outperforms all baselines on both datasets. The baselines have no mechanism to explicitly model stylistic differences as they are either mining discriminate patches over a subregion in label space (Singh et al.) or using quantized

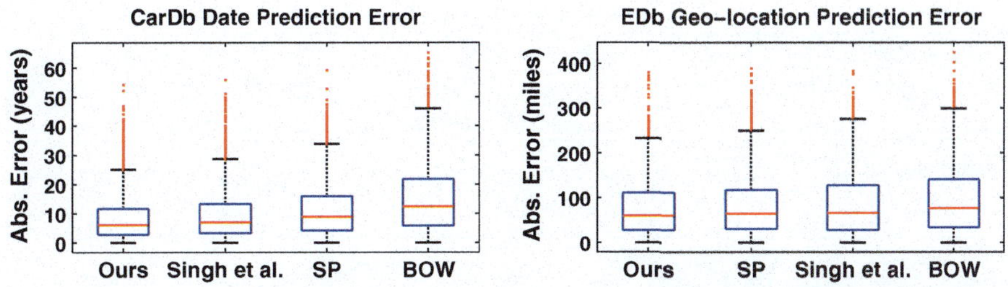

Fig. 2.7 *Box* plots showing date and location prediction error on the CarDb and EDb datasets, respectively. Lower values are better. Our approach models the subtle stylistic differences for each discovered element in the data, which leads to lower error rates compared to the baselines

Table 2.1 Mean absolute error on CarDb, EDb, and IMCDb for all methods

	Ours	Singh et al. [6, 30]	Spatial pyramid [17]	Bag-of-words
CarDb (years)	**8.56**	9.72	11.81	15.39
EDb (miles)	**77.66**	87.47	83.92	97.78
IMCDb (years)	**13.53**	15.32	17.06	18.65

The result on IMCDb evaluates cross-dataset generalization performance using models trained on CarDb. Lower values are better

local features (BOW and SP) that result in loss of fine detail necessary to model subtle stylistic changes. Without explicitly making connections over space/time, the baselines appear to have difficulty telling apart signal from noise. In particular, we show substantial improvement on CarDb, because cars exhibit more pronounced stylistic differences across eras that require accurate modeling. The stylistic differences in architecture and vegetation for EDb are much more subtle. This makes sense, since the geographic region of interest only spans about 530 miles along the U.S. east coast. Still, our method is able to capture more of the stylistic differences to produce better results. Note that chance performance is around 19 years and 113 miles for CarDb and EDb, respectively; all methods significantly outperform chance, which shows that stylistic patterns correlated with time/location are indeed present in these datasets.

Figure 2.8 shows some discovered correspondences. Notice the stylistic variation of the car parts over the decades (e.g., windshield) and the change in amount/type of vegetation from north to south (e.g., trees surrounding the houses). In Fig. 2.9 we visualize the learned styles of a few style-aware regressors on CarDb by averaging the most confident detected instances of each predicted decade.

Fig. 2.8 Example correspondences on CarDb (*top*) and EDb (*bottom*). Notice how a visual element's appearance can change due to change in time or location

Fig. 2.9 In each row, we visualize the styles that a single style-aware regressor has learned by averaging the predictions for each decade

2.4.2 *Cross-Dataset Generalization Accuracy*

Recent work on dataset bias [35] demonstrated that training and testing on the same type of data can dramatically overestimate the performance of an algorithm in a real-world scenario. Thus, we feel that a true test for an algorithm's performance should include training on one dataset while testing on a different one, whenever possible.

To evaluate cross-dataset generalization performance, we take the models trained on CarDb and test them on IMCDb. The third row in Table 2.1 shows the result. The error rates have increased for all methods compared to those on CarDb (with BOW now almost at chance level!). Overall, IMCDb is more difficult as it exhibits larger appearance variations due to more significant changes in scale, viewpoint, and position of the cars. CarDb, on the other hand, is a collection of photos taken by car enthusiasts, and thus, the cars are typically centered in the image in one of a few canonical viewpoints. Note also that the gap between BOW and SP is smaller compared to that on CarDb. This is mainly because spatial position is ignored in BOW while it is an important feature in SP. Since the objects' spatial position in IMCDb is more varied, SP tends to suffer from the different biases. Since our generic detectors are scale- and translation-invariant, we generalize better than the baselines. Singh et al. is also scale- and translation-invariant, and thus, shows better performance than BOW and SP. Still, our method retains a similar improvement over that baseline.

2.4.3 *Detailed Comparisons to Singh et al. [30]*

In this section, we present detailed comparisons to Singh et al. [30], which is similar to our method but does not capture the style-specific differences.

2.4.3.1 Robustness to Number of Detectors

Figure 2.10 (left) plots the geo-location prediction error as a function of the number of detectors on EDb for the two methods (the curve averages the error rates over five runs; in each run, we randomly sample a fixed number of detectors, and corresponding style-aware models for ours, among all 315 detectors to train the image-level SVR). Our approach outperforms the baseline across all points, saturating at a much lower error rate. This result demonstrates that when the visual patterns in the data change subtly, we gain a lot more from being style-aware than being discriminative.

We also analyze how well our models generalize across the label space. Using generic detectors initialized only with the visual patterns discovered within a specific decade (which results in 10 detectors), we train the corresponding style-aware regression models. We then use their outputs to train the image-level regressor. Across all eight different decade initializations, we find our final mean prediction error rates to be quite stable (∼10 years). This shows our approach's generalizability and robustness to initialization.

2.4.3.2 Visual Consistency of Correspondences

We next evaluate the quality of our discovered correspondences to that of Singh et al. using a purity/coverage plot. Purity is the % of cluster members that belong to the same visual element and coverage is the number of images covered by a given cluster. These are standard metrics used to evaluate discovery algorithms, and Singh et al. already showed superior performance over common feature/region clustering approaches using them. Thus, we feel it is important to evaluate our approach using the same metrics. We randomly sample 80 test images (10 per decade) from CarDb, and randomly sample 15 generic detectors and 15 discriminative detectors for ours and the baseline, respectively. We fire each detector and take its highest scoring detected patch in each image. We sort the resulting set of detections in decreasing detection score and ask a human labeler to mark the inliers/outliers, where inliers are the majority of high-scoring detections belonging to the same visual element. Using these human-marked annotations and treating each set of detections as a cluster, we compute average purity as a function of coverage.

Figure 2.10 (right) shows the result. We generate the curve by varying the threshold on the detection scores to define cluster membership and average the resulting purity scores (e.g., at coverage = 0.1, purity is computed using only the top 10 % scoring detections in each cluster). Both ours and the baseline produce high purity when the clusters consist of only the highest scoring detections. As more lower scoring instances are included in the clusters, the baseline's purity rates fall quickly, while ours fall much more gracefully. This is because the baseline is trained to be discriminative against visual elements from other time periods. Thus, it succeeds in detecting corresponding visual elements that are consistent within the same period, but cannot generalize outside of that period well. Our detectors are trained to be generic and thus able to generalize much better, maintaining high purity with increased coverage.

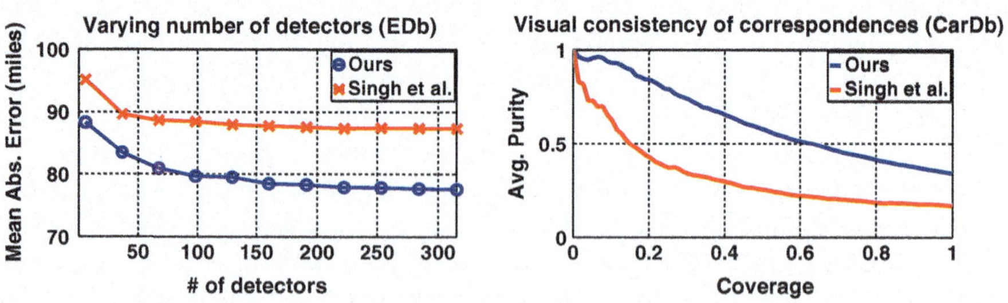

Fig. 2.10 Absolute prediction error rates when varying the number of detectors (lower is better), and visual consistency of correspondences measured using purity and coverage (higher is better)

Fig. 2.11 Examples of accurate (**a**) and inaccurate (**b**) date predictions on CarDb. *GT* ground-truth year, *Pred* predicted year, and *Error* absolute error in years

2.4.4 Qualitative Predictions

We next show qualitative examples of accurate and inaccurate date predictions on CarDb in Fig. 2.11a, b, respectively. Some common errors, as shown in Fig. 2.11b, are due to the car having atypical viewpoint (first example), resembling a vehicle from a different decade (second example), being too small in the image (third example), or being uncommon from that decade (fifth example).

2.4.5 Fine-Grained Recognition

Finally, the idea of first making visual connections across a dataset to create a "closed world," and then modeling the style-specific differences is applicable to several other domains. As one example, we adopt our method with minor modifications to the task of fine-grained recognition of bird species, where the labels are discrete.

Specifically, we first mine recurring visual elements that repeatedly fire inside the foreground bounding box (of any bird category) and not on the background (cf. style-sensitive clusters). We take the top-ranked clusters and train generic unsupervised bird-part detectors. Then, given the correspondence sets produced by each detector, we train 1-vs-all linear SVM classifiers to model the style-specific differences (cf. style-aware SVRs). Finally, we produce an image-level representation, pooling the maximum responses of the detectors and corresponding classifier outputs in a spatial pyramid. We use those features to train image-level 1-versus-all linear SVM classifiers, one for each bird category.

We evaluate classification accuracy on the CUB-200-2011 dataset [37], which is comprised of 11,788 images of 200 bird species, using the provided bounding box annotations. We compare to the state-of-the-art methods of [1, 2, 11, 40, 41], which are all optimized to the fine-grained classification task. The strongly supervised

Table 2.2 Fine-grained recognition on CUB-200-2011 [37]

	Ours	Zhang et al. [40]	Berg and Belhumeur [1]	Zhang et al. [41]	Chai et al. [2]	Gavves et al. [11]
Mean accuracy (%)	41.01	28.18	56.89	50.98	59.40	62.70
Supervision	Weak	Strong	Strong	Strong	Weak	Weak

methods [1, 40, 41] use ground-truth part annotations for training, while the weakly supervised methods (ours and [2, 11]) use only bounding box annotations.

Table 2.2 shows mean classification accuracy over all 200 bird categories. Even though our method is not specialized for this specific task, its performance is respectable compared to existing approaches (some of which employ stronger supervision), and even outperforms [40] despite using less supervision. We attribute this to our generic detectors producing accurate correspondences for the informative bird parts (see Fig. 2.12), allowing our style-specific models to better discriminate the fine-grained differences.

Fig. 2.12 Example correspondences on CUB-200-2011

2.5 Conclusion

In this chapter, we presented a novel approach that discovers recurring mid-level visual elements and models their *visual style* in appearance-varying datasets. By automatically establishing visual connections in space and time, we created a "closed world" where our regression models could focus on the stylistic differences. We demonstrated substantial improvement in date and geo-location prediction over several baselines that do not model style.

Source code, data, and additional results are available on our project webpage: http://www.cs.ucdavis.edu/~yjlee/iccv2013.html

Acknowledgments We thank Olivier Duchenne for helpful discussions. This work was supported in part by Google, ONR MURI N000141010934, and the Intelligence Advanced Research Projects Activity (IARPA) via Air Force Research Laboratory. The U.S. government is authorized to reproduce and distribute reprints for governmental purposes notwithstanding any copyright annotation thereon. Disclaimer: The views and conclusions contained herein are those of the authors and should not be interpreted as necessarily representing the official policies or endorsements, either expressed or implied, of IARPA, AFRL or the U.S. government.

References

1. Berg T, Belhumeur P (2013) POOF: part-based one-vs-one features for fine-grained categorization, face verification, and attribute estimation. In: CVPR
2. Chai Y, Lempitsky V, Zisserman A (2013) Symbiotic segmentation and part localization for fine-grained categorization. In: ICCV
3. Chen CY, Grauman K (2011) Clues from the beaten path: location estimation with bursty sequences of tourist photos. In: CVPR
4. Cristani M, Perina A, Castellani U, Murino V (2008) Geolocated image analysis using latent representations. In: CVPR
5. Dalal N, Triggs B (2005) Histograms of oriented gradients for human detection. In: CVPR
6. Doersch C, Singh S, Gupta A, Sivic J, Efros AA (2012) What makes Paris look like Paris? In: SIGGRAPH
7. Duan K, Parikh D, Crandall D, Grauman K (2012) Discovering localized attributes for fine-grained recognition. In: CVPR
8. Faktor A, Irani M (2012) Clustering by composition unsupervised discovery of image categories. In: ECCV
9. Farrell R, Oza O, Zhang N, Morariu V, Darrell T, Davis L (2011) Birdlets: subordinate categorization using volumetric primitives and pose-normalized appearance. In: ICCV
10. Fu Y, Guo G-D, Huang T (2010) Age synthesis and estimation via faces: a survey. TPAMI
11. Gavves E, Fernando B, Snoek C, Smeulders A, Tuytelaars T (2013) Fine-grained categorization by alignments. In: ICCV
12. Grauman K, Darrell T (2006) Unsupervised learning of categories from sets of partially matching image features. In: CVPR
13. Hays J, Efros A (2008) Im2gps: estimating geographic information from a single image. In: CVPR
14. Kalogerakis E, Vesselova O, Hays J, Efros A, Hertzmann A (2009) Image sequence geolocation with human travel priors. In: ICCV
15. Kim G, Xing E, Torralba A (2010) Modeling and analysis of dynamic behaviors of web image collections. In: ECCV

16. Knopp J, Sivic J, Pajdla T (2010) Avoiding confusing features in place recognition. In: ECCV
17. Lazebnik S, Schmid C, Ponce J (2006) Beyond bags of features: spatial pyramid matching for recognizing natural scene categories. In: CVPR
18. Lee YJ, Efros AA, Hebert M (2013) Style-aware mid-level representation for discovering visual connections in space and time. In: ICCV
19. Lee YJ, Grauman K (2009) Foreground focus: unsupervised learning from partially matching images. In: IJCV, vol 85
20. Lee YJ, Grauman K (2011) Object-graphs for context-aware visual category discovery. In: TPAMI
21. Li L-J, Su H, Xing E, Fei-Fei L (2010) Object bank: a high-level image representation for scene classification and semantic feature sparsification. In: NIPS
22. Malisiewicz T, Efros A (2009) Beyond categories: the visual memex model for reasoning about object relationships. In: NIPS
23. Palermo F, Hays J, Efros AA (2012) Dating historical color images. In: ECCV
24. Parikh D, Grauman K (2011) Relative attributes. In: ICCV
25. Payet N, Todorovic S (2010) From a set of shapes to object discovery. In: ECCV
26. Raptis M, Kokkinos I, Soatto S (2012) Discovering discriminative action parts from mid-level video representations. In: CVPR
27. Rastegariy M, Farhadi A, Forsyth D (2012) Attribute discovery via predictable discriminative binary codes. In: ECCV
28. Schindler G, Brown M, Szeliski R (2007) Cityscale location recognition. In CVPR
29. Shrivastava A, Singh S, Gupta A (2012) Constrained semi-supervised learning using attributes and comparative attributes. In: ECCV
30. Singh S, Gupta A, Efros AA (2012) Unsupervised discovery of mid-level discriminative patches. In: ECCV
31. Sivic J, Russell B, Efros A, Zisserman A, Freeman W (2005) Discovering object categories in image collections. In: ICCV
32. Sivic J, Zisserman A (2003) Video Google: a text retrieval approach to object matching in videos. In: ICCV
33. Smola A, Schlkopf B (2003) A tutorial on support vector regression. Technical report, Statistics and Computing
34. Tenenbaum J, Freeman W (2000) Separating style and content with bilinear models. Neural Comput 12(6)
35. Torralba A, Efros AA (2011) Unbiased look at dataset bias. In: CVPR
36. Wah C, Branson S, Perona P, Belongie S (2011) Multiclass recognition part localization with humans in the loop. In: ICCV
37. Wah C, Branson S, Welinder P, Perona P, Belongie S (2011) The caltech-UCSD birds-200-2011 dataset. Technical report
38. Yang S, Bo L, Wang J, Shapiro L (2012) Unsupervised template learning for fine-grained object recognition. In: NIPS
39. Yao B, Khosla A, Fei-Fei L (2011) Combining randomization and discrimination for fine-grained image categorization. In: CVPR
40. Zhang N, Farrell R, Darrell T (2012) Pose pooling kernels for sub-category recognition. In: CVPR
41. Zhang N, Farrell R, Iandola F, Darrell T (2013) Deformable part descriptors for fine-grained recognition and attribute prediction. In: ICCV

Chapter 3
Where the Photos Were Taken: Location Prediction by Learning from Flickr Photos

Li-Jia Li, Rahul Kumar Jha, Bart Thomee, David Ayman Shamma, Liangliang Cao and Yang Wang

Abstract In this chapter, we explore the characteristics of geographically tagged Internet photos and determine their location based on the visual content. We develop a principled machine learning model to estimate geographical locations of photos by modeling the relationship between location and the photo content. To build reliable geographical estimators, it is important to find distinguishable geographical clusters in the world. These clusters cover general geographical regions not limited to just landmarks. Geographical clusters provide more training samples and hence lead to better recognition accuracy. We develop a framework for geographical cluster estimation, and employ latent variables to estimate the geographical clusters. To solve this estimation problem, we propose to build an efficient solver to find the latent clusters. We illustrate detailed qualitative results obtained from beaches photos taken at different continents. In addition, we show significantly improved quantitative results over other approaches for recognizing different beaches using the Flickr beach dataset as validation.

L.-J. Li (✉)
Yahoo! Research, Sunnyvale, USA
e-mail: lijiali@yahoo-inc.com; lijiali@cs.stanford.edu

R.K. Jha
University of Michigan, Ann Arbor, USA
e-mail: rahuljha@umich.edu

B. Thomee · D.A. Shamma
Yahoo! Research, San Francisco, USA
e-mail: bthomee@yahoo-inc.com

D.A. Shamma
e-mail: shamma@yahoo-inc.com

L. Cao
IBM Watson Research, New York, USA
e-mail: liangliang.cao@gmail.com

Y. Wang
University of Manitoba, Winnipeg, Canada
e-mail: ywang@cs.umanitoba.ca

A.R. Zamir et al. (eds.), *Large-Scale Visual Geo-Localization*,
Advances in Computer Vision and Pattern Recognition,
DOI 10.1007/978-3-319-25781-5_3

3.1 Introduction

Photo community sites such as Flickr and Instagram host a massive amount of publicly shared photos that are rich with annotations, metadata, and location coordinates (geo-tag) information. Researchers have come to leverage such community annotations to organize and search these large-scale photo corpora—through advancements in social computing, computer vision, and machine learning.

Modern cameras and camera phones are GPS enabled and generate photos that are automatically geo-tagged; this enables much research that seeks to understand more about locations and photography. Geographic information of user generated photos is associated with a two-dimensional vector, latitude and longitude, representing a unique location on the earth. There is great need for such geographic information; many people have high interest in not only the place they live but also other interesting places around the world. Geographic annotation is also desirable when reviewing the travel and vacation images. For example, when a person becomes interested in a nice photo, she or he may want to know where exactly it is. Moreover, if one plans to visit a place, she or he may want to find out the points of interest nearby. Recent studies suggest that geo-tags expand the context that can be employed for image content analysis by adding extra information about the subject or environment of the image. Unfortunately, a large amount of online photos do not have geo-information available for such purpose. One emerging line of research is to estimate the geographical locations even when they are not provided.

However, estimating the geo-location of images is a challenging task. Pioneering work by Hays and Efros [1] and Crandall et al. [2], have been able to locate only a quarter of the test images subject to a rough region (approximately 750 km) near their true location. At the metropolitan scale, visual feature-based annotations perform no better than chance. While it is difficult to estimate the exact location at which a photo was taken, photos belong to a coarser location in terms of geographical clusters share similar properties, representing the characteristics of the location. It is not difficult to observe that photos of Alcatraz Islands would represent topics related to the lighthouse and prison cells, whereas photos taken at Time Square reflect topics such as the skyscrapers and busy streets. The goal of our approach is to find meaningful geographical clusters corresponding to different geographical regions. It is relatively easier to curate more training samples to build more reliable classifiers for predicting the geographical clusters than exact geo-information. The use of geographical clusters can provide the wisdom of the crowd for many applications. If we can correctly assign geo-locations to image, we will be able to produce tourist maps using geographical annotation techniques [3]. We can also compare the distribution of different topics, such as cars, food, or landscapes in the world [4]. However, in practice, it is not easy to find meaningful geographical clusters. Country borders that separate the geographical regions are too coarse for large countries but too fine for small ones. In this article, we propose to predict meaningful geographical clusters using efficient clustering approach built upon the latent SVM learning framework.

Note that geographical cluster prediction is different from landmark recognition [5]. A landmark usually corresponds one view or one subject with a unique appearance, while a coarse location such as Waikiki Beach may be constituted by multiple aspects including water, boats, people dresses, buildings, and plants. Moreover, a landmark is usually limited to a point on the earth, while a beach usually covers a region. It is often inaccurate and also unnecessary to estimate the exact GPS coordinates whereas estimating coarse location of photos can lead to more accurate result.

Since the photos at specific location carry unique information about it, an interesting application is to estimate the location by learning from the large-scale photos. In this chapter, we exercise the idea by focusing on the problem of predicting where the beach photos were taken.

To address the problem of beach photo location prediction, an important component in beach photo localization is to find meaningful *geographical clusters* corresponding to different beaches. The use of geographical clusters benefits the problem of localization in two aspects: On the training stage, geographical clusters provide more training samples and hence lead to better recognition accuracy; on the testing stage, estimation of the most possible region for each query photo will be relatively easier than the estimation of GPS coordinates, while the information of geographical cluster will be good enough for trip planning and photo organization applications. We propose an approach that considers image clustering and model learning in a single unified framework. Earlier versions of this work were introduced in [6, 7].

3.2 Related Work

Geo-location related research has made promising progress in the last 10 years. By associating geo-tagged information with an image, one might be able to infer the the image location from the visual content. There has been a growing body of visual research community investigating geographical information for image understanding [4, 8–19]. Many applications are inspired by one of the information technology research mission proposed by Bush [20] to build a personal Memex which can record everything a person sees and hears, and quickly retrieve any item on request. With the vast amount of photos shared on online photo sharing communities, it is possible to aggregate information from one user's album or even the albums from a large number of users, so that group wisdom can be mined from these media. The results of geo-driven Memex can be used in a number of applications, and hence opens many possibilities in location based services. Naaman et al. [8] showed that location-based metadata can help to suggest candidate identity labels in photo organization. Joshi and Luo [12] proposed to explore the geographical information systems (GIS) database using geo-tags in a neighborhood region. Yu and Luo [11] propose another way to leverage nonvisual contexts including location and seasonal information, both of which can be obtained through picture metadata automatically, to improve the accuracy of object region recognition. In [17], the authors explore satellite images corresponding to picture location data and investigate their novel use to recognize the picture-taking environment, as if through a third eye above the object. Agarwal et al. [21] developed an exciting system with a 500 core cluster which matches and reconstruct 3D scene from extremely large number of photographs collected from Flickr. The results shows that 3D reconstruction algorithms can scale with the size of the problem and the amount of available computation. Cao et al. [19] proposes to build tourism recommendation system using web photos with GPS information. Ji et al. [22] consider the problem of mining geographical information from web blogs and forums together with images. Quack et al. [16] develops a system for linking images from community photo collections to relevant Wikipedia articles.

In order to build the geo-driven Memex, it requires huge amount of geo-tagged information, which is still not practical given the fact that a significant portion of Flickr photos do not have related geographical information associated with them. Toward tackling this challenge, one line of research is devoted to estimating the geographical information from generic images. Hays and Efros [1] are among the first to consider the problem of estimating the location of a single image using its visual content alone. They collect millions of geo-tagged Flickr images. Using a comprehensive set of visual features, they employ nearest neighbor search in the reference set to estimate the location of the image. Motivated by Hays and Efros, Gallagher et al. [23] propose to combine the visual content of image and its related textual tags to estimate the geographical locations of images. Their results show that textual tags perform better than visual content and the combination of textual and visual information outperforms either alone. Cao et al. [10] also recognize the effectiveness of tags in estimating the geo-locations. They propose a novel model named logistic canonical correlation regression which explores the canonical correlations between geographical locations, visual content, and community tags. Unlike Hays and Efros [1], they argue that it is difficult to estimate the exact location at which a photo was taken and propose to estimate only the coarse location. Similarly, Crandall et al. [2] only estimate the approximate location of a test photo. Using SVM classifiers, a test image is geo-located by assigning it to the best cluster based on its visual content and annotations. In a recent work, Zhen et al. [5] built a web-scale landmark recognition engine named "Tour the world" using 20 million GPS-tagged photos of landmarks together with online tour guide web pages. The experiments demonstrate that the engine can achieve satisfactory recognition performance with high efficiency.

However, recognizing the location of a non-landmark image reliably still remains a challenging problem. Due to tremendous variety of views, subjects, activities, objects and environmental conditions, visual information based classifiers only perform comparable to chance for the non-landmark locations. In our approach, we propose a principled approach to discover "geographical clusters" to build classifiers. The use of geographical clusters benefits the problem of localization in two aspects: On the training stage, geographical clusters provide more training samples and hence lead to better recognition accuracy; on the testing stage, estimation of the most probable region for each query photo will be relatively easier than the estimation of exact GPS coordinates, while the information of geographical cluster will be good enough for trip planning and photo organization applications. Our approach is motivated by the success of latent structure learning in object detection [24] and max-margin clustering [25]. In the object detection approach proposed by Felzenszwalb et al. [24], the locations of object parts are unknown, and are treated as latent (hidden) variables in a learning framework called the *latent SVM*. A latent SVM is an extension of regular SVMs to handle latent variables—one that is is semi-convex and the training problem becomes convex once latent information is specified for the positive examples. This leads to an iterative training algorithm which alternates between fixing latent values for positive examples and optimizing the latent SVM objective function. Similar ideas can also be found in Xu et al. [25] which finds maximum margin hyperplanes through data. In our approach, we treat the geographical clusters of training images as hidden labels, and develop a principled learning method that recognizes the geographical clusters of images. More recently, Choi et al. [26] did a very exciting experiment and compared the performance of human and computers for the task of geo-placing. Although the overall performance of humans' multimodal video location estimation is better than current machine learning approaches, the difference is quite small.

3.3 Photos as Identities for Locations

Photos at specific locations exhibit unique image content properties. Figure 3.1 illustrates randomly sampled Flickr photos at Times Square, Brooklyn Bridge at New York and the Bund at Shanghai. We observe that the photos carry distinct information at different locations. At Times Square, the photos contain skyscrapers and billboards, providing a modern city view at the location. Photos at the Bund show the historic buildings and busy streets with lots of vehicles reflecting the characteristics of this location as the mixture of history and modern. Photos at the Brooklyn Bridge illustrate a comprehensive view of water front, bridge appearance from different view points and activities of ships/boats close to the bridge location.

Fig. 3.1 Sample photos at Times Square at New York, the Bund at Shanghai and Brooklyn Bridge at New York

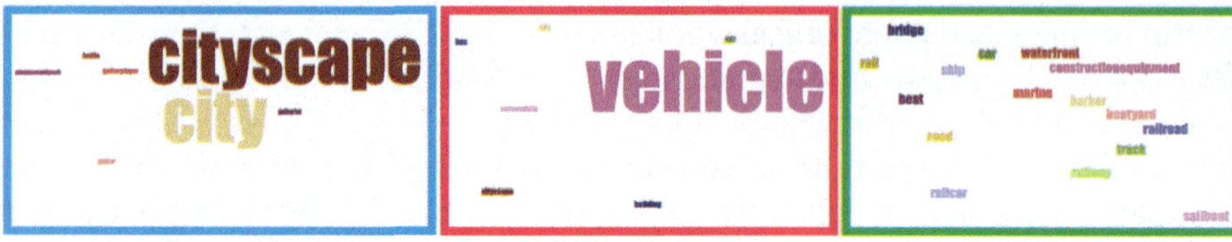

Time Square, New York   The Bund, Shanghai   Brooklyn Bridge, New York

Fig. 3.2 The most popular concepts discovered at Times Square at New York, the Bund at Shanghai and Brooklyn Bridge at New York respectively

We study the concepts related to the photos within 5 km of the interested location too. The most frequent concepts are showed in Fig. 3.2. Figure 3.2 demonstrates that photos indeed carry distinct visual content and the concepts related to the photos reflect the distinct visual characteristics too.

3.4 Estimate Where a Beach Photo Was Taken

In this section, we propose our algorithm for finding meaningful clusters from beach photos with geo-tags. These clusters can be then used to predict where a beach photo was taken.

3.4.1 Geo-location Regularized Clustering

Our method is inspired by the max-margin clustering (MMC) [25]. Naively applying MMC to our dataset is problematic, since MMC is a generic clustering algorithm and does not take into account of the geo-location information of the data. We propose an extension of MMC that clusters training images so that images in the same cluster are both visually similar and have close GPS locations.

We assume that we are given a training dataset with N instances. Each instance is in the form of (x_i, y_i), where x_i is the image feature of the i-th image, and y_i is its corresponding geo-location. Our goal is to cluster the training images into C groups in some sensible manner. We would also like to have a discriminative model that can assign an unseen image to one of the clusters. If we ignore the geo-location information y_i in the training data and only consider the image feature x_i, we can use standard clustering algorithms to partition the training images into C clusters. But now the challenge is how to incorporate the GPS location information into the clustering process.

Let us assume that the number of clusters is known as C. Clustering the training data is equivalent to assigning a binary vector z_i to each image x_i. Here z_i is a vector of length C, where its c-th component z_{ic} is defined as:

$$z_{ic} = \begin{cases} 1 & \text{if } x_i \text{ belongs to cluster } c \\ 0 & \text{otherwise} \end{cases} \quad (3.1)$$

Note that if z_i is observed on training data, we can use this information to learn a multiclass SVM classifier to assign the cluster membership of an unseen image by solving the following optimization problem:

$$\mathscr{P}(w^*) = \min_{w,\xi} \ \frac{1}{2}||w||^2 + C_1 \sum_i \xi_i \quad (3.2a)$$

$$\text{s.t. } w^\top \phi(x_i, z_i) - w^\top \phi(x_i, z) \geq \Delta(z_i, z) - \xi_i, \quad \forall i, \forall z \quad (3.2b)$$

where w is a vector of model parameter, $\phi(x_i, z_i)$ is a feature vector, ξ_i is the slack variable for handling soft margins in SVM classifiers. $\Delta(z_i, z)$ is a loss function that measures the penalty incurred on x_i by predicting its cluster to be z when the ground-truth cluster is z_i. Here we use the 0–1 loss defined as:

$$\Delta(z_i, z) = \begin{cases} 1 & \text{if } z_i \neq z \\ 0 & \text{otherwise} \end{cases} \quad (3.3)$$

Using our notation, $\mathscr{P}(w)$ is a function of w. For a given w (and corresponding $\{xi_i : \forall i\}$, the function value $\mathscr{P}(w)$ is obtained by plugging w and $\{\xi_i : \forall i\}$ in Eq. 3.2a.

An interesting fact about $\mathscr{P}(w)$ is that the optimal value of the objective $\mathscr{P}(w^*)$ has a specific meaning. In particular, it is the inverse of the SVM margin [25, 27]. In other words, let w^* be the solution to Eq. 3.2, we have $\mathscr{P}(w^*) = \frac{1}{\text{SVM margin}}$. This fact will turn out to be useful in the following when we need to consider z_i to be unobserved during training.

Now let us consider the case when $\{z_i : \forall i\}$ are not observed. First, we notice that the optimal solution $\mathscr{P}(w^*)$ in Eq. 3.2 now implicitly depends on $\{z_i : \forall i\}$. So from now on we will write $\mathscr{P}(w^*, \{z_i : \forall i\})$ to make it clearer. Note that $\mathscr{P}(w^*, \{z_i : \forall i\})$ is implicitly a function of $\{z_i\}$ as well. Now let us consider what might be an appropriate objective to optimize for choosing the labels of $\{z_i : \forall i\}$. As we mentioned above, for fixed values of $\{z_i : \forall i\}$, $\mathscr{P}(w^*, \{z_i : \forall i\})$ is the inverse of the SVM margin of the corresponding SVM. Then a natural objective is to find labels of $\{z_i : \forall i\}$ that minimize $\mathscr{P}(w^*, \{z_i : \forall i\})$, which in turn is equivalent to finding the maximum SVM margin. In other words, in order to find the optimal $\{z_i : \forall i\}$, we need to solve the following optimization problem:

$$\min_{\{z_i\}} \mathscr{P}(w^*, \{z_i : \forall i\}) \quad (3.4)$$

Note that in Eq. 3.4, the value of w^* implicitly depends on the values of $\{z_i\}$. For given values of $\{z_i\}$, the value of w^* is obtained by solving the optimization problem in Eq. 3.2.

Combing Eqs. 3.2 and 3.4, we get the following joint optimization problem that simultaneously solves for $\{z_i\}$ and w:

$$\mathscr{P}(w^*, \{z_i : \forall i\}) = \min_{w,\xi} \min_{\{z_i\}} \frac{1}{2}||w||^2 + C_1 \sum_i \xi_i \tag{3.5a}$$

$$\text{s.t. } w^\top \phi(x_i, z_i) - w^\top \phi(x_i, z) \geq \Delta(z_i, z) - \xi_i, \quad \forall i, \forall z \tag{3.5b}$$

$$\sum_c z_{ic} = 1, \quad \forall i \tag{3.5c}$$

$$z_{ic} \in \{0, 1\}, \quad \forall i, \quad \forall c \tag{3.5d}$$

The optimization problem in Eq. 3.5 tries to find $\{z_i : \forall i\}$ so that the resultant SVM achieves the maximum margin. The constraints in Eqs. 3.5c and 3.5d will make sure that z_i is a valid vector, i.e., it is a vector of all zeros with a single one for the corresponding cluster.

Unfortunately, without additional constraints or regularization, Eq. 3.5 has a degenerate solution—we can assign all training data to the same cluster and learn w to achieve arbitrarily large margin. In [25, 27], this problem is addressed by adding an additional constraint that tries to make sure that the clusters are balanced.

For our application, we have extra information (i.e., GPS locations) in addition to images. In the following, we will use this additional information to regularize Eq. 3.5. Intuitively, we would like the clusters to have the following property. If two images are close in terms of their geo-locations, they are more likely to be in the same cluster. One natural way to formalize this intuition is to solve the following optimization problem:

$$\mathscr{P}(w^*, \{z_i\} : \forall i) = \min_{w,\xi} \min_{\{z_i\}} \frac{1}{2}||w||^2 + C_1 \sum_i \xi_i \tag{3.6a}$$

$$+ C_2 \sum_i \sum_j (-|z_i - z_j| d_{ij}) \tag{3.6b}$$

$$\text{s.t. } w^\top \phi(x_i, z_i) - w^\top \phi(x_i, z) \geq \Delta(z_i, z) - \xi_i, \quad \forall i, \forall z \tag{3.6c}$$

$$\sum_c z_{ic} = 1, \quad \forall i \tag{3.6d}$$

$$z_{ic} \in \{0, 1\}, \quad \forall i, \quad \forall c \tag{3.6e}$$

where d_{ij} is the distance of two images x_i and x_j in terms of their geo-locations (which can be obtained from y_i and y_j).

Note that $|z_i - z_j| = 0$ if x_i and x_j are in the same cluster, $|z_i - z_j| = 1$ if they are not in the same cluster. So Eq. 3.6b will try to make the distance (in terms of GPS locations) between images in different clusters to be large.

The optimization problem in Eq. 3.6a can be solved using an iterative approach:

- Fix $\{z_i\}_{i=1}^N$, optimize over w and ξ.

$$\min_{w,\xi} \frac{1}{2}||w||^2 + C_1 \sum_i \xi_i \tag{3.7a}$$

$$\text{s.t. } w^\top \phi(x_i, z_i) - w^\top \phi(x_i, z) \geq \Delta(z_i, z) - \xi_i, \quad \forall i, \forall z \tag{3.7b}$$

- Fix w and ξ, optimize over $\{z_i\}_{i=1}^N$.

$$\min_{\{z_i\}} \sum_i \sum_j (-|z_i - z_j| d_{ij}) \tag{3.8a}$$

$$\text{s.t. } w^\top \phi(x_i, z_i) - w^\top \phi(x_i, z) \geq \Delta(z_i, z) - \xi_i, \quad \forall i, \forall z \tag{3.8b}$$

$$\sum_c z_{ic} = 1, \quad \forall i \tag{3.8c}$$

$$z_{ic} \in \{0, 1\}, \quad \forall i, \quad \forall c \tag{3.8d}$$

The first step (Eq. 3.7) of this iterative approach is straightforward since it is equivalent to solving a standard multiclass SVM problem. The second step (Eq. 3.8) is more challenging. It involves solving a combinatorial problem. One can shows that it can be reformulated as an integer program as follows:

$$\min_{\{z_i\}} \sum_i \sum_j \sum_c (-\gamma_{ijc} d_{ij}) \tag{3.9a}$$

$$\text{s.t. } w^\top \phi(x_i, z_i) - w^\top \phi(x_i, z) \geq \Delta(z_i, z) - \xi_i, \quad \forall i, \forall z \tag{3.9b}$$

$$-\gamma_{ijc} \leq z_{ic} - z_{jc} \leq \gamma_{ijc}, \quad \forall i, \quad \forall j, \quad \forall c \tag{3.9c}$$

$$\sum_c z_{ic} = 1, \quad \forall i \tag{3.9d}$$

$$z_{ic} \in \{0, 1\}, \quad \forall i, \quad \forall c \tag{3.9e}$$

If we relax the integral constraint (Eq. 3.9e) to a linear constraint $0 \leq z_{ic} \leq 1$, the optimization problem in Eq. 3.9 becomes a standard linear program and efficient solvers (e.g., Simplex algorithm) exist. However, the linear program is still too large for any off-the-shelf LP solvers. In the following section, we introduce a new formulation that is more amenable to efficient algorithms.

3.4.2 More Efficient Formulation

The main observation that enables our new formulation is the following. Suppose we know the cluster centers $\{g_c\}_{c=1}^C$ (in term of geo-locations), a natural way to solve our problem is to use the following optimization:

$$\mathcal{P}(w^*, \{z_i : \forall i\}) = \min_{w,\xi} \min_{\{z_i\}} \frac{1}{2}||w||^2 + C_1 \sum_i \xi_i \tag{3.10a}$$

$$+ C_2 \sum_i \sum_c (z_{ic}||y_i - g_c||^2) \tag{3.10b}$$

$$\text{s.t. } w^\top \phi(x_i, z_i) - w^\top \phi(x_i, z) \geq \Delta(z_i, z) - \xi_i, \quad \forall i, \forall z \tag{3.10c}$$

$$\sum_c z_{ic} = 1, \quad \forall i \tag{3.10d}$$

$$z_{ic} \in \{0, 1\}, \quad \forall i, \quad \forall c \tag{3.10e}$$

Note that Eq. 3.10b computes the distance (in term of geo-locations) between images and their corresponding cluster centers. When those cluster centers are known, the optimal clustering is obtained by choosing cluster membership that minimizes this distance (i.e., minimizing over $\{z_i : \forall i\}$).

Now the challenge is that the cluster centers $\{g_c : \forall c\}$ are also unknown. Using the same reasoning in Sect. 3.4.1, we can treat the cluster centers as yet another set of latent variables in the formulation. Similarly, we propose the following optimization problem for jointly learning w, $\{z_i : \forall i\}$ and $\{g_c : \forall c\}$

$$\mathcal{P}(w^*, \{z_i : \forall i\}, \{g_c : \forall c\}) = \min_{w,\xi} \min_{\{z_i\}} \min_{\{g_c\}} \frac{1}{2}||w||^2 + C_1 \sum_i \xi_i \tag{3.11a}$$

$$+ C_2 \sum_i \sum_c (z_{ic}||y_i - g_c||^2) \tag{3.11b}$$

$$\text{s.t. } w^\top \phi(x_i, z_i) - w^\top \phi(x_i, z) \geq \Delta(z_i, z) - \xi_i, \quad \forall i, \forall z \tag{3.11c}$$

$$\sum_c z_{ic} = 1, \quad \forall i \tag{3.11d}$$

$$z_{ic} \in \{0, 1\}, \quad \forall i, \quad \forall c \tag{3.11e}$$

and use the following iterative method to solve it:

- Fix $\{z_i\}_{i=1}^N$ and $\{g_c\}_{c=1}^C$, optimize over w and ξ: this step is equivalent to Eq. 1.7 and can be solved with standard SVM solvers. We use liblinear [28] for it.
- Fix w, ξ and $\{z_i\}_{i=1}^N$, optimize over $\{g_c\}_{c=1}^C$: this step involves solving the following optimization problem:

$$\min_{\{g_c\}} \sum_i \sum_c z_{ic}||y_i - g_c||^2 \tag{3.12}$$

To derive the optimal solution for g_c, let us define $\mathcal{L} = \sum_i \sum_c z_{ic}||y_i - g_c||^2$. Note that $\mathcal{L}$ is convex in g_c. By setting the derivative $\partial \mathcal{L}/\partial g_c$ to 0, we can derivate an analytical solution for the optimal g_c^*:

$$g_c^* = \frac{\sum_i z_{ic} y_i}{\sum_i z_{ic}} = \frac{\sum_{k \in \text{cluster(c)}} y_k}{|\text{cluster(c)}|} \tag{3.13}$$

where cluster(c) is the set of data points in the c-th cluster. In other words, the optimal value of the c-th cluster center g_c is the average of the geo-locations of images assigned (based on $\{z_i\}$) to this cluster.

- Fix w, ξ and $\{g_c\}_{c=1}^C$, optimize over $\{z_i\}_{i=1}^N$: this is equivalent to the following optimization problem:

$$\min_{\{z_i\}} \sum_i \sum_c z_{ic} ||y_i - g_c||^2 \tag{3.14a}$$

$$\text{s.t.} \sum_c z_{ic} = 1, \quad \forall i \tag{3.14b}$$

$$z_{ic} \in \{0, 1\}, \quad \forall i, \quad \forall c \tag{3.14c}$$

It is easy to show that we can independently solve z_i for every i in Eq. 3.14. For a fixed i, we need to solve the following problem:

$$\min_{z_{ic}} \sum_c z_{ic} ||y_i - g_c||^2 \tag{3.15a}$$

$$\text{s.t.} \sum_c z_{ic} = 1 \tag{3.15b}$$

$$z_{ic} \in \{0, 1\}, \quad \forall c \tag{3.15c}$$

Let us define $c^* = \arg\min_c ||y_i - g_c||^2$, i.e., c^* is the cluster closest to y_i. Then it is easy to show that the solution of Eq. 3.15 is $z_{ic^*} = 1$ and $z_{ic} = 0 : \forall c \neq c^*$. In other words, we simply assign z_i according to its closest cluster.

3.4.3 Dataset

To demonstrate the effectiveness of our approach, we apply it to predict where a beach photo was taken. A lot of beach photos with GPS tags are available on online photo sharing web sites. We test our approach on a dataset [6] containing 35k images downloaded from Flickr with the tags of "beach" or "coast". Each photo in this dataset is labeled by a two-dimensional GPS coordinate vector. 1.1K photos of this dataset is hand labeled for test with typical beach scenes, while removing photos with nontypical scenes such as personal portraits, and indoor pictures. The testing photos are evenly distributed in the world scale and there is no overlap between training and testing sets.

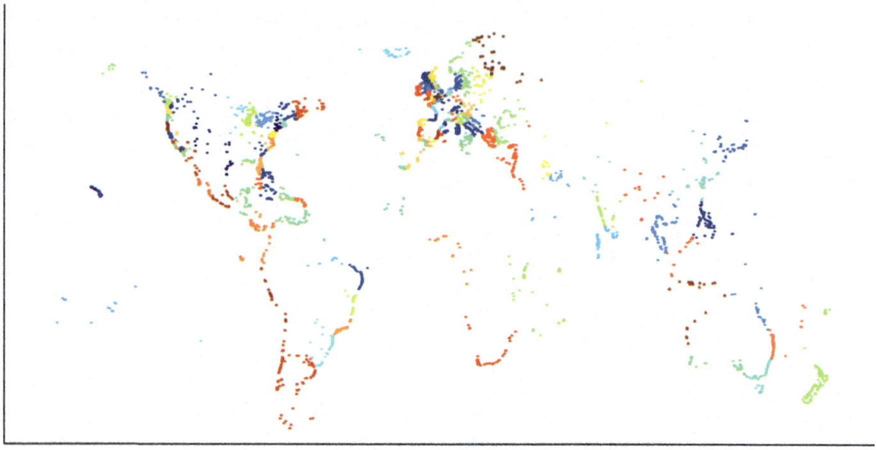

Fig. 3.3 Visualization of clustering training images using our method. Each *color* represents a different cluster

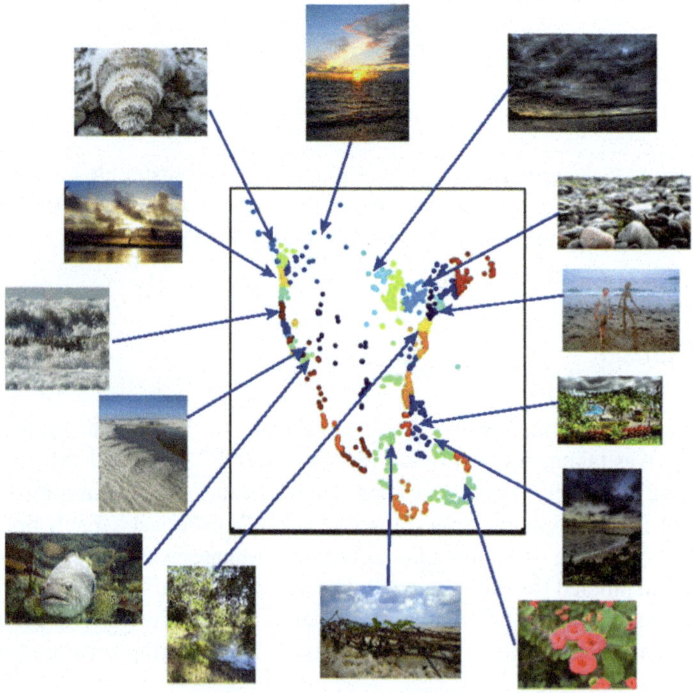

Fig. 3.4 Visualization of representative images for North America

3.4.4 Experiment Result

We use 34,558 images from the beach photo dataset [6] for training and 1185 images for testing. We use GIST features to represent the visual content of the images.

To illustrate the learned geographical clusters, we visualize them in Fig. 3.3. We plot the distribution of training images and their clusters in different colors. As we can observe in Fig. 3.3, photos are taken at geographically close locations are clustered together and labeled in the same color. The accurate clustering performance is largely attribute to the visual consistency of the photos at specific locations as well as the strength of the latent SVM for learning this consistency based upon the visual content representation.

In Figs. 3.4, 3.5, 3.6, 3.7, 3.8 and 3.9, we visualize some representative images in sample clusters. While photos taken at the geographically close locations are assigned to same clusters, photos sampled from different clusters exhibit diverse visual properties across different clusters. Take Fig. 3.4 as an example, the clusters of beaches reflect different topics at these beaches. Some of the beaches are more famous for their wild sea creatures and plants while other beaches are more popular for sunset/sunrise views. Such information is carried by the photos taken at these

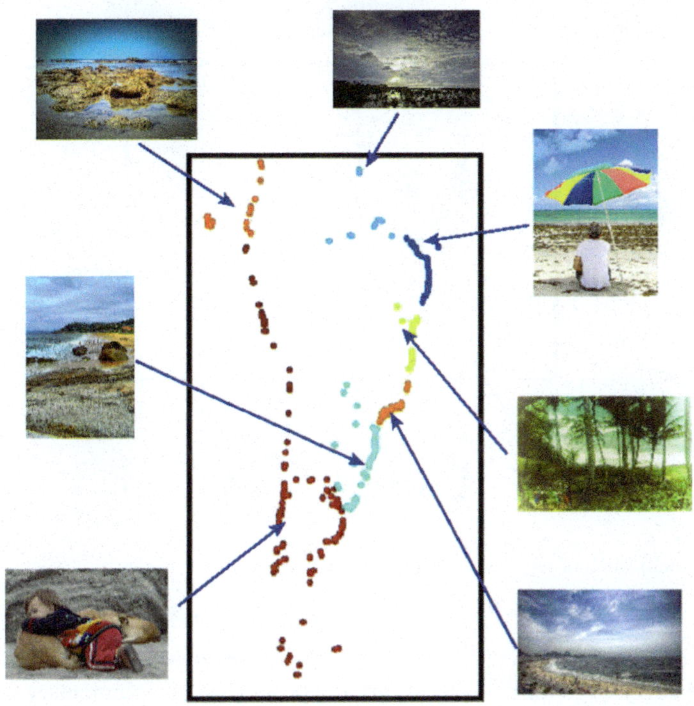

Fig. 3.5 Visualization of representative images for South America

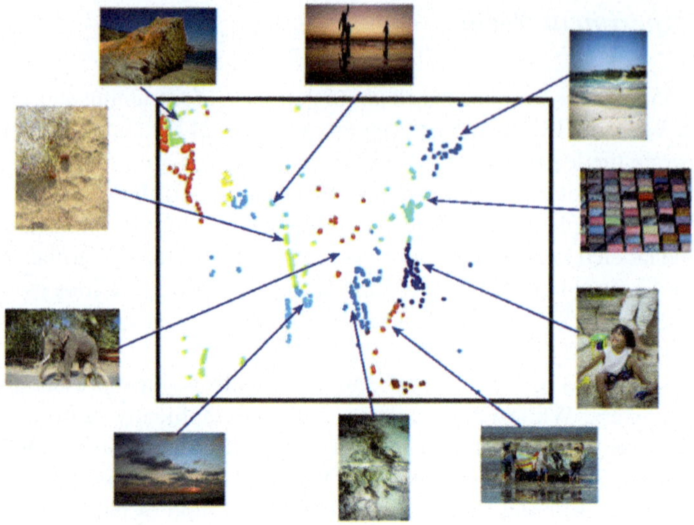

Fig. 3.6 Visualization of representative images for Asia

Fig. 3.7 Visualization of representative images for Europe

locations and uploaded to photo sharing web sites, which serves as the wisdom of the crowd knowledge for these locations and provides a complementary description to the locations for people who are interested in them.

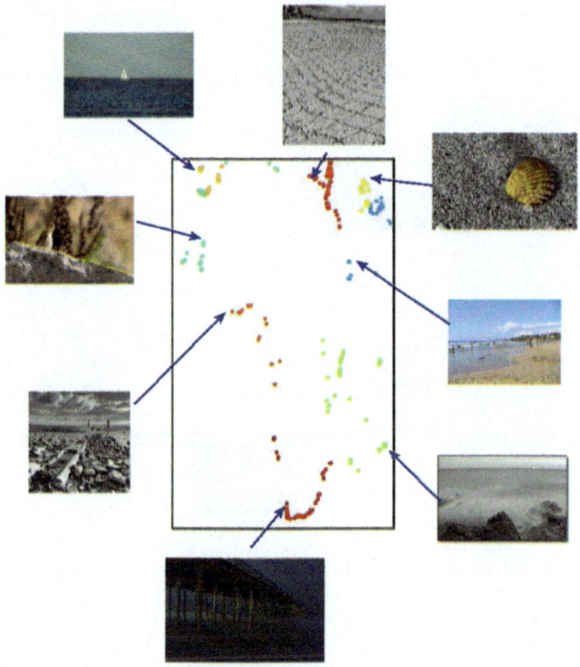

Fig. 3.8 Visualization of representative images for Africa

We also evaluate our approach quantitatively by comparing the recognition results of our method, random guess and the nearest neighbor method [1]. The accuracy reflects whether a testing photo is classified into the correct clusters. It is measured by the number of correctly classified test photos over the total number of test photos.

As shown in Fig. 3.10a, our system by learning from visual features works $19.3/0.7 = 27$ times better in recognition accuracy than random guess, and also demonstrates significantly better performance than the nearest neighbor method. Another intuitive way to evaluate our system is to check how far is the estimated location from its true GPS coordinates. We use the center of geographical cluster as the estimated location and compute the distance between estimation and GPS coordinates. As shown in Fig. 3.10b, about 17.5 % of testing photos are localized within a GPS coordinate neighborhood of 5°. Again, our approach shows a remarkable improvement over the nearest-neighbor-based approach [1].

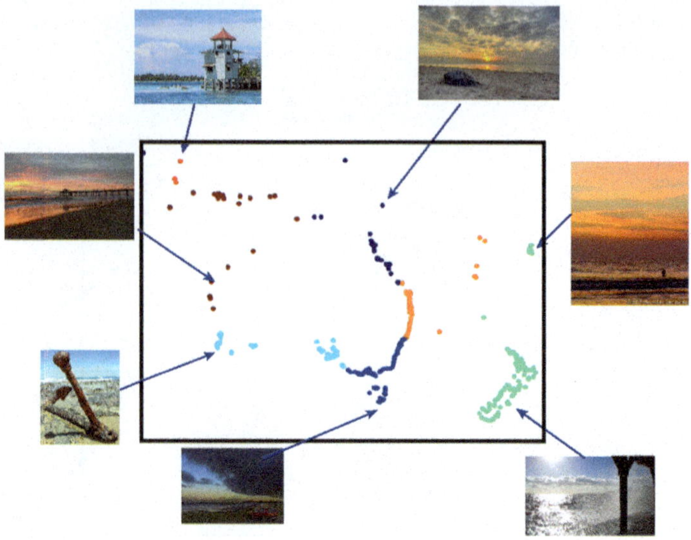

Fig. 3.9 Visualization of representative images for Australia

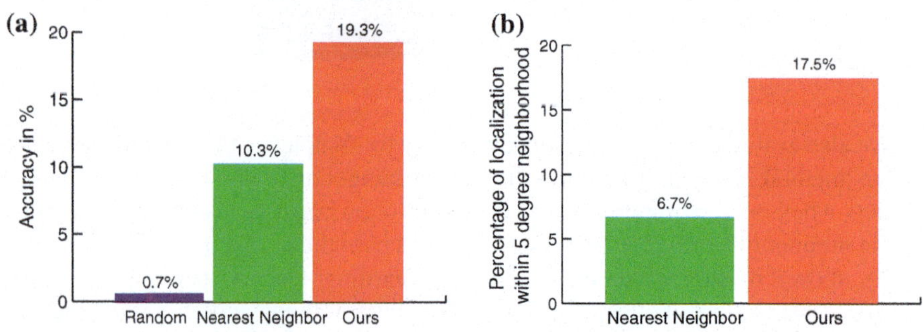

Fig. 3.10 Recognition (**a**) and localization accuracy (within $\leq 5^\circ$ neighborhood) (**b**) of different methods including nearest neighbor approach by [1] and our approach

3.5 Conclusion

The large-scale photos shared on online photo web sites make it possible to learn representative information from the crowd of wisdom. We have introduced a new framework for geographical cluster estimation by learning from the online photos. We develop a principled machine-learning model to estimate geographical locations of photos by modeling the relationship between location and the photo content. Our approach treats the geographical cluster of an image as a latent variable and model the relationship using the latent SVM approach. An efficient solver is designed for learning the latent geographical clusters. We show striking results on predicting the coarse location of where the photos were taken by recognizing different beaches using the beach dataset as validation. In the future, we would like to scale it to estimate other coarse locations over the world. We would like to incorporate more descriptive image representation such as that learned from convolutional neural network, which might capture the visual characteristics at locations better. We have shown that visual content is effective for predicting the geographical clusters. We would be interested to explore the potential of combining visual content with other multimodality data such as user tags, title of the photo, and social activities related to the photo.

References

1. Hays J, Efros AA (2008) Im2gps: estimating geographic information from a single image. In: IEEE conference on computer vision and pattern recognition
2. Crandall D, Backstrom L, Huttenlocher D, Kleinberg J (2009) Mapping the world's photos. In: International conference on world wide web, pp 761–770
3. Chen W, Battestini A, Gelfand N, Setlur V (2009) Visual summaries of popular landmarks from community photo collections. In: ACM international conference on Multimedia, pp 789–792
4. Yin Z, Cao L, Han J, Zhai C, Huang T (2011) Geographical topic discovery and comparison. In: Proceedings of the 20th international conference on world wide web. ACM, pp 247–256
5. Zheng Y, Zhao M, Song Y, Adam H, Buddemeier U, Bissacco A, Brucher F, Chua T, Neven H (2009) Tour the World: building a web-scale landmark recognition engine. In: IEEE conference on computer vision and pattern recognition
6. Cao L, Smith J, Wen Z, Yin Z, Jin X, Han J (2012) BlueFinder: estimate where a beach photo was taken. In: WWW
7. Wang Y, Cao L (2013) Discovering latent clusters from geotagged beach images. In: Advances in multimedia modeling. Springer, pp 133–142
8. Naaman M, Song Y, Paepcke A, Garcia-Molina H (2004) Automatic organization for digital photographs with geographic coordinates. In: International conference on digital libraries, vol 7. pp 53–62
9. Agarwal M, Konolige K (2006) Real-time localization in outdoor environments using stereo vision and inexpensive GPS. In: International conference on pattern recognition
10. Cao L, Yu J, Luo J, Huang T (2009) Enhancing semantic and geographic annotation of web images via logistic canonical correlation regression. In: Proceedings of the seventeen ACM international conference on multimedia, pp 125–134
11. Yu J, Luo J (2008) Leveraging probabilistic season and location context models for scene understanding. In: International conference on content-based image and video retrieval, pp 169–178
12. Joshi D, Luo J (2008) Inferring generic places based on visual content and bag of geotags. In: ACM conference on content-based image and video retrieval
13. Yuan J, Luo J, Wu Y (2008) Mining compositional features for boosting. In: IEEE conference on computer vision and pattern recognition
14. Kennedy L, Naaman M, Ahern S, Nair R, Rattenbury T (2007) How flickr helps us make sense of the world: context and content in community-contributed media collections. In: ACM conference on multimedia
15. Naaman M (2005) Leveraging geo-referenced digital photographs. PhD thesis, Stanford University
16. Quack T, Leibe B, Van Gool L (2008) World-scale mining of objects and events from community photo collections. In: ACM conference on image and video retrieval, pp 47–56
17. Luo J, Yu J, Joshi D, Hao W (2008) Event recognition: viewing the world with a third eye. In: ACM international conference on multimedia, pp 1071–1080

18. Schindler G, Krishnamurthy P, Lublinerman R, Liu Y, Dellaert F (2008) Detecting and matching repeated patterns for automatic geo-tagging in urban environments. In: IEEE conference on computer vision and pattern recognition
19. Cao L, Luo J, Gallagher A, Jin X, Han J, Huang T (2010) A worldwide tourism recommendation system based on geotagged web photos. In: International conference on acoustics, speech, and signal processing (ICASSP)
20. Bush V (1945) As we may think. The Atlantic Monthly
21. Agarwal S, Snavely N, Simon I, Seitz SM, Szeliski R (2009) Building rome in a day. In: International conference on computer vision
22. Ji R, Xie X, Yao H, Ma WY (2009) Mining city landmarks from blogs by graph modeling. In: ACM Multimedia, pp 105–114
23. Gallagher A, Joshi D, Yu J, Luo J (2009) Geo-location inference from image content and user tags. In: Workshop on internet vision
24. Felzenszwalb PF, Girshick RB, McAllester D, Ramanan D (2010) Object detection with discriminatively trained part based models. IEEE Trans Pattern Anal Mach Intell 32:1672–1645
25. Xu L, Neufeldand J, Larson B, Schuurmans D (2005) Maximum margin clustering. In Saul LK, Weiss Y, Bottou L (eds) Advances in neural information processing systems, vol 17. MIT Press, Cambridge, MA, pp 1537–1544
26. Choi J, Lei H, Ekambaram V, Kelm P, Gottlieb L, Sikora T, Ramchandran K, Friedland G (2013) Human vs machine: establishing a human baseline for multimodal location estimation. In: Proceedings of the 21st ACM international conference on multimedia, MM '13 pp 867–876
27. Xu L, Wilkinson D, Southey F, Schuurmans D (2006) Discriminative unsupervised learning of structured predictors. In: Proceedings of the 23th international conference on machine learning
28. Fan RE, Chang KW, Hsieh CJ, Wang XR, Lin CJ (2008) LIBLINEAR: a library for large linear classification. J Mach Learn Res

Chapter 4
Cross-View Image Geo-localization

Tsung-Yi Lin, Serge Belongie and James Hays

Abstract The recent availability of large amounts of geo-tagged imagery has inspired a number of data-driven solutions to the image geo-localization problem. Existing approaches predict the location of a query image by matching it to a database of geo-referenced photographs. While there are many geo-tagged images available on photo sharing and Street View sites, most are clustered around landmarks and urban areas. The vast majority of the Earth's land area has no ground-level reference photos available, which limits the applicability of all existing image geo-localization methods. On the other hand, there is no shortage of visual and geographic data that densely covers the Earth—we examine overhead imagery and land cover survey data—but the relationship between this data and ground-level query photographs is complex. In this chapter, we introduce a cross-view feature translation approach to greatly extend the reach of image geo-localization methods. We can often localize a query even if it has no corresponding ground-level images in the database. A key idea is to learn a mapping from ground-level appearance to overhead appearance and land cover attributes. This relationship is learned from sparsely available geo-tagged ground-level images and the corresponding aerial and land cover data at those locations. We perform experiments over a 1135 km^2 region containing a variety of scenes and land cover types. For each query, our algorithm produces a probability density over the region of interest.

T.-Y. Lin · S. Belongie
Cornell University, Ithaca, NY, USA
e-mail: tl483@cornell.edu

S. Belongie
e-mail: sjb344@cornell.edu

J. Hays (✉)
Brown University, Providence, RI, USA
e-mail: hays@cs.brown.edu

A.R. Zamir et al. (eds.), *Large-Scale Visual Geo-Localization*,
Advances in Computer Vision and Pattern Recognition,
DOI 10.1007/978-3-319-25781-5_4

4.1 Introduction

Consider the photos in Fig. 4.1. How can we determine where they were taken? One might try to use a image-based search (e.g., Google Images) to retrieve visually similar images. This will only solve our problem if we can find an instance-level match with a known location. This approach will likely succeed for famous landmarks, but not for the unremarkable scenes in Fig. 4.1. If instead of instance-level matching we match based on scene-level features, as in im2gps [9], we can sometimes get a coarse geo-location based on the distribution of similar scenes (e.g., this photo looks like many others in the American Southwest). Is this our best hope for geo-localizing photographs? Fortunately, there are no many other types of image understanding that can help constrain the location of a photograph. For instance, in the popular "View from Your Window" contest,[1] humans utilize a litany of geo-informative visual evidence related to architecture, climate, road markings, style of dress, etc. However, recognizing these properties and then mapping them to geographic locations is at the limit of human ability and beyond current computational techniques.

In this chapter, we expand the types of evidence used to estimate photo location by exploiting two previously unused geographic data sets—overhead appearance and land cover survey data. For each of these datasets, we must learn the relationship between ground-level views and the data. These datasets have the benefits of being (1) densely available for nearly all of the Earth and (2) rich enough such that the mapping from ground-level appearance is sometimes unambiguous. For instance, a human might be able to verify that a putative match is correct (although it is infeasible for a human to do exhaustive searches manually).

The im2gps approach of [9] was the first to predict image geo-location by matching visual appearance. Following this thread, several approaches posed the geo-localization task as an image retrieval problem, focusing on landmark images on the Internet [5, 13, 27]. Street View imagery has also been used for geo-localization [4, 20, 22]. Agarwal et al. [1] developed an approach to build 3D point cloud models of famous buildings automatically from publicly available images. Many techniques leverage such 3D information for more efficient and accurate localization [10, 14, 15, 21]. Note that all the above-mentioned approaches assume that at least one training image is taken from a similar vantage point as that of the query. A query image in an unsampled location has no hope of being accurately localized. Recently, Baatz et al. [2] presented an algorithm that does not require ground-level training images. They use a digital elevation model to synthesize 3D surfaces in mountainous terrain and match the contour of mountains extracted from ground-level images to the synthetic ridgelines. The paper extends the solution space of existing geo-localization methods from popular landmarks and big cities to mountainous regions. However, most photographs do not contain mountains. For our region of interest in this chapter, reasoning about terrain shape would not be informative.

[1] http://dish.andrewsullivan.com/vfyw-contest/.

Fig. 4.1 The above ground-level images were captured around Charleston, South Carolina within the region indicated in this satellite image. What can we determine about their geo-locations? In this work, we tackle the case for which ground-level training imagery at the corresponding locations is not available

Matching ground-level photos to aerial imagery and geographic survey data has remained unexplored despite these features being easily available and densely distributed on the surface of the earth. One major issue preventing the use of these features is that the mapping between them is very complex. For instance, the visual appearance of a building in ground-level versus an aerial view is very different due the extremely wide baseline, varying focal lengths, nonplanarity of the scene, different lighting conditions, weather, and season, and mismatched image quality.

Our approach takes inspiration from recent works in cross-view data retrieval that tackle problems such as webcam localization with satellite weather imagery [11], cross-view action recognition [16], and image–text retrieval [19, 23]. These approaches achieve cross-view retrieval by learning the co-occurrence of features in different views. The training data consists of corresponding features that describe the same content in multiple views and these techniques often involve solving a generalized eigenvalue problem to find the projection basis that maximizes the cross-view feature correlation. One example is kernelized canonical correlation analysis [7] which requires solving the eigenvalue problem of training kernel matrix. However, this approach is not scalable to very large datasets.

Inspired by [8, 18, 26], we propose a data-driven framework for cross-view image matching. To this end, we collected a new dataset that consists of ground-level images, aerial images, and land cover attribute images as the training data. At test time, we find the best-matching ground-level scenes and use their corresponding aerial and land cover features to guide the search in those domains. For example, a ground-level photo of a golf course by the water will hopefully match to similar scenes with known location, and the corresponding overhead views and land cover distributions for those scenes will be used to train a overhead and land cover classifier to predict the query scene's location. The contributions to our approach are that (1) we are able to geo-localize images without corresponding ground-level images in the database; (2) we build a new dataset for cross-view matching and leverage aerial imagery and land cover attributes; (3) we produce a dense prediction score over a substantial region (33.7 km $\times$ 33.7 km).

4.2 Dataset

For the experiments in this chapter we examine a 33.7 km $\times$ 33.7 km region around Charleston, South Carolina. This region exhibits great scene variety (urban, agricultural, forest, marsh, beach, etc.) as shown in Fig. 4.1. We take one ground-level image as query and estimate its location using a database of aerial images and land cover attributes. Our training data consists of triplets of ground-level images, aerial images, and attribute maps; see Fig. 4.4. To justify the need for this "cross-view" approach, Fig. 4.2a shows the ground-level image distribution in our training data. Because the ground-level images are sparsely distributed over the region of interest, ground-level retrieval methods in the vein of im2gps will fail when no nearby training images are close to query images as shown in Fig. 4.2b. In our database, 98.76 % of the space is not covered by any ground-level image if we conservatively assume that each image occupies 133 m x 133 m field of view. In such cases, the proposed method can leverage co-occurrence information in training triplets to match "isolated images" to aerial and attribute images in the map database. Note that the proposed method can be generalized nationwide or globally because the training data we use is widely available.

4.2.1 Ground-Level Imagery

We downloaded 6756 ground-level images from Panoramio. We do not apply any filters to the resulting images, even though many are likely impossible to geo-locate (e.g., close up views of flowers). The great scene variety of photographs "in the wild" makes the dataset extremely challenging for geo-localization.

4.2.2 Aerial Imagery

The satellite imagery is downloaded from Bing Maps. Each 256 $\times$ 256 image covers a 133 m $\times$ 133 m region. The image resolution is 0.52 m/pixel. We collect a total of 182,988 images for the map database. For each ground-level training example with

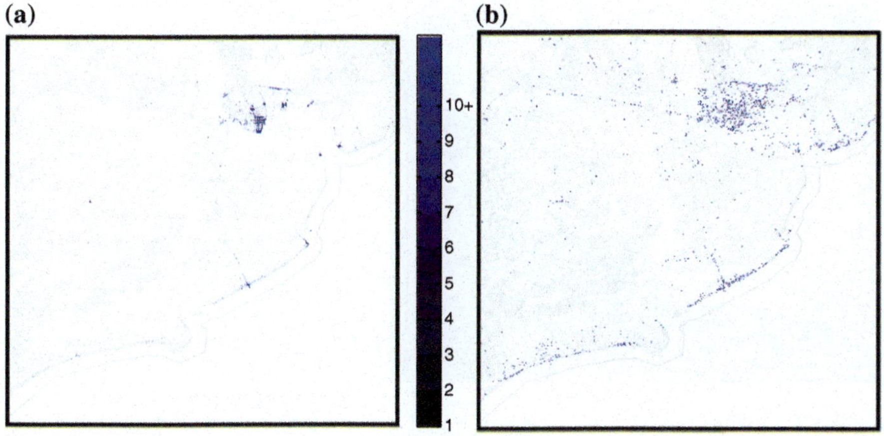

Fig. 4.2 The distribution of ground-level images from Panoramio within the region indicated in Fig. 4.1. **a** Most of the ground-level images are concentrated in a few highly populated areas for which ground-to-ground level matching will suffice. **b** Each point marks an "isolated" image for which no other ground-level image is within the same 133 m x 133 m grid cell. Such images must be localized using cross-view (ground-to-aerial) image matching

known location we crop the corresponding aerial imagery centered at that location. Unfortunately, we do not know the orientation of our ground-level training images so much of the content in the aerial crop which may not be represented in the ground-level photograph. For example, in Fig. 4.4 the ground-level photograph does not show any water but the aerial image does. This is not necessarily problematic—the system can correctly learn that this type of beach house scene should have water in its aerial view. However, since we cannot hope to discriminate photos based on orientation (a West-facing beach and an East-facing beach can look the same), we would prefer to be invariant to orientation. To this end, we rotate all aerial images such that their dominant orientation (the orientation with maximum gradient magnitude) is the same. This is analogous to keypoint feature representations like SIFT. When searching for matches at test time, we use a sliding window with half stride (67 m) on the region of interest.

4.2.3 *Land Cover Attribute Imagery*

The National Gap Analysis Program (GAP) conducted a nationwide geology survey to produce the USGS GAP Land Cover Data Set.[2] Figure 4.3 shows a view of the dataset for our region of interest. The dataset contains two hierarchical levels of classes including 8 general classes (e.g., Developed & Other Human Use, Shrubland & Grassland, etc.) and 590 land use subclasses (e.g., Developed/Medium Intensity,

[2]http://gapanalysis.usgs.gov/gaplandcover/.

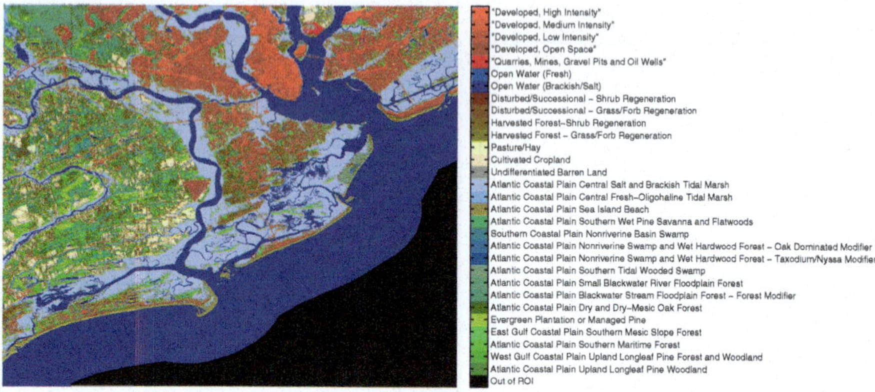

Fig. 4.3 Snapshot of the USGS GAP land cover attributes (available nationwide) for Charleston, SC

Fig. 4.4 Example "training triplet:" ground-level image, its corresponding aerial image and land cover attribute map

Salt and Brackish Tidal Marsh, etc.) Each pixel in the attribute map is assigned to one general class and one subclass. The land cover attribute resolution is 30 m/pixel. We crop 5 × 5 images for use in the training triplets and the scanning window search. We do not rotate the land cover maps as we did the aerial images because we will only use histogram representations which are already rotation invariant (Fig. 4.4).

4.3 Geo-localization via Cross-View Matching

In this section we discuss feature representation for ground, aerial, and attribute images and introduce several methods for cross-view image geo-location.

4.3.1 *Image Features and Kernels*

Ground and Aerial Image: We represent each ground image and aerial tile using four features: HoG [6], self-similarity [24], gist [17], and color histograms. The use of these features to represent ground-level photographs is motivated by their usage in scene classification tasks [25]. Even though these features were not designed to represent overhead imagery, we use them for that purpose because (1) it permits a simple "direct matching" baseline for cross-view matching described below and (2) aerial image representations have not been well explored in the literature, so there is no "standard" set of features to use for this domain. HoG and self-similarity local features are densely sampled, quantized, and represented with a spatial pyramid [12]. When computing image similarity, we use a histogram intersection kernel for HoG pyramids, χ^2 kernel for self-similarity pyramids and color histograms, and RBF kernel for gist descriptors. We combine these four feature kernels with equal weights to form the aerial feature kernel we use for learning and prediction in the following sections. All features and kernels are computed by code from [25].

Land Cover Attribute Image: Each ground and aerial image has a corresponding 5×5 attribute image. From this attribute image we build histograms for the general classes and subclasses and then concatenate them to form a multinomial attribute feature. We compare attribute features with a χ^2 kernel.

Our geo-location approach relies on translating from ground-level features to aerial and/or land cover features. The aerial and attribute features have distinct feature dimension, sparsity, and discriminative power. In the experiments section, we compare geo-location predictions based on aerial features, attribute features, and combinations of both.

4.3.2 Matching

In this section, we introduce two noval data-driven approaches—Data-driven Feature Averaging and Discriminative Translation—for cross-view geo-localization. We introduce three baseline algorithms for comparison: (1) im2gps [9], which can only match images within the ground-level view, (2) direct match, and (3) kernelized canonical correlation analysis. In the following section, x denotes the query image. The bold-faced $\mathbf{x}$ denotes ground-level images and $\mathbf{y}$ denotes aerial/attribute images of corresponding training triplets. $\mathbf{y}_{\text{map}}$ denotes the aerial/attributes images in the map database to be searched at test time. $k(.,.)$ denotes the kernel function.

im2gps: im2gps geo-localizes a query image by guessing the locations of the top k scene matches. im2gps or any image retrieval-based method makes no use of aerial and attribute information and can only geo-locate query images in locations with ground-level training imagery. We compute $k(x, \mathbf{x})$ to measure the similarity of the query image and the training images.

Direct Match (DM): In order to localize spatially isolated photographs, we need to match across views from ground level to overhead. The simplest method to match ground-level images to aerial images is just to match the same features with no translation, i.e., to assume that ground-level appearance and overhead appearance are similar. This is not universally true, but a beach from ground level and overhead do share some rough texture similarities, as do other scene types, so this baseline performs slightly better than chance. We compute similarity with:

$$\mathrm{sim}_{\mathrm{dm}} = \mathrm{k}(\mathrm{x}, \mathbf{y}_{\mathrm{map}}) \tag{4.1}$$

The direct match baseline cannot be used for ground-to-attribute matching, because those features sets are distinct.

Kernelized Canonical Correlation Analysis (KCCA): KCCA is a tool to learn the basis along the direction where features in different views are maximally correlated:

$$\max_{w_x \neq 0, w_y \neq 0} \frac{w_x^\top \Sigma_{xy} w_y}{\sqrt{w_x^\top \Sigma_{xx} w_x} \sqrt{w_y^\top \Sigma_{yy} w_y}} \tag{4.2}$$

where Σ_{xx} and Σ_{yy} represent the covariance matrices for ground-level feature and aerial/attribute feature in training triplets. Σ_{xy} represents the cross-covariance matrix between them. The optimization can be posed as a generalized eigenvalue problem and solved as a regular eigenvalue problem. In KCCA, we use the "kernel trick" to represent a basis as a linear combination of training examples by $w_{xi}^T = \alpha_i^\top \phi(\mathbf{x})$ and $w_{yi}^T = \beta_i^\top \phi(\mathbf{y})$. Here, w_{xi} indicates the basis of ith principle axis in high-dimensional space defined by mapping function $\phi(\mathbf{x})$. $\phi(\mathbf{x})$ is a Nxd data matrix where $\phi()$ maps data x into high-dimensional space. The vectors α_i and β_i are learned from training data by eigenvalue analysis.

We compute the cross-view correlation score of query and training images by summing over the correlation scores on the top d bases:

$$\mathrm{sim}_{\mathrm{kcca}} = \sum_{\mathrm{i}=1}^{\mathrm{d}} \alpha_{\mathrm{i}}^\top \mathrm{k}(\mathrm{x}, \mathbf{x}) \beta_{\mathrm{i}}^\top \mathrm{k}(\mathbf{y}, \mathbf{y}_{\mathrm{map}}) \tag{4.3}$$

We use the cross-view correlation as the matching score. KCCA has several disadvantages for large-scale cross-view matching. First, we need to compute the singular value decomposition for a nonsparse kernel matrix to solve the eigenvalue problem. The dimension of kernel matrix grows with the number of training samples. This makes the solution infeasible as training data increases. Second, KCCA assumes one-to-one correspondence between two views. But in our problem, it is common to have multiple ground-level images taken at the same geo-location. In this case, we need to throw away some training data to enforce the one-to-one correspondence.

Data-driven Feature Averaging (AVG): When images are similar in the ground-level view they also tend to have similar overhead views and land cover attributes. Based on this observation, we propose a simple method to translate ground-level to aerial and attribute features by averaging the overhead features of best matching ground-level scenes. Figure 4.5a shows the pipeline of AVG. We first find the k most similar training scenes as we do for im2gps. Then we average their corresponding

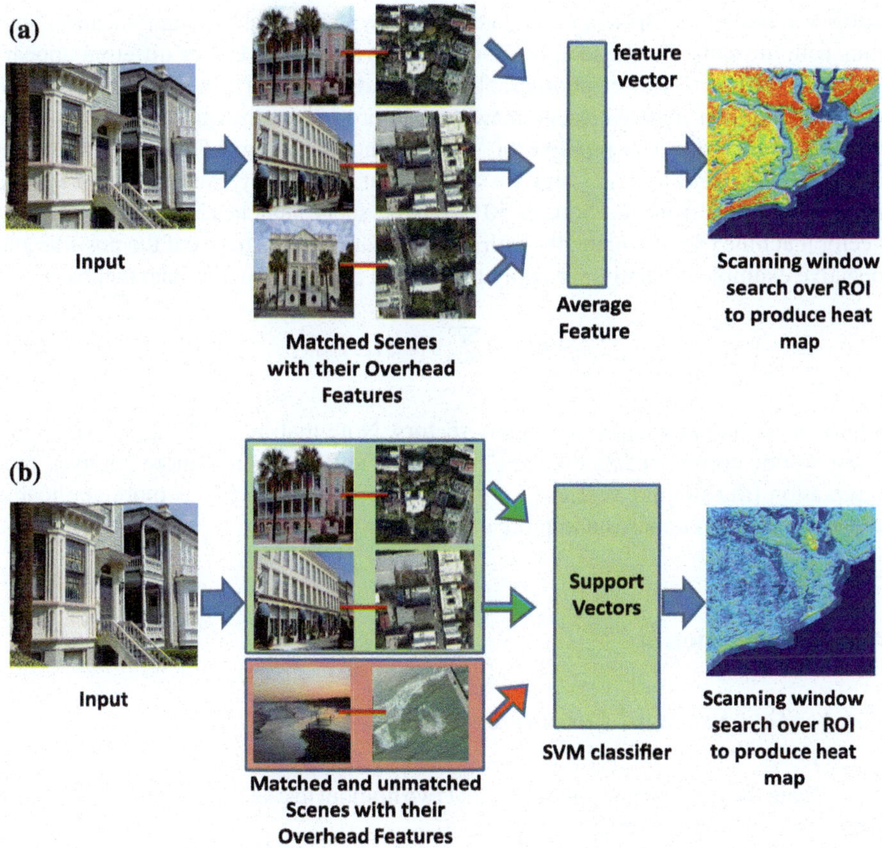

Fig. 4.5 Our proposed data-driven pipelines for cross-view matching. **a** The input ground-level image is matched to available ground-level images in the database, the features of the corresponding aerial imagery are averaged, and this averaged representation is used to find potential matches in the region of interest (ROI). **b** We train an SVM using batches of highly similar and dissimilar aerial imagery and apply it to sliding windows over the ROI

aerial or land cover features to form our predictions of the unknown features. Finally, we use the averaged features to match features in the map database:

$$\mathrm{sim_{avg}} = \mathrm{k}\left(\sum_{\mathrm{i}} \mathrm{y_i}, \mathbf{y}_{\mathrm{map}}\right), \quad \mathrm{i} \in \mathrm{knn}(\mathrm{x}, \mathbf{x}) \tag{4.4}$$

Discriminative Translation (DT): In the AVG approach, we only utilize the best scene matches to predict the ground truth aerial or land cover feature. But the *dissimilar* ground-level training scenes can also be informative—scenes with very different ground-level appearance tend to have distinct overhead appearance and ground cover attributes. Note that this can be violated when two photos at the same location (and thus with the same overhead appearance and attributes) have very different appearance (e.g., a macro flower photograph vs. a landscape shot). In order to capitalize on the benefits of negative scene matches, we propose a discriminative cross-view translation method. The pipeline of "DT" is shown in Fig. 4.5b. The positive set of DT is the same as AVG and we add to that a relatively large set of negative training samples from the bottom 50 % ranked scene matches. We train a support vector machine (SVM) on the fly with the aerial/attribute features of the positive and negative examples. Finally, we apply the trained classifier on the map database:

$$\mathrm{sim}_{\mathrm{dt}} = \sum_{\mathrm{i}} \alpha_{\mathrm{i}} \mathrm{k}(\mathrm{y}_{\mathrm{i}}, \mathbf{y}_{\mathrm{map}}) \tag{4.5}$$

where the α_i is the weight of support vectors. Note that unlike KCCA, DT allows many-to-one correspondence for training triplet and results in a more compact representation (the support vectors) for prediction. As a result, DT is more applicable to large training databases and more efficient for testing.

4.4 Experiments

4.4.1 Test Sets and Parameter Settings

To evaluate the performance, we construct two holdout test sets **Random** and **Isolated** from the ground-level image database.

Random: we randomly draw 1000 images from the ground-level image database. Some images in this set come from frequently photographed locations and thus the training images could contain instance-level matches. Quantitative results for this test set are shown in Fig. 4.7.

Isolated: We use 737 isolated images with no training images in the same 133 m × 133 m block. We are most interested in geo-localizing these isolated images because existing methods (e.g., im2gps) fail for all images in this set. Quantitative results for this test set are shown in Fig. 4.8.

For discriminative translation (DT), for each query we select the top 30 scene matches and 1500 random bad matches for training. We train the kernel SVM classifier with the libsvm package [3]. For KCCA, we enforce one-to-one correspondence by only choosing one ground-level image from multiple images at the same location in the training data. We project the data onto first 500 learned bases to compute the cross-view correlation.

4.4.2 Performance Metric

Each cross-view matching algorithm returns a score for every map location. Figure 4.6 visualizes the scores assigned to each location by our various algorithms and baselines for a particular query.

For quantitative evaluation, we sort the location scores of each query and then measure the rank assigned to the ground truth location. The second column of Figs. 4.7 and 4.8 show the fraction of query images where the ground truth location is ranked in the top 1 % of highest scoring map locations. For im2gps, we look at the top 30 scene matches for evaluation. Note that we use the same 30 matched images as our positive training examples for AVG and DT. That means the accuracy of im2gps indicates the fraction of query images that retrieve training examples at the ground truth location for AVG and DT.

4.4.3 Matching Performance

In this section, we evaluate the various geo-localization methods. The top rows of Figs. 4.7 and 4.8 show the performance of matching ground-to-aerial features. The performance of DT is better than AVG especially in the high precision, low recall regime. Figure 4.6e shows that the heat map of AVG is too uniform while Fig. 4.6g shows that the heat map of DT concentrates around the ground truth location. This suggests that the predicted features from AVG fail to learn a discriminative representation for each query. This is unsurprising, as discriminative methods have been successful across many visual domains. For instance, in object detection, averaging descriptors from objects in the same class may produce a new feature that is also similar to many different classes. On the other hand, DT can leverage the negative training data to predict the aerial feature that is only similar to the given positive examples in a discriminative way.

The middle rows of Figs. 4.7 and 4.8 show the performance of matching ground-to-attribute features. This approach is less accurate than matching to overhead appearance because attribute features are not discriminative enough by themselves. For instance, an attribute feature may correspond to many locations. Perhaps because the attribute representation is low dimensional and there are many duplicate instances in the database, DT does not perform as well as the simple AVG approach.

The bottom rows of Figs. 4.7 and 4.8 show the performance of matching based on predicting both aerial and attribute features. This approach performs better than each individual feature which suggests that the aerial and attribute features complement each other. In particular, we find that the best performance comes by combining with equal weight the attribute-based scores from the AVG method and the overhead appearance-based scores from the DT method. This "hybrid" method achieves 39.5 % accuracy for **Random** and 17.37 % for **Isolated**.

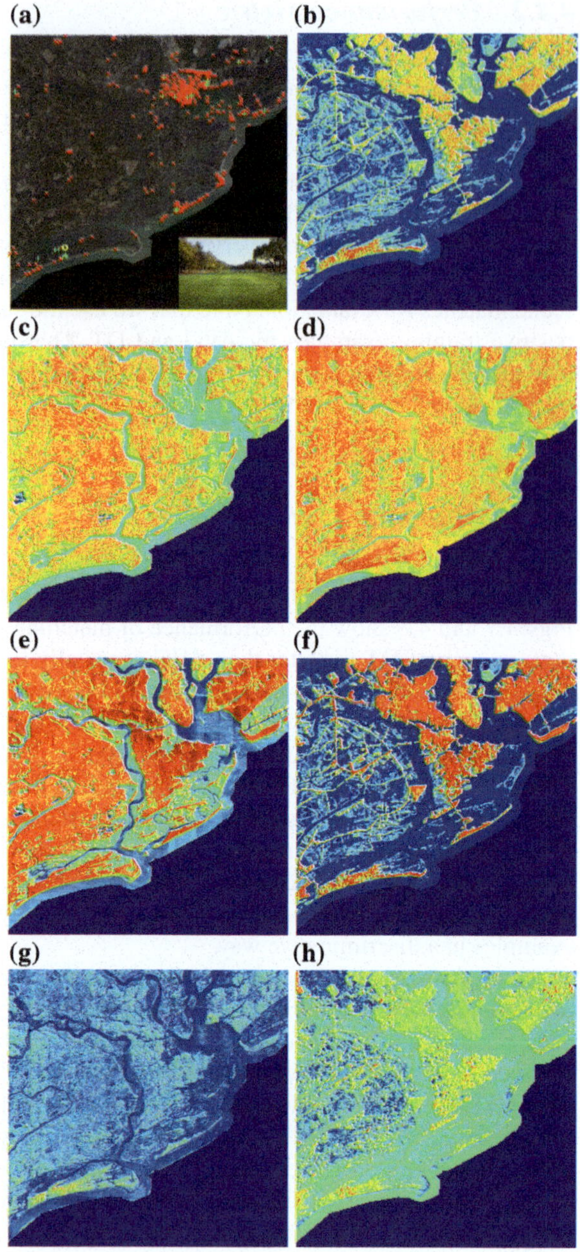

Fig. 4.6 **a** Shows the geo-location of query image (*yellow*), the best 30 scene matches (*green*) and 1500 bad scene matches (*red*). **b** The "hybrid" approach has the best performance in the experiment. (**c**–**d**) Direct matching (DM) and KCCA as baseline algorithms. (**e**–**h**) Feature averaging (AVG) and discriminative translation (DT) on aerial and attribute feature. AVG is more reliable with attribute features while DT is more accurate with aerial features

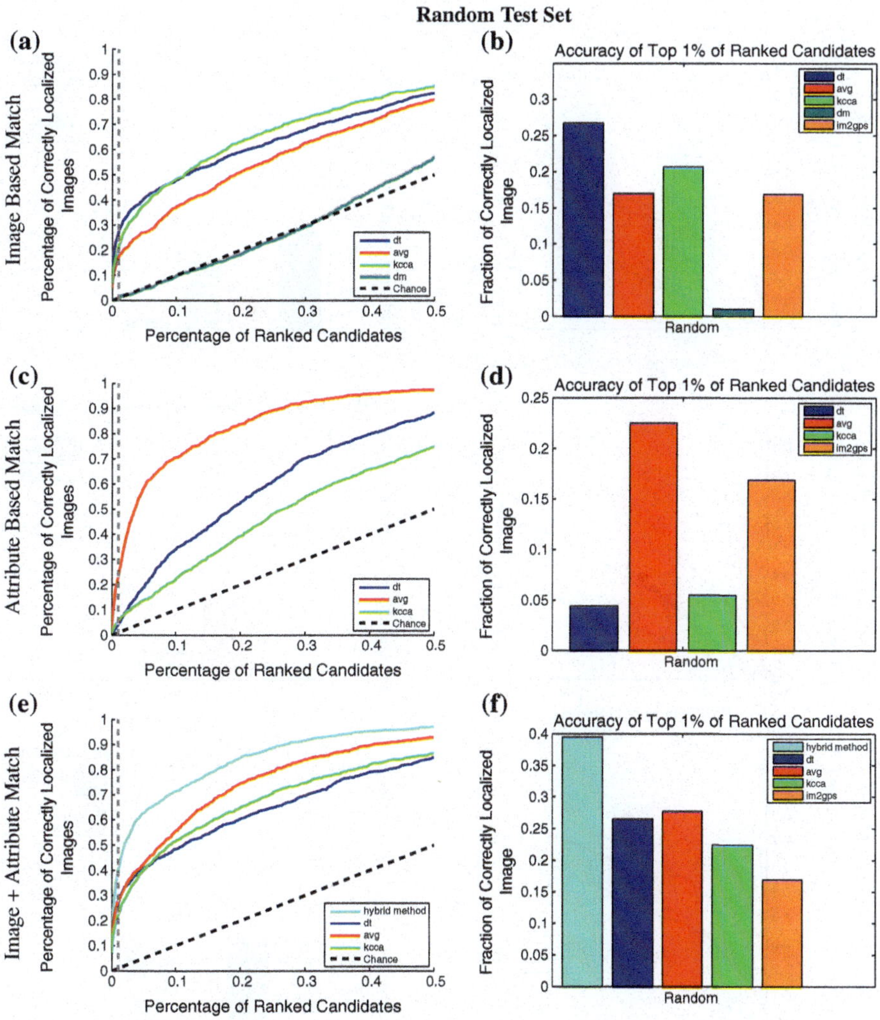

Fig. 4.7 Each *row* shows the accuracy of localization as a function of the fraction of retrieved candidate locations for "random" test cases, followed by a *bar* plot slice through the curve at the 1 % mark (rank 1830). (**a–b**) Image-based matching only, (**c–d**) land cover attribute-based matching only, (**e–f**) combined image and land cover attribute-based matching. im2gps is competitive on this test set because many of the queries do have nearby ground-level training images to match to

Figure 4.9 shows a random sample of successfully and unsuccessfully localized query images. Our system handles outdoor scenes better because they are more correlated to the aerial and attribute images. Photos focused on objects may fail

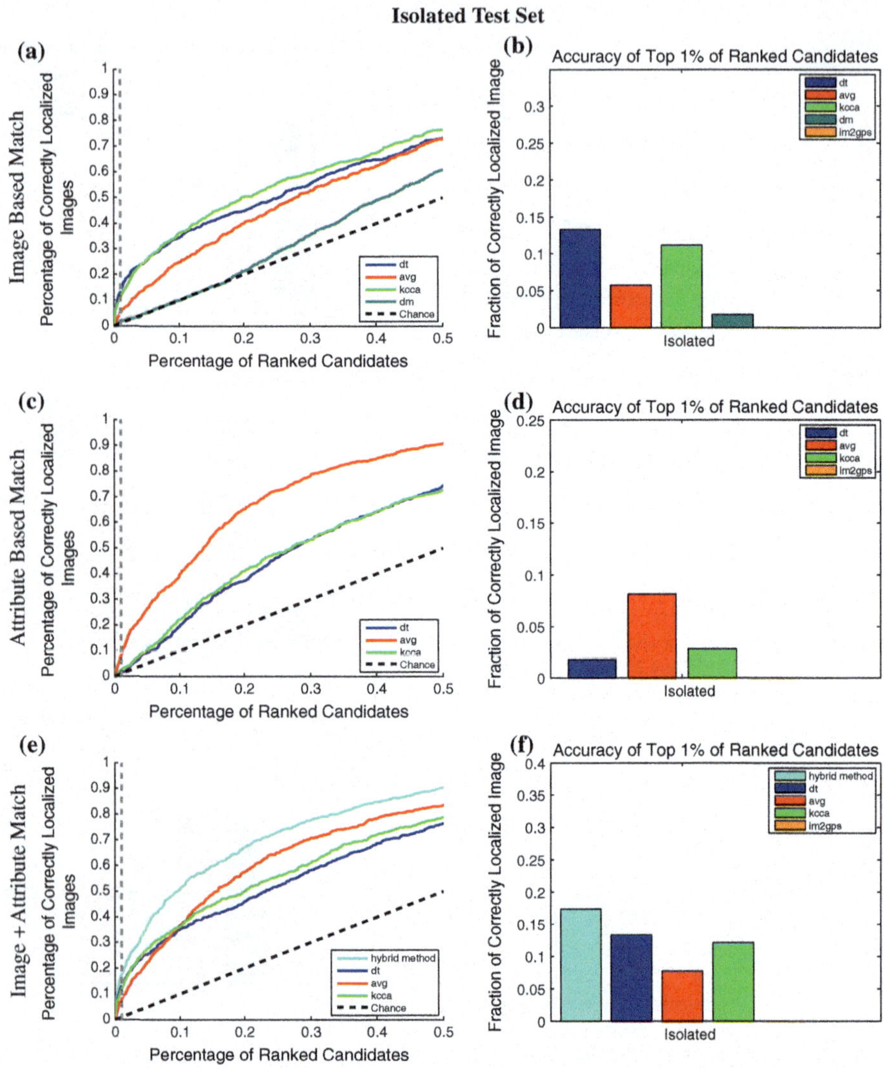

Fig. 4.8 Each *row* shows the accuracy of localization as a function of the fraction of retrieved candidate locations for test photos from isolated locations, followed by a *bar* plot slice through the curves at the 1 % mark (rank 1830). (**a**–**b**) Image-based matching only, (**c**–**d**) land cover attribute-based matching only, (**e**–**f**) combined image and land cover attribute-based matching. Note that the accuracy of im2gps is exactly zero for the isolated test set since these photos have no other nearby ground-level reference imagery. With our hybrid approach, we can determine the correct geo-location for 17.37 % of our query images when we consider the top 1 % best matching candidates

Fig. 4.9 Gallery of (**a**) successfully and (**b**) unsuccessfully matched isolated query images in our experiments. Photos that prominently feature objects tend to fail because the aerial and land cover features are not well constrained by the scene features

because the correlation between objects and overhead appearance and/or attributes is weak. Figure 4.10 visualizes query images, corresponding scene matches, and heat maps. We demonstrate that our proposed algorithm can handle a variety of scenes by showing three very different query examples.

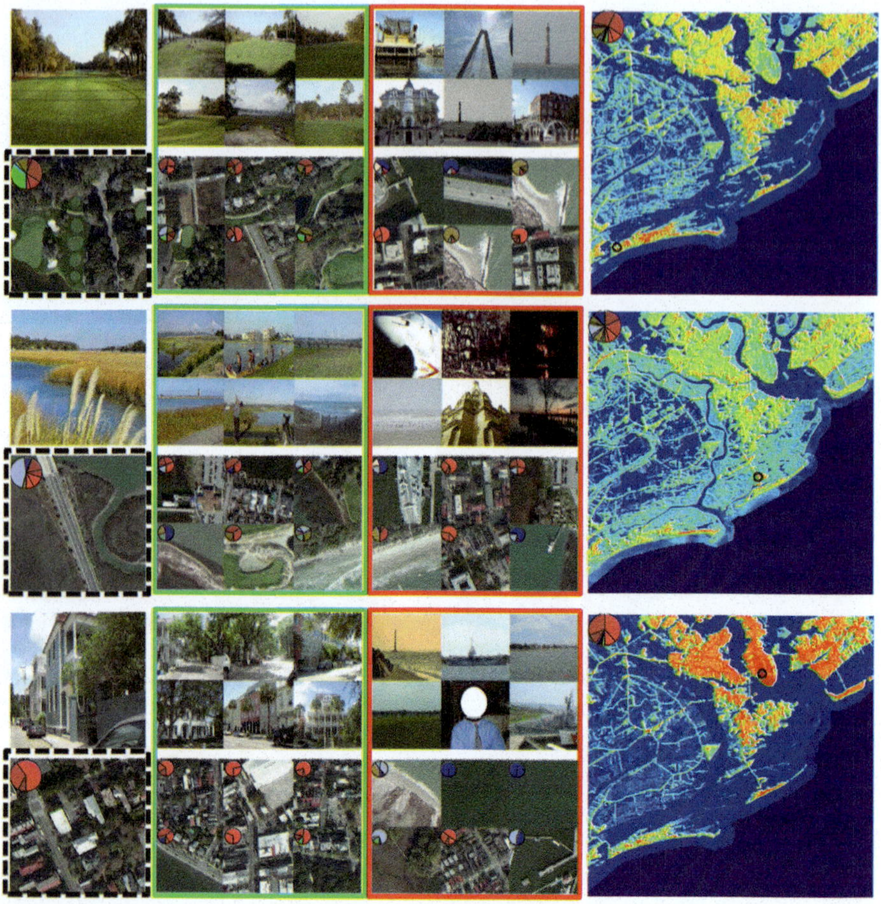

Fig. 4.10 *Left* input ground-level image (shown *above*) and corresponding satellite image and pie chart of attribute distribution (shown *below*, but not known at query time). *Middle* similar (in *green*) and dissimilar (in *red*) ground-level and satellite image pairs used for training the SVM in our discriminative translation approach. *Right* geo-location match score shown as a heat map. The ground truth location is marked with a *black circle*

4.5 Conclusion

In this chapter, we propose a cross-domain matching framework that greatly extends the reach of image geo-localization techniques. Most of the Earth's surface has no ground-level geo-tagged photos publicly (98 % of the area of our experimental region, even though it is part of the densely populated eastern coast of the United States) and traditional image geo-location methods based on ground-level image matching will fail for queries in such locations. Using our new dataset of ground-level, aerial, and ground cover attribute images, we quantified the performance of several baseline and novel approaches for "cross-view" geo-location. In particular, our "discriminative translation" approach in which an aerial image classifier is trained based on ground-level scene matches can roughly locate 17 % of isolated query images, compared to 0 % for existing methods. While the experiments in this chapter are at a modest scale (1135 km^2 region of interest), the approach scales up easily both in terms of data availability and computational complexity. Our approach is the first to use overhead imagery and land cover survey data to geo-locate photographs, and it is complementary with the impressive mountain-based geo-location method of [2] which also uses a widely available geographic feature (digital elevation maps).

Limitations: The lack of dense correspondence between the ground-level and aerial views at training time limits the types of algorithmic approaches we can use. Our proposed algorithms probably learn something along the lines of "this ground-level texture is correlated with this aerial texture and this land cover distribution" but there is no reasoning about the spatial layout of particular structures. A human verifying a cross-view match would use such information, e.g., "here is a house with a large tree to its left, a swimming pool to its right, and a circular driveway in the front."

Acknowledgments This research was supported by the Intelligence Advanced Research Projects Activity (IARPA) via Air Force Research Laboratory, contract FA8650-12-C-7212. The U.S. government is authorized to reproduce and distribute reprints for governmental purposes notwithstanding any copyright annotation thereon. Disclaimer: The views and conclusions contained herein are those of the authors and should not be interpreted as necessarily representing the official policies or endorsements, either expressed or implied, of IARPA, AFRL, or the U.S. government.

References

1. Agarwal S, Snavely N, Simon I, Seitz SM, Szeliski R (2009) Building Rome in a day. In: ICCV, 3
2. Baatz G, Saurer O, Köser K, Pollefeys M (2012) Large scale visual geo-localization of images in mountainous terrain. In: ECCV, 3, 12
3. Chang C-C, Lin C-J (2011) LIBSVM: a library for support vector machines. ACM Trans Intell Syst Technol 11
4. Chen D, Baatz G, Köser K, Tsai S, Vedantham R, Pylvanainen T, Roimela K, Chen X, Bach J, Pollefeys M, Girod B, Grzeszczuk R (2011) City-scale landmark identification on mobile devices. In: CVPR, 3
5. Crandall DJ, Backstrom L, Huttenlocher D, Kleinberg J (2009) Mapping the world's photos. In: WWW, 3
6. Dalal N, Triggs B (2005) Histograms of oriented gradients for human detection. In: CVPR, 7
7. Hardoon DR, Szedmak SR, Shawe-Taylor JR (2004) Canonical correlation analysis: an overview with application to learning methods. Neural Comput 16(12):2639–2664
8. Hays J (2009) Large scale scene matching for graphics and vision. Ph.D. thesis, Carnegie Mellon University, 3
9. Hays J, Efros A (2008) IM2GPS: estimating geographic information from a single image. In: CVPR, 2, 3, 7
10. Irschara A, Zach C, Frahm J-M, Bischof H (2009) From structure-from-motion point clouds to fast location recognition. In: CVPR, 3

11. Jacobs N, Satkin S, Roman N, Speyer R, Pless R (2007) Geolocating static cameras. In: ICCV, Oct 2007, 3
12. Lazebnik S, Schmid C, Ponce J (2006) Beyond bags of features: spatial pyramid matching for recognizing natural scene categories. In: CVPR, 7
13. Li X, Wu C, Zach C, Lazebnik S, Frahm J (2008) Modeling and recognition of landmark image collections using iconic scene graphs. In: ECCV, 3
14. Li Y, Snavely N, Huttenlocher DP (2010) Location recognition using prioritized feature matching. In: ECCV, 3
15. Li Y, Snavely N, Huttenlocher D, Fua P (2012) Worldwide pose estimation using 3D point clouds. In: ECCV, 3
16. Liu J, Shah M, Kuipers B, Savarese S (2011) Cross-view action recognition via view knowledge transfer. In: CVPR, 3
17. Oliva A, Torralba A (2001) Modeling the shape of the scene: a holistic representation of the spatial envelope. IJCV, 7
18. Ordonez V, Kulkarni G, Berg TL (2011) Im2text: Describing images using 1 million captioned photographs. In: NIPS, 3
19. Rasiwasia N, Costa Pereira J, Coviello E, Doyle G, Lanckriet G, Levy R, Vasconcelos N (2010) A new approach to cross-modal multimedia retrieval. In: ACM international conference on multimedia, 3
20. Roshan Zamir A, Shah M (2010) Accurate image localization based on Google maps street view. In: ECCV, 3
21. Sattler T, Leibe B, Kobbelt L (2011) Fast image-based localization using direct 2D-to-3D matching. In: ICCV, 3
22. Schindler G, Brown M, Szeliski R (2007) City-scale location recognition. In: CVPR, 3
23. Sharma A, Kumar A, Daumé III H, Jacobs DW (2012) Generalized multiview analysis: a discriminative latent space. In: CVPR, 3
24. Shechtman E, Irani M (2007) Matching local self-similarities across images and videos. In: CVPR, 7
25. Xiao J, Hays J, Ehinger KA, Oliva A, Torralba A (2010) SUN database: large-scale scene recognition from abbey to zoo. In: CVPR, 7
26. Zhang H, Berg AC, Maire M, Malik J (2006) SVM-KNN: discriminative nearest neighbor classification for visual category recognition. In: CVPR, 3
27. Zheng Y, Zhao M, Song Y, Adam H, Buddemeier U, Bissacco A, Brucher F, Chua T, Neven H, Yagnik J (2009) Tour the world: building a web-scale landmark recognition engine. In: CVPR, 3

Chapter 5
Ultrawide Baseline Facade Matching for Geo-localization

Mayank Bansal, Kostas Daniilidis and Harpreet Sawhney

Abstract Matching street-level images to a database of airborne images is hard because of extreme viewpoint and illumination differences. Color/gradient distributions or local descriptors fail to match forcing us to rely on the structure of self-similarity of patterns on facades. We propose to capture this structure with a novel "scale-selective self-similarity" (S^4) descriptor which is computed at each point on the facade at its inherent scale. To achieve this, we introduce a new method for scale selection which enables the extraction and segmentation of facades as well. We also introduce a novel geometric method that aligns satellite and bird's-eye-view imagery to extract building facade regions in a stereo graph-cuts framework. Matching of the query facade to the database facade regions is done with a Bayesian classification of the street-view query S^4 descriptors given all labeled descriptors in the bird's-eye-view database. We also discuss geometric techniques for camera pose estimation using correspondence between building corners in the query and the matched aerial imagery. We show experimental results on retrieval accuracy on a challenging set of publicly available imagery and compare with standard SIFT-based techniques.

5.1 Introduction

In this chapter, we propose a novel method for matching facade imagery from very different viewpoints—like from a low-flying aircraft and from a street-level camera. The scenario we address entails a database of preprocessed bird's-eye-view (BEV) images and street-view (SV) queries. Such images are characterized by large

M. Bansal (✉) · H. Sawhney
Center for Vision Technologies, SRI International, Princeton, NJ, USA
e-mail: mayank.bansal@sri.com; mayban@gmail.com

H. Sawhney
e-mail: harpreet.sawhney@sri.com

K. Daniilidis
Department of Computer and Information Science,
University of Pennsylvania, Philadelphia, PA, USA
e-mail: kostas@cis.upenn.edu

A.R. Zamir et al. (eds.), *Large-Scale Visual Geo-Localization*,
Advances in Computer Vision and Pattern Recognition,
DOI 10.1007/978-3-319-25781-5_5

differences in local appearance which render any comparison of bags of visual words infeasible. A visual comparison of this imagery even after rectification testifies to the hardness of the problem. Moreover, a vast majority of facades contain repetitive patterns which make correspondence estimation highly ambiguous. We rather have to rely on comparing the structures of the facade patterns and still account for any transformations between such structures.

The key idea is to avoid direct matching of features to solve this extreme case of wide-baseline matching. Thus, we formulate the problem as "embeddings" within each respective dataset (SV and BEV) so that large variations are incorporated within the structure of embeddings. This idea has not been explored before especially in the context of air-ground matching. We make the following contributions to the state of the art: (a) we introduce an approach for matching image regions with significant appearance, scale, and viewpoint variations based on a novel *Scale-Selective Self-Similarity* (S^4) feature that combines intrinsic scale selection with self-similarity descriptors, and (b) we demonstrate a novel system for matching street-level queries to a database of birds-eye views. We show experimental results on the retrieval accuracy from our technique and compare our performance with standard SIFT descriptors.

We approach the facade detection and matching problem from a combined statistical and structural viewpoint. While other approaches model the lattice structure explicitly [13], we capture the statistical self-similarity (or dis-similarity) of a local patch to its neighbors. By avoiding using a specific feature like SIFT, MSER, or line segments, we can capture this structure at any point—in implementation we do it on a randomly jittered grid. In addition, the self-similarity descriptor also captures the dis-similarity between neighboring elements ignored in lattice approaches but still observed, e.g., in [6]. The challenge with self-similarity is to capture the intrinsic local scale governed by the periodicity/generator group of a lattice. We estimate the scale by discovering the closest most salient repetition of a patch which can be centered anywhere. With the exception of [10], other approaches rely on the robustness of interest point or line segment detectors. Having obtained the intrinsic scale enables us to compute the scale-invariant S^4 descriptor and also allows us to **detect** facades as clusters of such points in space that have similar scale and descriptors. Similar descriptors are obtained from the query street-level image as well. At this point, instead of lattice or graph matching [6, 10], we apply a labeling approach that labels each query descriptor with the most probable facade label (cluster) in a naive-Bayes sense. This way, we match local lattice structures rather than global ones and the most likely closest database facade is obtained.

5.2 Scale-Selective Self-similarity Features

The viewpoint and appearance difference between oblique bird's-eye-view (BEV) and street-view (SV) imagery is too large to be captured by direct matching of descriptors like SIFT and MSER. Therefore, we propose to create a descriptor that captures the structure of repetition of patterns or more generally the relative similarity between local patches within facades. Instead of modeling the structure with a graph or lattice and relying on the robustness of the detection of their nodes, we define a new feature which we call the *Scale-Selective Self-Similarity* or S^4 feature. This feature improves upon the well-known self-similarity descriptor from Shechtman and Irani [17] by adding a SIFT-like scale normalization to allow characterization of the self-similar structure in a scale-invariant manner.

Using the same notation as [17], for a given pixel q, the local self-similarity descriptor d_q is computed as follows. A local image patch of width w_{ss} (e.g., 5 pixels) centered at q is correlated with a larger surrounding image region of radius r_{ss} (e.g., 40 pixels), resulting in a local internal 'correlation surface'. The correlation surface is then transformed into a binned log-polar representation which accounts for increasing positional uncertainty with distance from the pixel q, accounting, thus, for local spatial affine deformations.

Figure 5.1 shows a pair of (orthorectified) SV and BEV images of a facade that have been manually normalized to the same image scale, and compares how well their self-similarity descriptors match relative to their SIFT descriptors. The self-similarity descriptor at the center of the green ROI (local patch) is computed by correlating within the surrounding support region (blue ROI). The computed descriptors are noticeably quite similar even with the large appearance difference between the images themselves. In comparison, the SIFT descriptors computed using the same support region are dissimilar.

Scale Selection While it is clear that the inherent self-similar structure in building facades can serve as a good matching criterion, it is not clear how that structure can be matched if the building is seen at different scales. The basic self-similarity descriptor discussed above assumes a distance binning which is not scale invariant. To account for feature scale differences, Shechtman and Irani [17] suggest computing the self-similarity descriptors on a Gaussian image pyramid representation and then searching for the template object across all scales. For the purposes of retrieval, however, such an approach would not work. In particular, for building facades, capturing the self-similar structure at all scales will reduce the discriminability evident at the fundamental scale of the facade. Instead, we would like a SIFT-like normalization so that the descriptors between differently scaled buildings can still be matched. The repetitive structure of building facades provides one such normalization scale. However, building facades typically also exhibit *local* periodicity. While recovering this scale will serve the purpose of a valid normalizing scale, it may compromise on

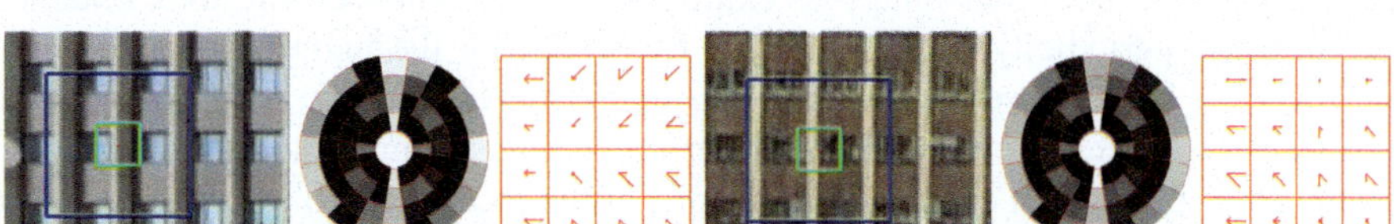

Fig. 5.1 Example of self-similarity and SIFT descriptors for corresponding facades from SV and BEV images respectively

the overall discriminability of the computed descriptor by (a) being too local, and (b) by being too dependent on the inherent image scale (the smallest scale structure will be lost first in a noisy query image).

In this work, we focus on recovering the *motif scale*. We define the motif scale at a pixel in the facade as the smallest wavelength at which any patch in this pixel's local neighborhood repeats. Defined this way, a local window scale would be ignored if it is not consistent with a few other window pixels in its neighborhood—thus making this scale robust against local pattern noise. This motif scale can be measured independently in both horizontal and vertical directions; in our implementation, we have only used the horizontal scale (denoted as λ_x), but the approach is symmetric with respect to using either of the two. Given the motif scale λ_x value at any pixel, the S^4 descriptor is defined as the self-similarity descriptor computed by setting the patch size w_{ss} to the estimated motif scale λ_x and the correlation radius to $r_{ss} = 2\lambda_x$.

Our approach for motif scale selection is based on the peaks in the autocorrelation surface in a local neighborhood surrounding a pixel. Consider a pixel (x, y) inside an image $\mathscr{I}$ exhibiting periodic structure and let λ_x be its scale along the x-direction. Now consider a small $w \times h$ patch of pixels around this pixel and correlate it with patches extracted at various offsets (r, θ) in a polar representation. To capture the correlations most relevant to the self-similarity descriptor, we measure the correlation profile using the following SSD measure. Let $\mathscr{J}(s, t) = \mathscr{I}(x + s, y + t)$, then:

$$q(r, \theta) = \sum_{t_y=-\frac{h}{2}}^{\frac{h}{2}} \sum_{t_x=-\frac{w}{2}}^{\frac{w}{2}} (\mathscr{J}(t_x, t_y) - \mathscr{J}(t_x + r\cos(\theta), t_y + r\sin(\theta)))^2 \tag{5.1}$$

Then, the correlation profile $p_{(x,y)}(r)$ is computed by integrating the scores $q(r, \theta)$ in a 20° lobe ($\theta_0 = 10°$) around the horizontal direction:

$$p_{(x,y)}(r) = \exp\left(-\frac{1}{2\theta_0 + 1} \sum_{\theta=-\theta_0}^{\theta_0} q(r, \theta)\right) \tag{5.2}$$

where the subscript (x, y) makes explicit the fact that the profile was obtained by correlating the patch around pixel (x, y). The angular integration provides robustness against image distortions and orthorectification errors. The value of r is varied such that $r \in \{1, \ldots, S_{\max}\}$, where $S_{\max}$ is a predefined maximum scale value we expect the structure in the input image to exhibit. The correlation profile thus obtained captures the periodicity of the structure by producing the highest correlation for $r \in \{\lambda_x, 2\lambda_x, \ldots\}$. However, depending on the starting location (x, y), the correlation profile can exhibit peaks at r values which are nonintegral multiples of λ_x. This will be the case if the patch contains a sub-motif of the facade which is locally periodic at a higher frequency. The illustration in Fig. 5.2 depicts this happening for the green and blue profiles obtained from the (black) 1-D signal. The wavelength of both these curves is smaller than the motif scale λ_x by our definition above. To alleviate this

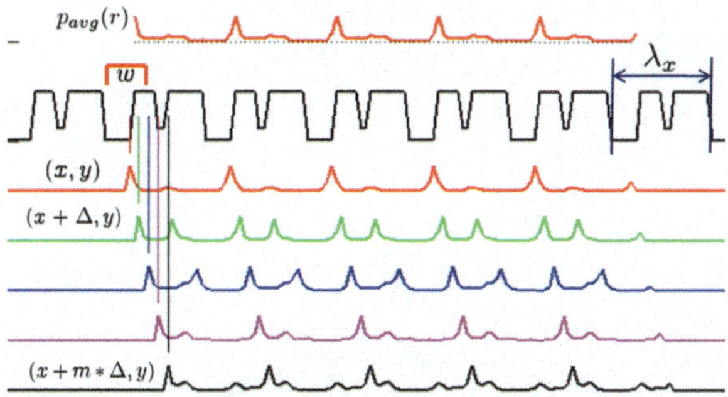

Fig. 5.2 Scale selection. To determine the scale λ_x of the (*black*) 1D signal in the second row, if we autocorrelate a patch of width w, we get one of the profiles shown in *rows 3–7* depending on the starting offset. However, for a poor offset choice (*green* and *blue curves*), one can get comparable peaks in the correlation profile for scale values $<\lambda_x$ making it difficult to extract the correct scale. Integrating across these profiles, however, resolves this issue and results in a well-defined profile $p_{\text{avg}}(r)$ shown in the *first row*. The high peaks now correspond to the correct wavelength λ_x

issue, we compute multiple correlation profiles by varying the starting offset in an interval $\mathcal{O} = \{(x, y), (x + 1, y), (x + 2, y), \ldots, (x + m, y)\}$. The maximum offset $(x + m, y)$ is set so that the patch around it covers the structure at the maximum scale S_{max} from the starting position, i.e., $m + w/2 \geq S_{\text{max}}$. The correlation profiles are combined into a single profile $p_{\text{avg}}(r)$ by integrating across the offsets, i.e., $p_{\text{avg}}(r) = \sum_{o \in \mathcal{O}} p_o(r)$. This removes the higher frequency peaks in the individual profiles, leaving only the peaks corresponding to the actual wavelength λ_x as depicted in Fig. 5.2. Furthermore, the scale estimation becomes robust to the choice of the patch dimensions w and h assuming that these are chosen to be reasonable w.r.t. the number of pixels occupied by typical building regions in the image. In our implementation, we have fixed these to be 13 pixels which allows us to detect facades for fairly small building areas.

To be robust against shallow peak responses, we measure a peakness measure around each peak in the profile $p_{\text{avg}}(r)$ and prune peaks which are shallower than a threshold t_{peak}. This threshold is set empirically by running the scale estimator on textureless and non-repetitive structures. From the locations of the remaining peaks, the scale value λ_x can be readily obtained by a discrete Fourier transform. In the absence of any peaks the underlying structure is labeled aperiodic (assigned scale *zero*)—this removes most of the non-facade pixels and serves as an effective building detection mechanism.

5.3 Facade Extraction and Segmentation

We now describe our general approach for extracting building facade regions which is applicable to both BEV and SV images. The key idea is to exploit the self-similar structure of building facades: orthorectify the image, compute motif scales at sampled

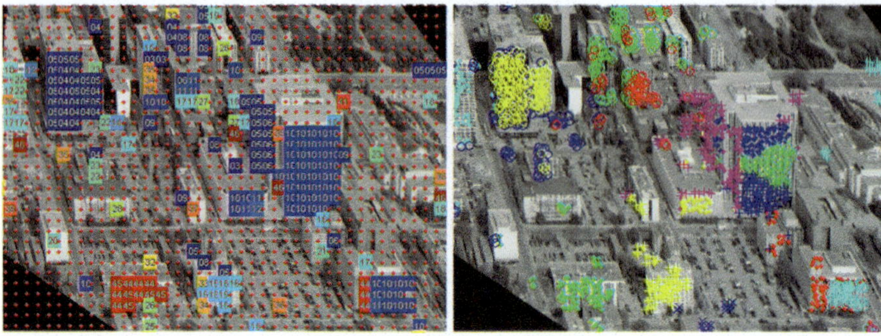

Fig. 5.3 Facade extraction and segmentation. Rectified BEV images showing, *left* the selected horizontal scales with *red dots* at the locations assigned zero scale value, and *right* cluster assignments after K-means

locations in the given image, compute S^4 descriptors at the computed scales and then cluster the descriptors to group similar structures together.

Motif scale computation In the rectified image, we sample a grid of pixel locations every $\sigma_f = 5$ pixels apart and add uniformly random spatial jitter of amplitude $\sigma_f/2$ at each sample location. This jitter allows us to capture a good sampling of the feature distributions expected from this facade structure at the matching stage. At each sample location, we compute the motif scales using the approach discussed in Sect. 5.2. An example result at this stage is shown in the left half of Fig. 5.3. Note that the scale selection has removed the nonbuilding areas almost completely by labeling them with a *zero* scale value (shown as red dots in the figures). Also note the wide range of motif scales seen across buildings stressing the importance of proper scale selection. At this point, we need a way to segment out individual facades into disjoint groups so that a matching approach can predict labels at the building level.

Facade Segmentation At each sample location, we compute the S^4 descriptor ($n_\theta = 20$ angular bins and $n_r = 4$ distance bins) by setting the patch size w_{ss} to the estimated motif scale λ_x and the correlation radius to $r_{ss} = 2\lambda_x$. Now, we perform K-means clustering in this S^4 feature space using L_1 norm as our distance measure. To avoid descriptor grouping across different buildings, we penalize clustering of descriptors which were sampled from far-off locations by adding a penalty term driven by the spatial distance between the sampled feature locations. The desired number of clusters N is set as follows. We manually mark the boundaries of a small number of buildings (5 in our case) in each BEV image and initialize $N = N_0$. Now, we iteratively run K-means with decreasing value for N as long as the following invariant is maintained: clusters on the marked buildings are contained within the marked boundaries. At the end of this process, we obtain a clustering that has the fewest number of clusters within each building and does not merge two different buildings into a single cluster (note that this is not guaranteed for unmarked buildings in general, but due to the descriptor-based grouping, we have not seen any merging of separate buildings into a single cluster in our experiments). For our test BEV set, we typically obtain 1–3 clusters per facade after this procedure. The right half of Fig. 5.3 shows an example of the clusters obtained after K-means clustering. Figure 5.4 shows another example of the full facade segmentation pipeline from a BEV image.

Notation In the following, we will denote the S^4 descriptor vectors obtained from the entire set of BEV imagery by words $\mathscr{V} = \{v_1, v_2, \ldots, v_m\}$, the cluster labels as $\mathscr{C} = \{c_1, c_2, \ldots, c_N\}$ and the labeling function mapping each word to its cluster assignment by the function $\mathscr{L} : \mathscr{V} \rightarrow \mathscr{C}$.

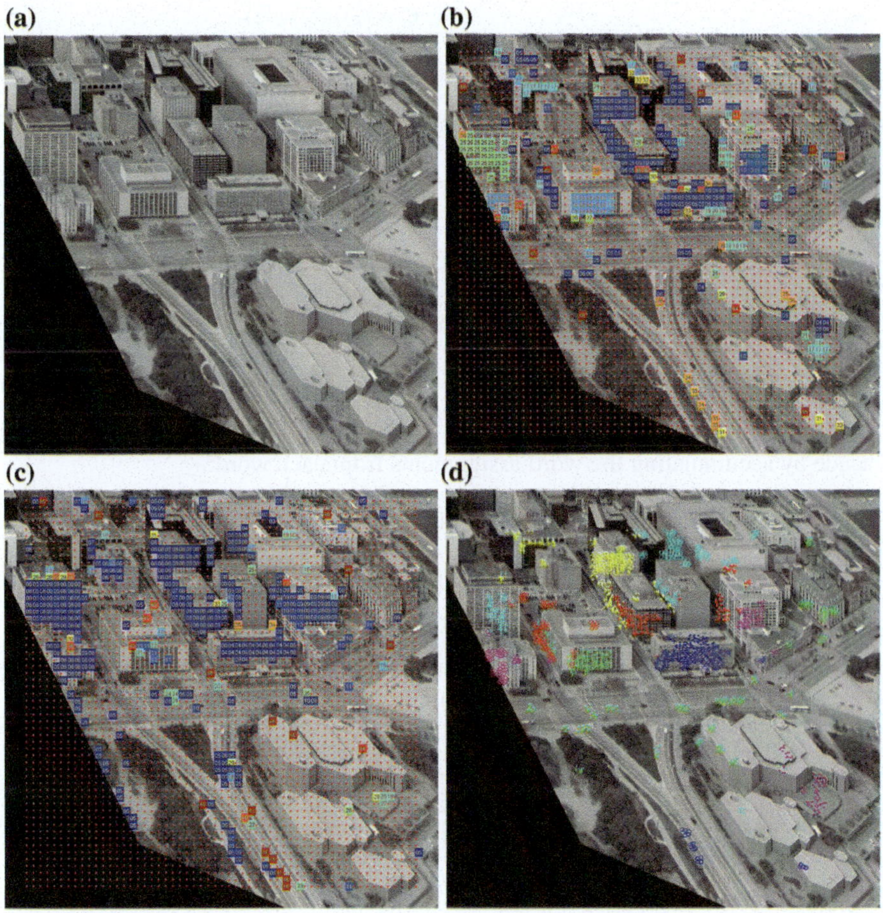

Fig. 5.4 Scale extraction and oversegmentation of facades from rectified BEV image. **a** BEV image, **b** horizontal scales, **c** vertical scales, **d** K-means clusters

5.4 Facade Matching

Given a query street-view image, we would like to retrieve facades from our BEV database that match the dominant facade(s) in the query. Section 5.5.3 and Fig. 5.7 illustrate the key steps in our SV-to-BEV matching pipeline. After orthorectification, motif scale selection and S^4 descriptor computation, we obtain a set of descriptor vectors $\mathscr{W} = \{w_1, w_2, \ldots, w_n\}$ from the query. For each of these words, we would like to estimate the probability $p(C = c_k|w_i)$ of being assigned to one of the clusters c_k in $\mathscr{C}$. The problem of finding the closest cluster label for each word w_i can be formulated in a Bayesian settings as follows. By Bayes' theorem,

$$p(C = c_k|w_i) = \frac{p(w_i|C = c_k)p(C = c_k)}{\sum_{j=1}^{N} p(w_i|C = c_j)p(C = c_j)} \tag{5.3}$$

For each word w_i, we estimate the likelihoods $p(w_i|C = c_k)$ by kernel density estimation using a Gaussian kernel $\mathscr{K}(w_i, v_j)$ with wavelength parameter $\sigma_{\mathscr{K}}$. The likelihood is then computed as:

$$p(w_i|C = c_k) = \frac{1}{|c_k|} \sum_{\mathscr{L}(v_j)=c_k} \mathscr{K}(w_i, v_j) \tag{5.4}$$

where $|c_k|$ denotes the cardinality of cluster k. The prior probability $p(C = c_k)$ is simply set from the sample proportions: $p(C = c_k) = \frac{|c_k|}{m}$. For each word w_i, we estimate the MAP estimate of the label by choosing the label k with the maximum a posteriori probability: $\mathscr{L}(w_i) = \arg\max_k p(C = c_k|w_i)$. Given the above word assignments, we can now compute the most probable label for the entire query facade by accumulating the word assignments from each word:

$$f(k) = \sum_i \delta(\mathscr{L}(w_i) = c_k) \tag{5.5}$$

$$\mathscr{L}(\mathscr{W}) = \arg\max_k \{f(k) \,|\, k = 1, \ldots, N\} \tag{5.6}$$

where $\delta(.)$ is the indicator function. The label $\mathscr{L}(\mathscr{W})$ identifies a cluster $c^* \in \mathscr{C}$ which, by construction of the clustering algorithm, identifies a single BEV facade.

Algorithm 1: BEV processing

1. Orthorectify BEV image using vanishing points.
2. Compute motif scale λ_x at a jittered grid of pixel locations on the BEV.
3. Compute S^4 descriptors v_i at locations with nonzero scales.
4. Cluster S^4 descriptors v_i using K-means to obtain label-set $\mathscr{C}$ and labeling function $\mathscr{L}$.

Algorithm 2: SV processing

1. Orthorectify SV image using vanishing points.
2. Compute motif scale λ_x at a jittered grid of pixel locations on the SV.
3. Compute S^4 descriptor-set $\mathscr{W} = \{w_j\}$ at locations with nonzero scales.
4. Compute labels $\mathscr{L}(w_j)$ using Eq. 5.3.
5. Best matching BEV facade: Facade containing cluster $\mathscr{L}(\mathscr{W})$ (Eq. 5.6).
6. Top matching facade set: For threshold t, return facades containing clusters k s.t. $f(k) > t$ (Eq. 5.5).

5.5 Experiments and Results

Algorithm Parameters In Table 5.1, we list all the parameter settings we used in our implementation. The scale estimation process was found robust against different choices of patch size parameters w and h. $S_{\max}$ was set to a number greater than the maximum horizontal building scale for our BEV dataset (manually eyeballed). The S^4 values for n_θ and n_r were set the same as in [17].

BEV and SV Imagery Datasets Our dataset comprises of BEV imagery (2000 × 1500 pixels) downloaded using Microsoft's Bing service for an area approximately 2 km × 1.2 km in size (Fig. 5.5a) in downtown Pittsburgh, PA, USA. This dataset is challenging due to a large number (approx. 40) of buildings and very similar facade patterns. This dataset also covers a much larger area than used in related works in air-ground-based localization, e.g., 440 m × 440 m in [5]. Street-view images downloaded using Panoramio, Flickr, Google Street-View(screenshots), and Microsoft Bing's Streetside(screenshots) were used as queries. For ground-truth purposes, only

Table 5.1 Parameter settings

w	h	$S_{\max}$	σ_f	w_{ss}	r_{ss}	n_θ	n_r	N_0	$\sigma_{\mathscr{K}}$
13 px	13 px	48 px	5 px	λ_x	$2\lambda_x$	20	4	100	2.5

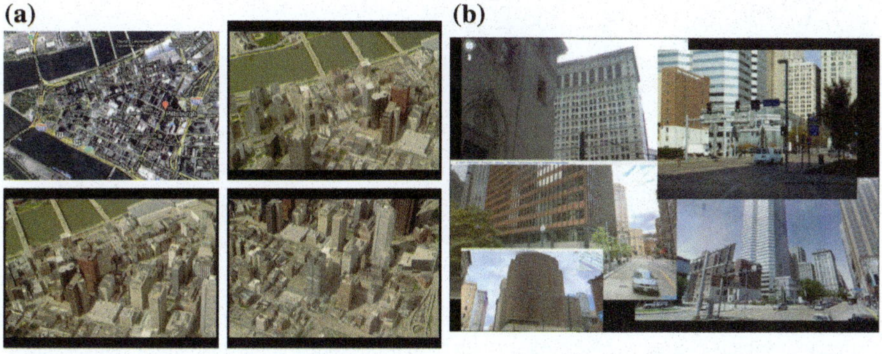

Fig. 5.5 Pittsburgh dataset. **a** Satellite coverage and sample BEVs, **b** sample queries

the SV imagery with geo-tags or visually identifiable facade correspondence (with the BEV) was retained.

Imagery Rectification We rectify BEV to an orthographic view aligned with the dominant city-block direction. Similarly, the SV imagery is rectified to an orthographic view of the dominant facade in the scene using the Geometric Parsing-based vanishing point estimation approach and code [4, 18]. For the BEV imagery, we adapted the approach to handle the high density of lines detected in these images. Using the estimated vanishing points corresponding to the two facade axes, we obtain pairs of extremal rays in the horizontal and vertical directions. Intersecting these gives us four corners of a quadrilateral which are then used to estimate a rectification homography using the approach in [11]. This transformation warps the facade to be fronto-parallel and also corrects the aspect ratio between the horizontal and vertical directions.

5.5.1 *Scale Selection Results*

To characterize our scale selection algorithm, we selected a test set of 10 building facades extracted from the Pittsburgh BEV dataset. We manually measured the ground-truth horizontal scale(s) for each facade and compared them to those estimated by our approach. Since we densely estimate these scale values over the facade, we computed a histogram of the estimated scale values and the normalized histogram values are shown as the blue circles (with radii proportional to the histogram values) in the bubble plot of Fig. 5.6. The red pluses denote the ground-truth scale values—multiple in cases where the facade exhibits more than one motif scale. The comparison shows the accuracy of our scale estimation and the presence of very few outliers.

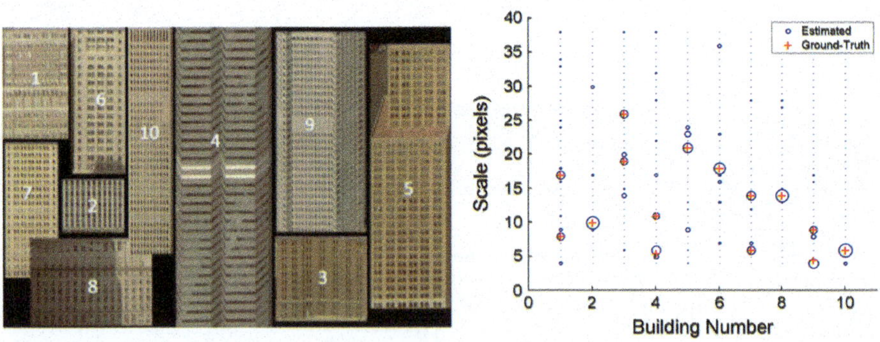

Fig. 5.6 Evaluation of scale estimation accuracy for a test set of 10 building facades from the BEV imagery. Densely estimated scale values from each facade are used to compute a normalized histogram which is plotted as *blue circles* (with radii proportional to the histogram values); the *red pluses* denote ground-truth scale values

Table 5.2 Facade detection performance

Scene	TP rate (%)	# Buildings	# FPs
BEV-1	86	29	8
BEV-2	91	33	3
BEV-3	86	21	5

5.5.2 Facade Detection Evaluation

Table 5.2 shows results from our facade detection algorithm. For each BEV scene, we looked at the computed horizontal scales—points with nonzero scale values are treated as potential facades. We quantify the performance as follows: for each building facade, if at least 50 % of its visible area was assigned a nonzero scale, then we count it as a true detection. If in any 4×4 sub-grid of sampled locations not on a building facade, at least 25 % are assigned a nonzero scale, then we count it as a false-positive.

5.5.3 SV-to-BEV Matching

Figure 5.7 illustrates our typical query SV processing pipeline. The algorithmic steps are outlined in Algorithm 2. First, the query SV image is orthorectified [4, 18]. Next, the motif scale computation algorithm described in Sect. 5.3 is employed to compute the horizontal scales on a uniform grid. S^4 descriptors are then computed at locations with nonzero scale producing the query word set $\mathscr{W} = \{w_1, w_2, \ldots, w_n\}$. The facade matching algorithm in Sect. 5.4 is used to label each word with a cluster label.

Figure 5.8a shows the retrieval performance of our approach (along with a comparison with SIFT—details in Sect. 5.5.4) with a query set of 79 images including 33 true negatives, i.e., buildings which were either not part of the BEV database or were significantly occluded. The query set contains challenging images with significant uncorrected image distortions, urban clutter, and varied zoom range. A third of these images are high-resolution pictures from Flickr and Panoramio and the remaining are low-resolution screenshots from Google Street-View and Bing Streetside. A few samples from the query set are shown in Fig. 5.5b. For generating the ROC curves, instead of using the most probable label from Eq. 5.6 directly, we treat the vector of frequency of each label $f(k) = \sum_i \delta(\mathscr{L}(w_i) = c_k)$ as a probability distribution. Then, to get a point on the ROC curve, we pick a value between 0.0 and 1.0 and select all the labels with probabilities higher than this value. This becomes our retrieval set which is compared with the ground-truth facade set to compute the TP and FP rates in the usual manner.

Figure 5.10 shows few examples of the top three retrieval matches on representative (screen-captured) Google street-view queries. From the amount of perspective

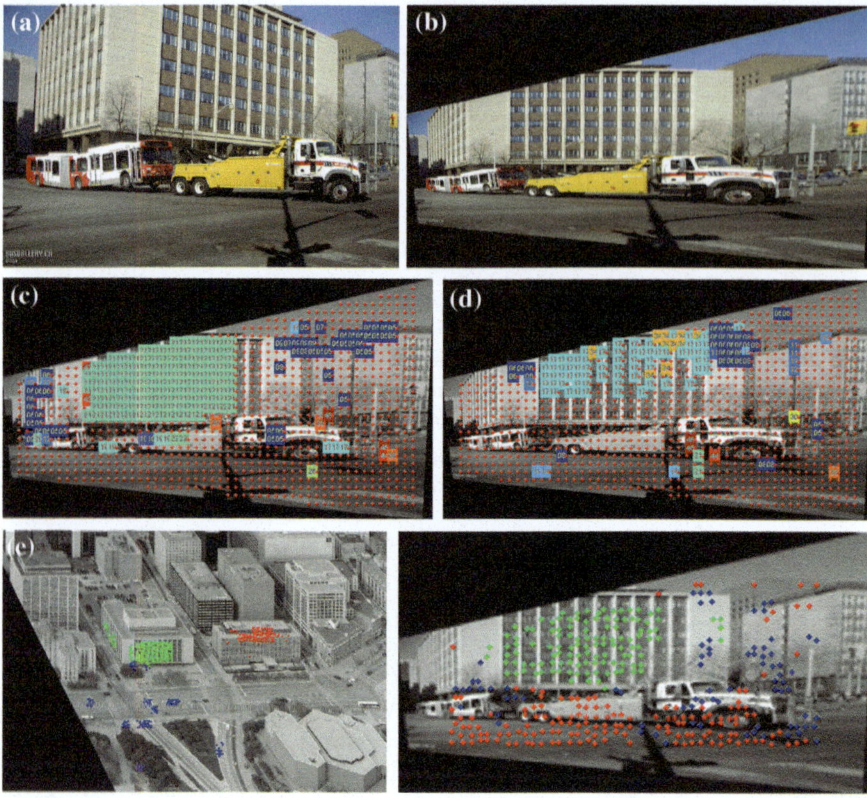

Fig. 5.7 Example street-view processing. **a** Query street-view, **b** orthorectified street-view, **c** street-view horizontal scales, **d** street-view vertical scales, **e** matching result with BEV with correspondingly matching clusters shown in *same colors*

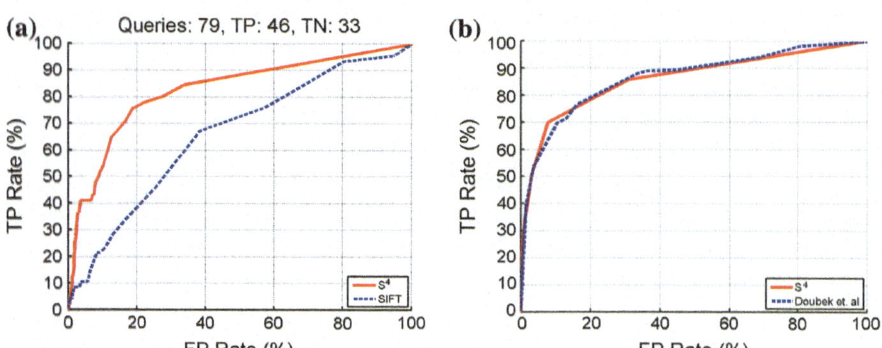

Fig. 5.8 **a** ROC curve for BEV-to-SV matching on Pittsburgh dataset, **b** SV-to-SV matching performance on the "Pankrac+Marseilles" public dataset

(and distortion) in the SV imagery, it is clear that features like MSER and SIFT would hardly find any correspondences. The bottom row shows some problem cases due to the scale difference between the SV and BEV images or due to the global rectification of the BEV image—as expected, the descriptors from non-fronto-parallel facades do not match well to fronto-parallel rectified SV imagery.

5.5.4 Comparison with SIFT Features

Given the prevalence of SIFT features in wide-baseline matching literature, we present experimental comparison of its performance with our approach. To avoid any bias against SIFT due to perspective distortions (and to preclude comparison with SIFT variants like A-SIFT), we extract SIFT features on orthorectified BEV and orthorectified SV imagery. Next, we use the building clusters found using our S^4-based algorithm and perform an assignment of the SIFT features to these clusters using a nearest-neighbor association on pixel coordinates thus discarding any features on nonbuilding background clutter. The Bayesian classification from Sect. 5.4 is used on the SIFT clusters to retrieve matching facades for the query images and the quantitative results are shown in the ROC in Fig. 5.8a which illustrates that we achieve significant improvement in performance using S^4 features instead of SIFT features.

5.5.5 Street-View to Street-View Matching

We present results of our approach applied to the public dataset "Pankrac+ Marseilles"[1] of SV-only images. This dataset contains 106 images of approx. 30 buildings from Pankrac, Prague, and Marseille appearing in more than one image, number of appearances ranges from 2 to 6. Figure 5.8 shows the performance of our approach on this dataset and compares it with the best performance shown by [8]. Their approach uses detection of repeated lattice tiles followed by appearance features on the detected tile pattern as a means to match between facades in SV imagery and it is not surprising that we come close to their results using the self-similarity descriptor. In fact, the SV-to-SV matching problem does not entail the same challenges as the SV-to-BEV matching we address here. In the former, once the projective distortions are removed, one can still achieve good performance using direct feature matching across images because the appearance does not look as dissimilar as in the case of SV versus BEV. A qualitative snapshot of our results is shown in Fig. 5.9 (Fig. 5.10).

[1] http://cmp.felk.cvut.cz/data/repetitive.

Fig. 5.9 SV-to-SV matching examples on the "Pankrac+Marseilles" public dataset. *First column* shows the query followed by the top three retrieved results. The *red*, *green* and *blue points* denote the subset of features in the query which match best with each of three top contenders respectively. The *black points* match some other images from the database. The database had only one matching candidate in the case of the second example and the other retrieval results have the closest matching facade structure to the query

5.5.6 Camera Pose Estimation

Facade matching is in itself good enough to localize the SV image within a constrained visibility zone defined by the facadc. However, for precise localization of the SV camera we compute the 6 DOF pose of the camera to establish the efficacy of our method. We have developed two different algorithms for this task and we briefly describe them below.

Manual Correspondence Algorithm In this method, we manually identify 7 point correspondences between the SV and BEV image in the structure surrounding the matched facade. These correspondences are used to estimate the fundamental matrix F [9] between the SV and BEV images. The epipole of the BEV image, as computed from F, then corresponds to the SV camera location in the BEV coordinate system.

Automatic RANSAC Algorithm In this method, we use purely geometric constraints to simultaneously estimate both the correspondences as well as the camera pose. We use the plane+parallax methodology for this problem. We use the recovered matching facade to estimate a homography between the BEV and SV imagery and then use this homography to measure the parallax corresponding to any point correspondence between the SV and BEV imagery. We employ constraints from the SAT imagery to make this problem more tractable. We use a few corners from the SV image and building top corner correspondence (between BEV and SAT imagery) to enforce the parallax geometry and recover the SV camera location. The top corner correspondence itself can be extracted from the facade extraction phase where we explicitly detected the top edge of each building when using the stereo-based algorithm.

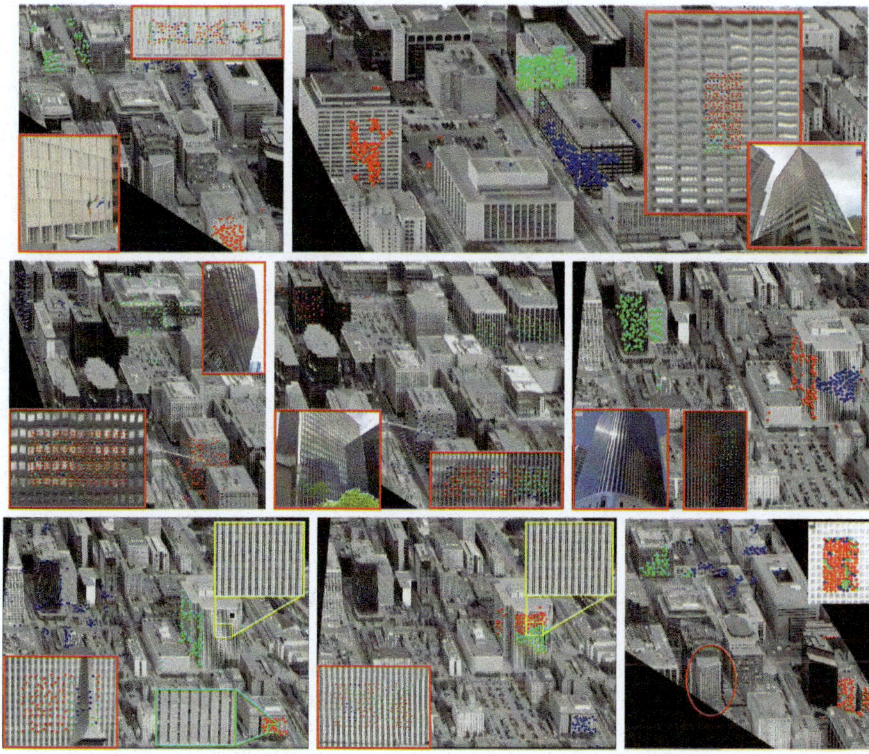

Fig. 5.10 Qualitative matching results. The main tiles show rectified BEV images. The *insets* show the original and rectified query street-view facades. On the *rectified inset*, the *colored points* are a subset of the words $w_1, w_2, \ldots, w_n$ with the top three most frequent recovered labels $\mathscr{L}(w_i)$ shown as *red*, *green* and *blue points* respectively; similarly *colored points* in the BEV image are words v_j which belong to these three clusters. *Top two rows* some examples of correct retrieval. *Bottom row* Mismatched result (*left*) due to missing fine structure in the street-view image that was seen in the BEV image. Correct matching result for another street-view image which shows the fine structure is shown in the *middle*. The *bottom right* example shows a problem scenario where the non-fronto-parallel facade in the BEV causes it to be mismatched due to the difference in the self-similarity structure presented by the descriptors

The SV camera location in the BEV image is mapped to absolute lat-long coordinates using the ground plane correspondence with the SAT imagery. Finally, the metric (cms/pixel) information in the SAT image is used to estimate the camera focal length which can be used in conjunction with any knowledge about the CCD array dimensions to establish the camera field of view as well. The camera look-at direction is also estimated using the metric information available from the SAT imagery by a simple trigonometric calculation. Figure 5.11 shows the localization results obtained for three query images using the manual correspondence algorithm.

Fig. 5.11 Localization results. The *top row* shows the SV images, and the *bottom row* shows the estimated and ground-truth camera locations

5.6 Facade Extraction Using Satellite Imagery

In Sect. 5.3, we described an approach to extract facade regions from BEV imagery by clustering S^4 descriptors. However, this technique would not be able to distinguish between nearby buildings if they exhibit similar facade pattern. We have developed a geometric technique [2] to extract building facades from BEV imagery by aligning it with satellite (SAT) imagery when available.

Given a BEV image and a SAT image for the same region of interest, first, we align the ground plane between the SAT and BEV imagery. Thus, BEV images can be rectified with respect to the ground plane with canonical axes (N–S, E–W) aligned. Then, we match building outlines extracted from SAT imagery with the corresponding outlines in the rectified BEV images. Subsequently, we use the identified building outlines to find the roofs of buildings thus identifying the facades. This allows extraction of orthorectified building facades from the BEV from which S^4 features can then be extracted and matched using the matching algorithms described before. In the following, we describe in sequence the above algorithmic steps in more detail.

Imagery Alignment Given the set of SAT and BEV image tiles (see examples in Fig. 5.12) and the mapping of their pixel coordinates to lat-long coordinates, we can warp the BEV images to the SAT coordinate system. To compute the warping transformation, we approximate it as a projective transformation between pixels in SAT and BEV—thus approximating the Earth's surface within each tile as a flat plane. Using the computed transformations, we warp each of the images to the SAT image coordinate system. As a result, the ground plane gets aligned well in all the images as shown in Fig. 5.13. To aid further processing, we also compute the dominant city block direction in the SAT imagery and rotate this image before warping the other

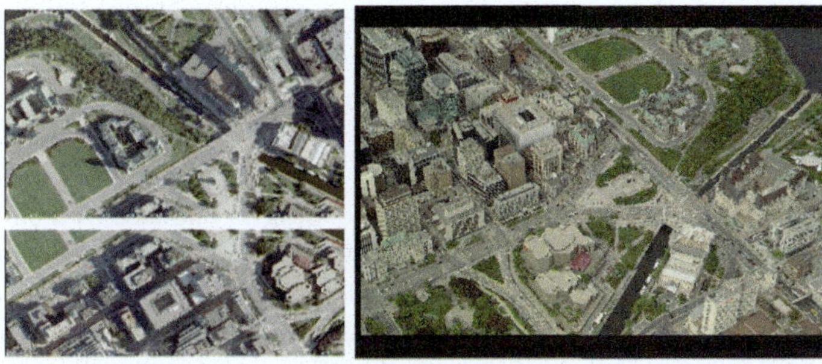

Fig. 5.12 Sample SAT tiles (*left*) and BEV imagery (*right*) from Ottawa, Canada

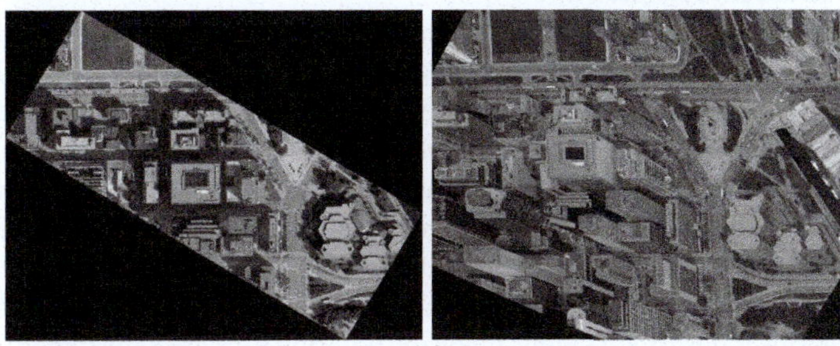

Fig. 5.13 SAT image rotated to align city-block direction with the x-axis (*left*) and the corresponding BEV image automatically aligned to the SAT image w.r.t. the ground plane using the geo-coordinate information (*right*)

images to its coordinate system. This renders most of the buildings parallel to the scan lines in the image—a feature which will be exploited in further processing.

After initial imagery rectification, we extract regions from the BEV imagery corresponding to building facades. To ensure least distortion, we concentrate only on the facade planes which face the heading direction of the particular BEV image. Since the SAT imagery was previously rotated to align the city blocks with the image scan lines, we can now restrict our attention to facade planes whose 2D projections are horizontal in the SAT images.

Vertical Vanishing Point Estimation In the ground-aligned BEV imagery, lines along the vertical (gravity) vanishing direction can be seen to be convergent. Before extracting affine corrected facades, we first rectify the BEV imagery so that these lines are rendered parallel. We detect canny edges in the BEV image and then group these edges into line segments. Lines along horizontal and vertical directions correspond to city block axes and can hence be rejected. From the remaining line segments, a RANSAC-based process then determines the inlier set of lines that intersect at the required vanishing point.

Image Rectification Given the computed vanishing point, we now rectify the BEV image by mapping this vanishing point to a point at infinity (in particular to $v_x = [1, 0, 0]'$), thus making the building edges parallel. This rectifying transformation is a projective warp which is computed by a method similar to the epipolar rectification method described in [9]. Due to the choice of v_x, the building facade edges in the rectified BEV become parallel to the image scan lines.

SAT Edge Extraction To extract building facades from BEV, we start by detecting building contours in the overhead SAT imagery. The contours need to be detected as chains of line segments, each segment corresponding to one face of a building. We developed an iterative algorithm to extract these line segments from an initial canny edge-detector processed SAT image. Briefly, the algorithm links edges into edge-chains based on proximity and then fits line segments to these edge-chains, splitting wherever the deviation of the edges from the fitted line segment becomes greater than a threshold. Consistent line segments are merged into longer line segments and the overall process is iterated a few times.

Facade ROI Search From the line segments extracted in the SAT imagery, we keep only the segments along the dominant facade direction in the BEV. Using the ground plane homography between SAT and BEV, we warp these segments into the rectified BEV image coordinate system. These segments then map to approximately the bottom of the buildings in the BEV image because the transformation corresponds to the ground plane. In the rectified BEV imagery, the gravity vanishing direction is aligned with the scan lines and therefore the tops of these buildings can be found by sliding the mapped line segments horizontally (Fig. 5.15). Our algorithm to determine the building tops is described below. Once the building tops are determined, we obtain the coordinates of the four corners of each facade which can then be mapped back to the original (unrectified) BEV imagery for high-resolution texture retrieval. For each facade, we crop the texture from the original BEV imagery and then warp it into a rectilinear coordinate system. Figure 5.14 shows an example of this process where a few of the facades are extracted and rectified to their orthographic representations.

Computation of Building Tops using GC Given the nature of the rectified BEV imagery, the top of each building can be determined as a translation $\delta(s)$ for each segment s projected to the building bottom. We formulate this problem as a Graph

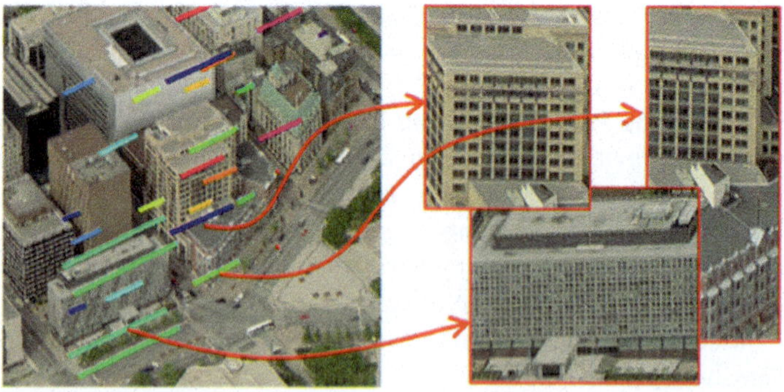

Fig. 5.14 Example of the facade extraction process. **a** Detected building tops and bottoms, and **b** extracted facade tiles

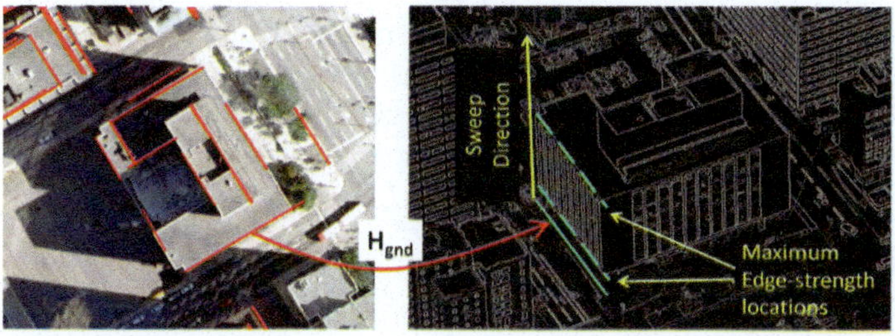

Fig. 5.15 Building top search. Line segments extracted from the SAT imagery are projected to the canny edge map of BEV where a sweep along the gravity direction is expected to give a maximal point at the top edge of the building

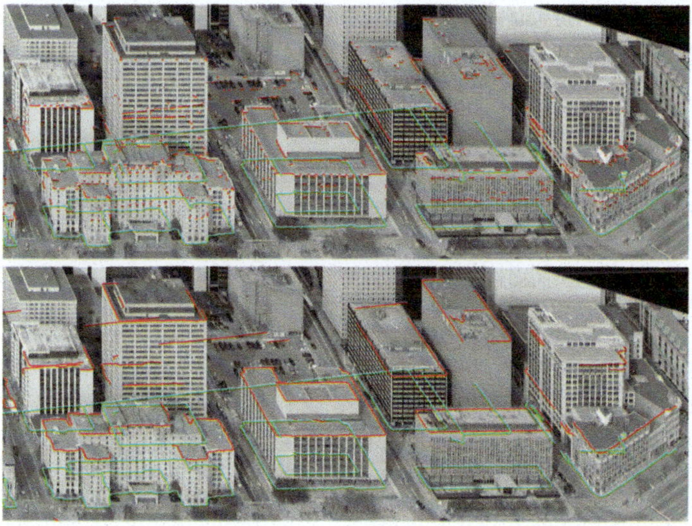

Fig. 5.16 Effect of graph-cuts optimization. The *green edges* are the SAT edges directly projected to this view and they lie in the ground plane. The *red edges* are the estimated building top edges. The *top row* shows the estimates obtained by picking the maximum score for each edge pixel independently; the *bottom row* shows these estimates refined by the GC optimization

Cut (GC) optimization of an objective function that consists of the usual data and smoothness costs. The data cost for a line segment is strictly a function of the hypothesized translation and is computed by measuring the average edge strength in the rectified BEV image when the line segment is translated by this value. Thus, when the segment lands on the top of a building, we incur a lower cost due to the high edge strength. To ensure smoothness in the translation values for connected line segments, we add a smoothness cost that penalizes difference in translation values for line segments that are spatially close to each other at their endpoints. For the typical polygonal chains of line segments that we detect for each building, the smoothness cost enforces a strong constraint that the entire building top be at a single translation and avoids the problem of local optima occurring at the numerous edges in the middle of the building facade. Figure 5.16 shows an example of how this optimization approach helps the building extraction process.

5.7 Related Work

In the discussion of related work, we emphasize two main aspects: **detection** of facades/lattices and **matching**. Chung et al. [6] extract MSER regions in multiple scales which are then clustered w.r.t. similarity. Local histograms of gradient similarity, area ratio, and configuration entropy are used to build adjacency matrices which are matched by using a spectral approach comparing only the graph structure. The commonality with our approach is that we never use any direct comparison of appearance across images. On the other hand, their query and model graph structures have to match globally while our approach uses the statistics of the edges of these graphs represented by the self-similarity descriptor and hence exploits the redundancy in features better. Moreover, the self-similarity descriptor is more general and implicit than the concatenation of several neighborhood descriptions (HoG, area ratio, entropy). Park et al. [13] model the lattice discovery as a multi-target tracking problem using Mean-Shift Belief Propagation. Candidates for lattice vertices are interest points that are obtained through clustering. Hays et al. [10] randomly select regions and search for their repetition in two directions in their immediate neighborhood. Lattice discovery is formulated as a graph matching problem with higher order constraints that model the lattice structure of the region repetitions. The advantage of [10, 13] is that they can deal with deformed lattices in the detection step while almost all other approaches including ours remove projective and sometimes affine distortions using vanishing points and ratio constraints. Schindler et al. [16] detect lattices by mapping quadruples of SIFT features to the projective basis and checking the consistency of the rest of the points with respect to this basis. They combine multiple 2D-to-3D pattern correspondences and recover the camera orientation and location as an intersection of the family of solutions obtained using each correspondence.

Bansal et al. [2, 3] established the feasibility of matching highly disparate street-view images to aerial image databases to precisely geo-localize SV images without the need for GPS or camera metadata. Doubek et al. [8] match the similarity of repetitive patterns by comparing the grayscale tiles, the peaks in color histogram, and the sizes of the two lattices. In [15], corners are extracted and grouped according to consistency with the geometric transformations corresponding to the generators of the lattice. Kosecka and Zhang [11] extract rectangle projections by grouping line segments according to vanishing point consistency. Using [20] they match a query street-view image to a database of geo-tagged street-view images using wide-baseline matching. In [7, 14], a query street-view image is again matched to a database of street-view images and then used to compute the camera pose. They assume the query image camera internal parameters to be known and use a pyramid to match at multiple scales using geometric consistency. In [19], a viewpoint normalization of planar patches is followed by SIFT computation of the rectified patch. Matei et al. [12] used a LIDAR scan of the environment to create a DEM which is rendered exhaustively from multiple locations and viewpoints. Features extracted from these renderings are matched against query features to generate candidate camera locations. Bansal and Daniilidis [1] employ a DEM as the starting point of their geo-localization setup as well. However, instead of rendering a priori in a quantized pose space, they extract sparse PointRay features which are composed of a pair of building corner and edge direction, and are thus purely geometric. They describe a stratified algorithm that uses the two-point minimal algorithm to compute candidate query poses without any appearance matching. We close our discussion with Cham et al. [5] where omnidirectional views are matched to building outline maps by detecting the tallest vertical corners of the buildings which are matched through 2D to 1D projection.

5.8 Conclusion

We have been able to match query street-level facades to airborne imagery under challenging viewpoint and illumination variation by introducing a novel approach of selecting the intrinsic facade motif scale and modeling facade structure through self-similarity. Using the motif scale, we extract and segment lattice-like facades and construct scale-invariant S^4 descriptors. We localize queries by classifying descriptors, thus matching to facades with semi-local lattice consistency.

References

1. Bansal M, Daniilidis K (2014) Geometric urban geo-localization. In: The IEEE conference on computer vision and pattern recognition (CVPR)
2. Bansal M, Sawhney HS, Cheng H, Daniilidis K (2011) Geo-localization of street views with aerial image databases. In: Proceedings of the 19th ACM international conference on multimedia, vol 16
3. Bansal M, Daniilidis K, Sawhney H (2012) Ultra-wide baseline facade matching for geo-localization. In: Computer vision–ECCV 2012. Workshops and demonstrations. Springer, Berlin
4. Barinova O, Lempitsky V, Tretiak E, Kohli P (2010) Geometric image parsing in man-made environments. In: ECCV
5. Cham T, Ciptadi A, Tan W, Pham M, Chia L (2010) Estimating camera pose from a single urban ground-view omnidirectional image and a 2D building outline map. In: CVPR
6. Chung Y, Han T, He Z (2010) Building recognition using sketch-based representations and spectral graph matching. In: ICCV
7. Cipolla R, Robertson D, Tordoff B (2004) Image-based localisation. In: Proceedings of 10th international conference on virtual systems and multimedia
8. Doubek P, Matas J, Perdoch M, Chum O (2010) Image matching and retrieval by repetitive patterns. In: ICPR
9. Hartley R, Zisserman A (2003) Multiple view geometry in computer vision. Cambridge University Press, New York

10. Hays J, Leordeanu M, Efros A, Liu Y (2006) Discovering texture regularity as a higher-order correspondence problem. In: ECCV
11. Kosecka J, Zhang W (2005) Extraction, matching, and pose recovery based on dominant rectangular structures. In: CVIU, vol 100. Elsevier, pp 274–293
12. Matei BC, Vander Valk N, Zhu Z, Cheng H, Sawhney HS (2013) Image to lidar matching for geotagging in urban environments. In: WACV (2013)
13. Park M, Brocklehurst K, Collins R, Liu Y (2009) Deformed lattice detection in real-world images using mean-shift belief propagation. TPAMI 31(10):1804–1816
14. Robertson D, Cipolla R (2004) An image-based system for urban navigation. In: BMVC
15. Schaffalitzky F, Zisserman A (1999) Geometric grouping of repeated elements within images. In: Shape, contour and grouping in computer vision
16. Schindler G, Krishnamurthy P, Lublinerman R, Liu Y, Dellaert F (2008) Detecting and matching repeated patterns for automatic geo-tagging in urban environments. In: CVPR
17. Shechtman E, Irani M (2007) Matching local self-similarities across images and videos. In: CVPR
18. Tardif J (2009) Non-iterative approach for fast and accurate vanishing point detection. In: ICCV
19. Wu C, Clipp B, Li X, Frahm J, Pollefeys M (2008) 3d model matching with viewpoint-invariant patches (vip). In: CVPR
20. Zhang W, Kosecka J (2006) Image based localization in urban environments. In: Third international symposium on 3D data processing, visualization, and transmission

Part II
Semantic Reasoning Based Geo-localization

Identifying the geo-location based on high-level and semantic cues

High-level semantic features, such as architectural style, vegetation type, or language of signs, contain a large amount of geographically discriminative cues. Comprehensive databases of geo-referenced semantic information, for instance Geographical Information Systems (GIS), OpenStreetMap, or Wikimapia, are also widely available to public. For these reasons, it is feasible and beneficial to incorporate semantics in the geo-localization process. In the previous part of the book, we discussed the techniques that perform geo-localization using massive geo-referenced datasets but with no particular notion of semantics. In this part, we investigate how high-level semantic cues, in particular urban area concepts and landmarks, can be leveraged in the geo-localization process.

Chapter 6
Semantically Guided Geo-location and Modeling in Urban Environments

Gautam Singh and Jana Košecká

Abstract The problem of localization and geo-location estimation of an image has a long-standing history both in robotics and computer vision. With the advent of availability of large amounts of geo-referenced image data, several image retrieval approaches have been deployed to tackle this problem. In this work, we will show how the capability of semantic labeling of both query views and the reference dataset by means of semantic segmentation can aid (1) the problem of retrieval of views similar and possibly overlapping with the query and (2) guide the recognition and discovery of commonly occurring scene layouts in the reference dataset. We will demonstrate the effectiveness of these semantic representations on examples of localization, semantic concept discovery, and intersection recognition in the images of urban scenes.

6.1 Introduction

The task of geo-location of a query image is typically posed as an image matching problem where it is assumed that a reference dataset consisting of geo-tagged images is provided. Several location recognition algorithms use repeatable local features like SIFT [16] along with spatial verification methods to match scenes [20, 30]. To cope with the large amount of features that are typically computed for an image, the bag of visual words (BOW) models [2, 26] utilize clustering of similar descriptors to produce a vocabulary of features which are then used in efficient retrieval strategies [18, 20]. City scale localization by improving the quality of the computed vocabulary was studied in [1, 23]. Authors in [9] computed a vocabulary where visual words which are considered nondiscriminative for geographic location are discarded. In the work of [7], the authors proposed a data-driven method for computing the coarse geograph-

G. Singh (✉) · J. Košecká
George Mason University, Fairfax, VA, USA
e-mail: gsinghc@cs.gmu.edu

J. Košecká
e-mail: kosecka@cs.gmu.edu

A.R. Zamir et al. (eds.), *Large-Scale Visual Geo-Localization*,
Advances in Computer Vision and Pattern Recognition,
DOI 10.1007/978-3-319-25781-5_6

ical location of an image using a collection of global features including GIST [19] and color histograms. More recently, [3] focused on finding geo-informative patches which correspond to visual elements like balconies, windows, and street signs in a discriminative clustering framework. While these patches elements are not directly applied to geo-location, they are used for related tasks like mapping architectural patterns to geographic locations and visual correspondences across different cities, e.g., street lamps in Paris and London. In this work, we propose to utilize the semantic segmentation of scenes for geo-location and related tasks. The presence (and consequently also their absence) of semantic categories like building, ground, trees at different spatial locations in an image characterizes its semantic layout which can be used to enhance BOW techniques for geo-location. This semantic layout is also used for the association of metadata with geographic locations like the presence of intersections.

The problem of semantic segmentation requires simultaneous segmentation of an image into regions and categorization of all image pixels. With the development of methods for integration of object detection techniques, with various contextual cues and top-down information as well as advancements in inference algorithms used to compute the optimal labeling, semantic segmentation has been an active research problem in computer vision. A majority of the current approaches are formulated using a graphical model over image sites (the basic image element to label) which captures the dependencies between image sites. Conditional random fields [13] (CRFs) are the most popular variant of such approaches. These methods differ in the sites that they aim to label (e.g., individual pixels, square patches, superpixels) and also in the dependencies between the sites. In the work of TextonBoost [24], pixel-level label evidence is gathered by combining appearance, shape, and context-based features. These methods have been extended to use superpixels [5] which are computationally more efficient for inference purposes. Researchers in [10] proposed the use of higher order CRFs in a hierarchical framework which allowed the integration of features at different levels (pixels and superpixels). Other works look at incorporating object co-occurrence statistics [11, 21] including spatial offset based priors [6] and results from object detectors [12]. We will describe how the availability of information about the semantic labels present in a scene can enhance appearance-based geo-location and enable us to infer additional semantic concepts about the environment, e.g., highways, inner-city streets.

The remainder of the chapter is organized as follows. We describe our approach for semantic segmentation of scenes in Sects. 6.2.1 and 6.2.2. We then introduce the informative *semantic label descriptor* that summarizes the semantic layout of a scene in Sect. 6.2.3. The large-scale StreetView dataset used in our experiments is described by Sect. 6.3.1. Our first contribution is the acquisition of a topological map over an urban environment using semantic labeling is presented in Sect. 6.3.2. We also show the ability to learn semantic concepts like the presence of intersections inside a city in Sect. 6.3.3. Having shown the ability to associate meta-information with geographic locations by using semantic segmentation, we then present its direct application to geo-location displaying improvements in appearance-based geo-location by utilizing the semantic similarity between street scenes in Sect. 6.3.4. We conclude with a discussion in Sect. 6.4.

6.2 Approach

We begin with a discussion of the framework for computing the semantic layout of street scenes. This is followed by an introduction to the method to summarize the semantic layout in an informative feature which is then applied to different tasks like topological mapping, intersection detection, and improving appearance based location recognition.

6.2.1 Semantic Segmentation Formulation

We formulate the semantic labeling of an image segmented into superpixels. The output of the semantic segmentation is a labeling $\mathbf{L} = (l_1, l_2, \ldots l_S)^\top$ with hidden variables assigning each superpixel s_i a unique label, $l_i \in \{1, 2, \ldots, L\}$, where L is the total number of the semantic categories and S is the number of superpixels in the image. The posterior probability of a labeling $\mathbf{L}$ given the observed appearance feature vectors $\mathbf{A} = [\mathbf{a}_1, \mathbf{a}_2, \ldots, \mathbf{a}_S]$ computed for each superpixel can be expressed as:

$$P(\mathbf{L}|\mathtt{A}) = \frac{P(\mathtt{A}|\mathbf{L})\,P(\mathbf{L})}{P(\mathtt{A})}. \tag{6.1}$$

The labeling $\mathbf{L}$ can estimated as a Maximum A Posteriori Probability (MAP),

$$\underset{\mathbf{L}}{\operatorname{argmax}}\, P(\mathbf{L}|\mathbf{A}) = \underset{\mathbf{L}}{\operatorname{argmax}}\; P(\mathbf{A}|\mathbf{L})\,P(\mathbf{L}). \tag{6.2}$$

where $P(\mathbf{A}|\mathbf{L})$ is the observation likelihood and $P(\mathbf{L})$ is the joint prior. In order to compute the observation likelihood, a Naive Bayes assumption of independence between appearance features given the labels is made and it yields

$$P(\mathbf{A}|\mathbf{L}) \approx \prod_{i=1}^{S} P(\mathbf{a}_i|l_i). \tag{6.3}$$

where $P(\mathbf{a}_i|l_i)$ is the label likelihood and models the correlation between the appearance and semantic label for superpixel s_i. We compute this likelihood using a boosting classifier, the details of which are described in Sect. 6.2.2. While many approaches to semantic segmentation model the joint prior $P(\mathbf{L})$ using a pairwise smoothness term in a Markov random field (MRF), we forgo the use of an MRF in our approach and utilize only the observation likelihood for a faster semantic segmentation of an image.

6.2.2 Superpixels and Features

Our approach for semantic labeling is based on using a single bottom-up segmentation of the image where the superpixels are characterized with a variety of features including color, texture, location, and perspective cues. The labeling is performed using boosting classifiers which automatically compute feature relevance in the trained classifier. The semantic labels we consider are five commonly occurring semantic categories in street scenes—*ground, sky, building, car, tree*. The proposed semantic segmentation approach is closely related to [8, 27]. The labeling is done on superpixels obtained by the color-based oversegmentation scheme proposed in [4]. This segmentation algorithm typically generates large irregular regions of different sizes (see Fig. 6.1a).

Since we are interested in learning the coarse semantic layout of the urban environment, we use both geometric as well as appearance features to capture the statistics of individual regions. The choice of features has been adopted from [8] where each superpixel is characterized by location and shape (position of the centroid, relative position, number of pixels, and area in the image), color (mean of RGB and HSV values and histograms of hue and saturation), texture (mean absolute response of the filter bank of 15 filters and histogram of maximum responses) and perspective cues computed from long linear segments and lines aligned with different vanishing points. We use the publicly available code provided by authors of [8] for computing these features. In addition to the above features, we endow each superpixel region

Fig. 6.1 **a** Image from StreetView dataset. **b** Segmentation obtained using the graph-based method of [4]. Superpixel boundaries marked by *red color*. **c** Semantic labeling result for the given image. Color code for labels in *top row*

Table 6.1 List of features computed for the image superpixels

Color
RGB color mean
HSV values
Hue histogram (5 bins)
Saturation histogram (3 bins)
Texture
Histogram over LM filter [15] response
Mean absolute LM filter response
Location and shape
Mean normalized x and y
Normalized x and y, 10th and 90th percentile
Normalized y w.r.t. estimated horizon
Segment location w.r.t. horizon
Normalized superpixel area
Perspective cues
Lines: Number of line pixels
Lines: Percent of (nearly) parallel lines
Line intersections: 8 bin histogram over orientations
Line intersections: Percent right/above image center
Vanishing points: Number of line pixels with vertical VP membership
Vanishing points: Percent of line pixels with vertical VP membership
Vanishing points: Number of line pixels with horizontal VP membership
Vanishing points: Position of vertical VP w.r.t. segment center
Gradient histogram
Dense SIFT [16] histogram with 100 clusters

with a histogram of SIFT descriptors computed densely at each image location and quantized into 100 clusters. The entire feature vector is of 194 dimensions. Table 6.1 summarizes the features computed for each image region.

In order to compute the label likelihood, $P(\mathbf{a}_i|l_i)$ in Eq. (6.3), we use discriminative boosting classifiers [22]. Within the boosting framework, we use decision trees as the weak learners since they automatically provide feature selection. We learn separate classifiers for each of the five classes and this is done in a one versus all fashion. During testing, the separate classifiers are run on the individual feature vectors of the superpixels of an image and output confidence scores. The class with the maximum confidence score is assigned to be the superpixel's label. In our implementation, each strong classifier has 15 decision trees and each of the decision trees has 8 leaf nodes (using the implementation of [8]). An example of the obtained semantic layout is shown in Fig. 6.1c.

Table 6.2 Per pixel and average per class accuracy on 320 StreetView image dataset

System	Global	Average
Zhang-ECCV10 [29]	88.4	80.4
Zhang-CVPR11 [28]	93.2	73.1
Boosting	94.4	81

Table 6.3 Comparison of category level accuracy on 320 StreetView image dataset

System	Building	Car	Ground	Sky	Tree
Zhang-ECCV10 [29]	89.1	56.4	89.6	97.1	69.7
Zhang-CVPR11 [28]	95.3	40.5	96	92.5	41.4
Boosting	96.4	68.3	94.4	97.2	48.9

We annotated a dataset of 320 side and 90 frontal views where each pixel of an image is assigned one of the five classes or *void* if it does not fall into any of the categories. Two separate models are learned, one using the dataset of side views and the other using the frontal views. This is because while the classes may have similar appearance across side or frontal views (e.g., trees are generally green in color), they may not necessarily share the same geometric properties in the two different views. As an example, in a frontal view the buildings are generally observed on the sides with the ground/road in the middle while that is not the case in the side views where they appear in fronto-parallel views. To evaluate the performance of the boosting classifier and compare it to state-of-the art systems, we use the dataset of 320 side views. The classifier was trained using one-half of the dataset for training and the other half is used for testing.

The results for the boosting classifier and its comparison to the approach of supervised label transfer [29] and nonparametric scene parsing [28] methods on this dataset are provided in Tables 6.2 and 6.3. Here, the global accuracy is the per pixel accuracy and the average accuracy is the average of the five individual category accuracies. It is observed that the boosting classifier outperforms the other state-of-the-art systems on this dataset and therefore, we use this classifier through all our experiments for the semantic labeling of an image in this work. Some examples of the semantic labeling results can be found in Fig. 6.2. The illustrated examples include both frontal and side views.

6.2.3 *Semantic Label Descriptor*

The semantic labeling of an image provides a means of spatially aggregating the semantic information in the image. For example, the semantic labeling of a highway scene will typically be devoid of buildings while streets inside a city have high-rise

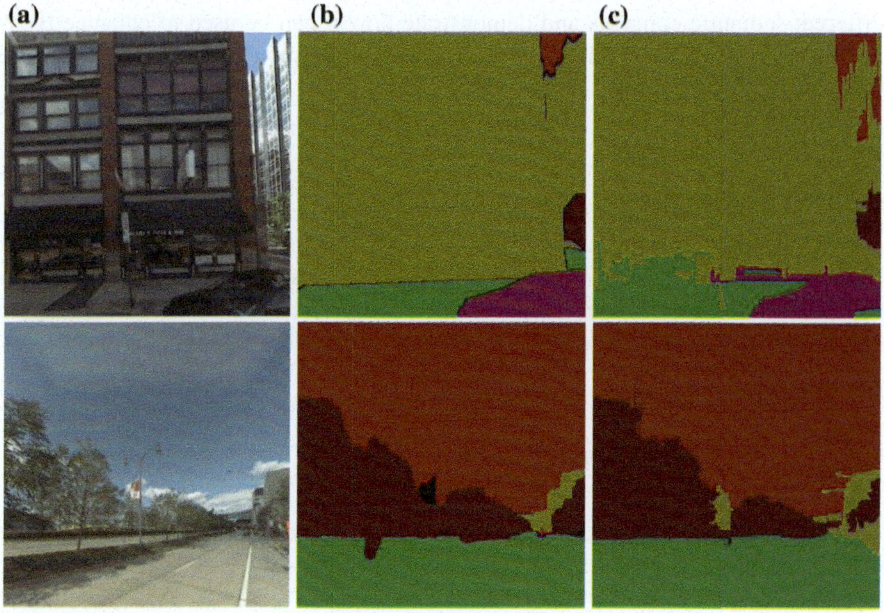

Fig. 6.2 Examples of semantic labeling of Streetview images. The *top row* shows a *side view* while the *bottom row* visualizes a *frontal view*. From *left* to *right*, **a** Input image **b** Ground truth annotation **c** Predicted labeling

buildings. We propose to exploit the scene labeling to obtain a richer understanding of the urban environment. To summarize the semantic information in the labeled image, we introduce the *semantic label descriptor*. This descriptor captures the basic underlying structure of the image and can help divide images into sets of visually and semantically similar images. This is done through encoding the spatial distribution of semantic categories in the image.

For a given image I, we divide I into a uniform $n_k \times n_k$ grid. We compute the semantic labeling of the image and within each cell of the $n_k \times n_k$ grid, we compute the distribution for each of the five semantic categories using the number of individual pixels in that grid cell which have been assigned that class. This results in a five-bin histogram for a single grid cell. The class distribution values for each cell are normalized so that they sum to one. The histograms for the n_k^2 grid cells are concatenated together resulting in a feature vector of length $5 \times n_k^2$. A high value for n_k will capture the details of the layout more precisely but be prone to classification errors while a low value for n_k would be less sensitive to errors in the labeling. In the experiments of this work, we use $n_k = 4$ resulting in a 80-dimensional semantic label descriptor. This semantic label descriptor is used to characterize the semantic layout of street scenes. In our experiments, we show that it provides evidence of different semantic concepts and demonstrate how it can be used to enhance BOW methods for localization and associate meta data like the presence of intersections with geographic locations.

6.3 Experiments

We now provide a description of the street scenes dataset used in our experiments followed by an evaluation of our proposed approach for using semantic labeling for improving appearance-based geo-location and topological mapping on this dataset.

6.3.1 Dataset Description

For the street scene imagery, we use StreetView™ panoramas acquired by a 360° field of view LadyBug multi-camera system. The sequence consists of 12,556 panoramas acquired from a run in an urban environment and has been previously used in loop detection experiments [17, 25].

A single panorama is obtained by warping the radially undistorted perspective images onto the sphere assuming one virtual optical center. The sphere is backprojected into a quadrangular prism to get a piecewise perspective panoramic image (see Fig. 6.3). Our panorama is composed of four perspective images covering 360° horizontally and 127° vertically. The system includes a top camera as well, but it is discarded as it does not provide much information. The panorama is represented by 4 views (front, left, back and right) each covering 90° horizontal FOV as seen in Fig. 6.3. We discard the bottom part of all views which contains parts of the vehicle acquiring the panoramas.

Fig. 6.3 A panoramic piecewise perspective image used in our experiments; composed of *four parts* (from *left* to *right*)—*left*, *front*, *right* and *back views*

6.3.2 Clustering Topology

The proposed semantic label descriptor is now used to cluster locations in the sequence based on their semantic similarity. While evaluating the performance of our boosting classifier (Tables 6.2 and 6.3), we had used half of the 320 labeled images for training and the second half for testing. When computing the semantic layout of the entire sequence of 12,000 views, the classifier is trained using all 320 side views and run on the left- and right-side views of each location. The classifier trained using the 90 frontal views is run on the front and back view of each location. Locations for which the ground truth labels are available were excluded from the sequence labeling exercise. The resulting semantic layouts for the four views are then converted into the semantic label descriptor as described in Sect. 6.2.3. They are then concatenated together to form a location descriptor for each individual location. Since each individual semantic layout results in a 80-dimensional descriptor, the dimensionality of the location descriptor is 320 (using the four views). This process of generating a descriptor from its individual images that characterizes a location's semantic structure is visualized in Fig. 6.4.

We then use the generated location descriptors for clustering the entire sequence. We perform k-means clustering and use cosine similarity between the descriptors instead of Euclidean distance. A color-coded visualization of the clustering output over the StreetView sequence is shown in Fig. 6.5. One may note that the highway section in the bottom part of the sequence is assigned to a single cluster as the

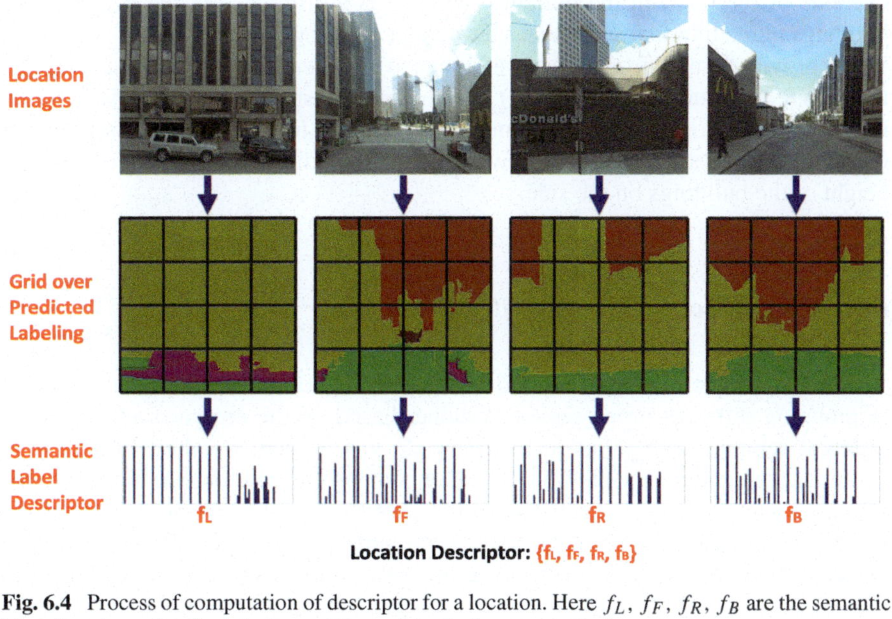

Fig. 6.4 Process of computation of descriptor for a location. Here f_L, f_F, f_R, f_B are the semantic label descriptors for the *left*, *front*, *right*, and *back views* respectively

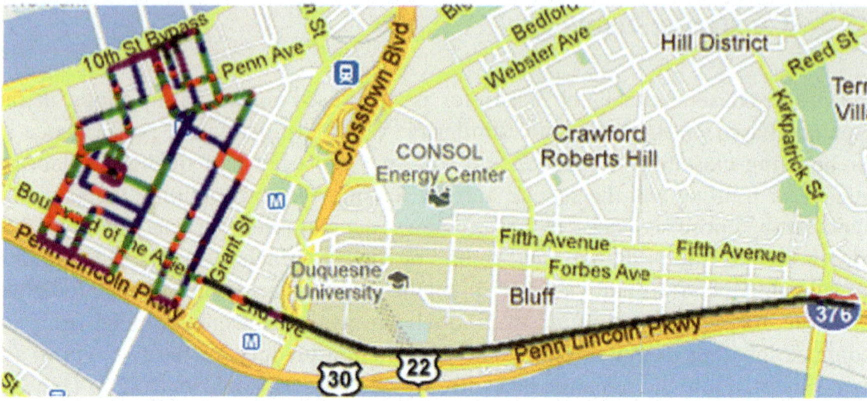

Fig. 6.5 Visualization of the k-means clustering for the entire StreetView sequence. Different colors distinguish the cluster assignments for individual locations (best viewed in *color*)

scene typically contains only sky and road with a few vehicles visible on the road. In Fig. 6.6, we look in more detail at the non-highway portion of the sequence. It is interesting to note how cluster assignments often change at block intersections signifying a change in the semantic structure of the street scenes.

In Fig. 6.7, we provide the average frontal view for a cluster when we set $k = 6$. It can be noted that the different clusters capture distinct semantic structures. For example, the top row has clusters for areas on highways or with buildings on only one side of the road. In the bottom row, there is a difference in the height of the buildings indicating that some areas have taller buildings than others.

We provide sample frontal views assigned to a cluster characterizing such layouts in Fig. 6.8. For example, column-(a) is a cluster corresponding to areas with buildings on one side of the road, column-(c) corresponds to highways, column-(d) has buildings on both sides of the street while column-(b) differs from column-(d) in the height of the buildings on the side.

6.3.3 *Intersection Classification*

The semantic label descriptor introduced in the previous section was instrumental in grouping different urban regions together based on the presence and layout of different semantic categories in the scene. In this section, we show how to infer additional semantic concepts from the attained image representation. In urban environments which can be described as networks of roads and intersections, it is useful to be able to classify a particular view as an intersection or not. The capability of detecting intersections often provides useful prior information of presence of addi-

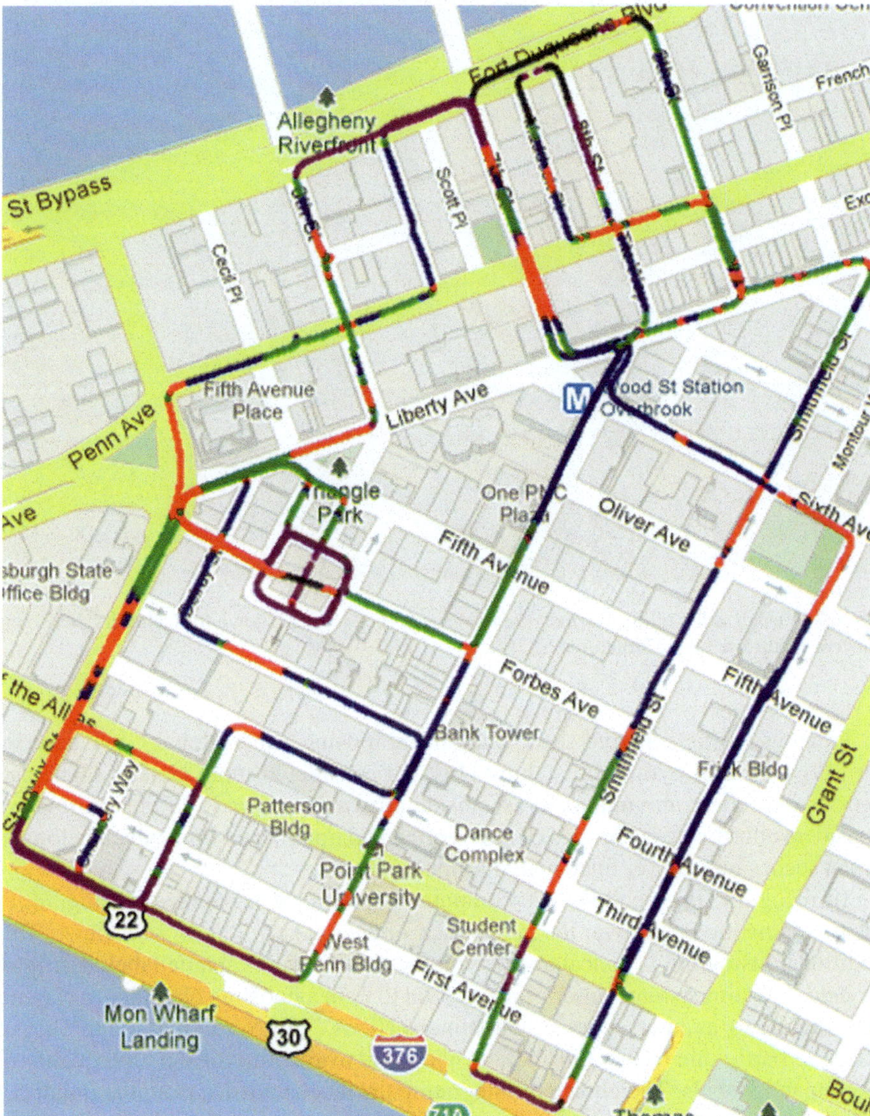

Fig. 6.6 Cluster visualization for the non-highway portion of the StreetView sequence. The highway section has been removed in this visualization. Note the *black color* assigned to the highway in Fig. 6.5 is missing except for an area next to the riverfront which is similar to the highway and lacks buildings on either side

tional semantic concepts, such as pedestrian crossings, stoplights, traffic lights, etc. Intersections also correspond to locations where navigations decisions can be made and hence are of interest for automated driving systems.

Fig. 6.7 Visualization of the average *frontal view* for each cluster (shown for 6 clusters). Different clusters capture semantic structures

Previous works explored scene classification using either global GIST descriptor [19] or spatial pyramid matching [14] and considered more general scene categories like coast, mountain, forest, inside city, and highway. In our setting, we consider subordinate categories of intersection and nonintersection, which belong to urban scenes but vary in finer spatial semantic layout.

To recognize intersections, we compute an additional normalized histogram of the five semantic labels over the middle part of the image width for side views. This additional histogram is concatenated with the side view's semantic label descriptor to yield a 85-dimensional descriptor and used to train a boosting classifier to classify the side views as intersections or nonintersections. This very simple approach is effective partly due to the 360° field of view and availability of the high quality of the semantic labels. The choice of integrating the label statistics from the middle of the side view is motivated by the distinguished appearance of intersections in inner city environments and also the fact that they typically appear at an angle from the main direction of travel. To visualize this intuition, we have computed for the side views (perpendicular to the direction of travel), for each pixel, the probability of a label occurring at that pixel at intersections and nonintersections in Fig. 6.9. Based on this observation, an additional histogram is computed over 70 % of the middle part of each side view.

The 320 side views dataset was annotated for the intersection classifier experiments. Each of the 320 images is manually labeled as an intersection or a nonintersection. This resulted in a set of 250 nonintersection and 70 intersection views. The

Fig. 6.8 Cluster examples. Each *column* shows the average *frontal image* for a cluster in the *top row* followed by some sample *frontal views* from locations assigned to that cluster

intersection descriptor is computed for all the 320 side views and another boosting classifier is trained using the resultant 320 descriptors. This boosting classifier has 5 decision trees and each of the decision trees has 4 nodes. This boosting classifier is now run on only the side views of the entire dataset. Locations which contributed images to the training of the intersection classifier were excluded from the test stage.

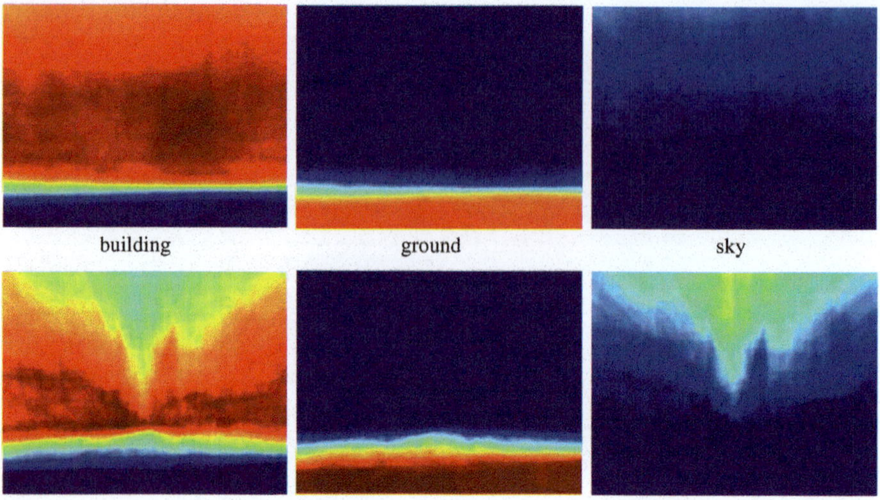

Fig. 6.9 *Top* Probability maps for each label occurring at a pixel at nonintersection side images. *Bottom* Probability maps for each label occurring at a pixel at intersection side images. *Red* indicates a high probability while *blue* indicates a low probability

If both the left and right side views of a location are classified as an intersection by the classifier, the location is categorized as an intersection. Otherwise the location is categorized as a nonintersection.

A visualization of our results is provided in Fig. 6.10. It can be observed that our intersection classifier successfully predicts intersection at many of the major intersections. A human annotator marked 79 unique areas of the sequence as intersections in the city. The intersection classifier correctly predicted an intersection for 63 of the 79 marked intersections for a recall rate of 79.7 % indicating the effectiveness of our approach. A successful detection implies that at least two locations within 10 m of an intersection were classified as an intersection by the classifier. The system also has a high precision with only 7 nonintersection locations in the large-scale sequence classified as intersections.

6.3.4 Location Recognition

We present experiments for using the semantic label descriptor to aid location recognition. In the previous sections, we showed the ability to induce a topological map and discover intersections by summarizing the semantic layout of a street scene. We now propose to exploit this summarized information about a scene for improving visual location recognition.

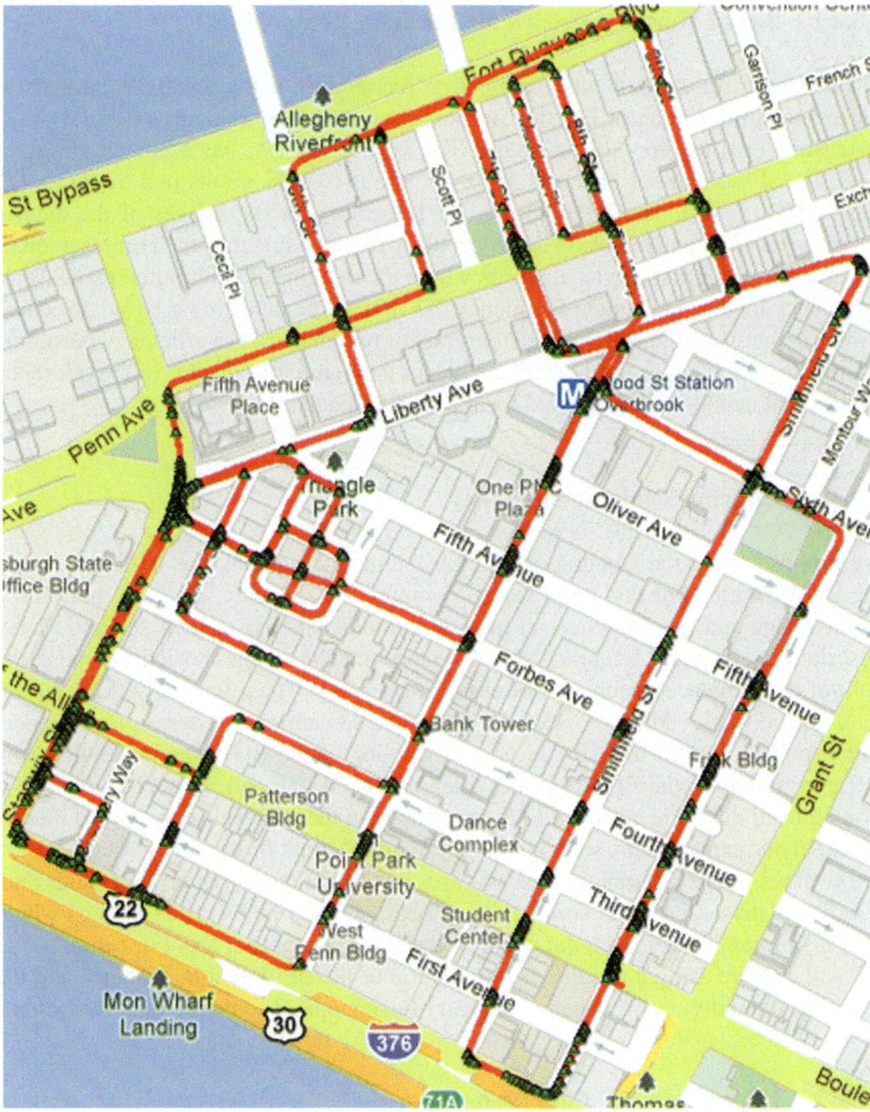

Fig. 6.10 Results for intersection recognition. Locations marked with *green icons* were predicted as intersections by our system

We use a bag of words (BOW) approach [2, 26], a method which has been previously applied in city scale localization using street side images [23]. However, the BOW approach ignores the context provided by scenes and we propose to improve its performance by using the semantic similarity between scenes.

6.3.4.1 Bag of Words Approach for Location Recognition

We first perform experiments for location recognition using the BOW model for retrieving similar images. For this purpose, we use a hierarchical vocabulary [18] over SIFT [16] features for an image. The vocabulary tree has a branching factor k of 10 and the number of levels M in the tree is 6. This results in a tree with k^M = 1,000,000 leaf nodes. We compute SIFT features for the images of the dataset which are then quantized using the vocabulary tree. An image is then represented as a weighted histogram of visual words where the weights are computed using Term Frequency Inverse Document Frequency (TF-IDF). The matching is based on cosine similarity between the weighted histograms for two images. For efficiency purposes, an inverted file index is used which decreases the number of images in the reference dataset that need to be considered for matching a query.

6.3.4.2 Utilizing Semantic Similarity

We propose to use the labeling obtained by semantic segmentation to help improve the BOW results. For this purpose, we compute the semantic similarity between a query image and the reference set images as the distance over their semantic label descriptors. Here, the reference set of images are the images which compose the split of the sequence which are not used for querying. The reference set images are then ranked in ascending order of the computed semantic label descriptor distance. These ranks are used in a simple post-processing of the BOW model results. Any retrieved image in the BOW output which is ranked low with respect to a query is pruned from the retrieval set and the next image in the retrieval set is moved into its place. In our experiments, we consider the semantic rank threshold of 1000, i.e., if a retrieval set image's semantic rank with respect to the current query is greater than 1000, it is discarded. This simple post-processing step helps prune images which may significantly vary from the query in the semantic label space. For future work, we will explore using a weighted combination of the BOW feature distance and the semantic label descriptor distance instead of using a hard semantic rank threshold.

6.3.4.3 Location Recognition Performance

For preparing the reference and query datasets, we utilize the GPS information provided with the dataset. Each StreetView panorama is provided with GPS coordinates specifying the latitude and longitude of that location. Using the GPS coordinates, the individual distance between all the locations in the sequence is calculated. A location which has a previously visited location within a threshold distance of 10 m is considered a revisited location. In order to avoid considering immediately preceding locations as revisits, we discard the previous 25 frames for a location so that views collected within a short timespan of each other are not considered as revisits. The first visit to a location in the StreetView sequence is added to the reference set and

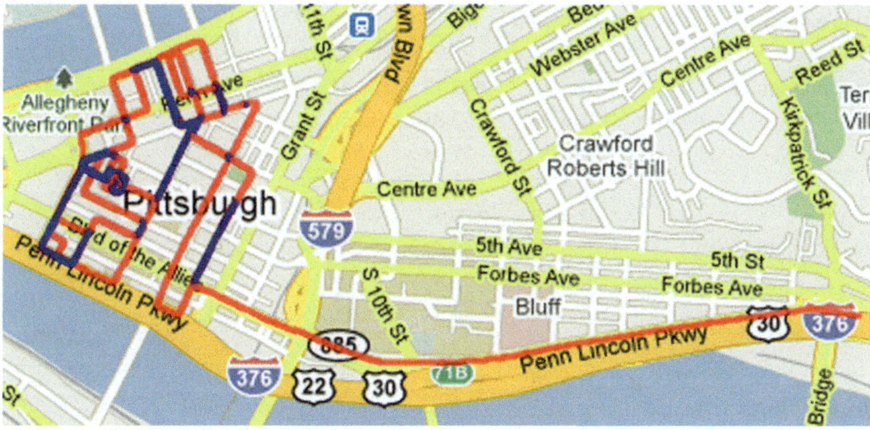

Fig. 6.11 Google Maps visualization of the location recognition dataset. *Blue points* represent the query images and *red points* are the locations of the reference set images

all further visits which occur at different time intervals in the future now compose the set of query images. This provides a reference set of 36,776 images from 9,194 locations and a query set of 13,448 query images from 3,362 locations. A Google Map visualization of the reference and query dataset is shown in Fig. 6.11.

An inverted file index is built over the quantized SIFT descriptors of the images of the reference set and then used to compute a retrieval set of images for each query image. We evaluate the location recognition performance as the percent of query images for which an reference set image located within a certain distance was successfully retrieved. In addition to considering the first retrieved image for a query, we also consider retrieval sets of different sizes in our evaluation. This is done since the first match is not necessarily the correct location and post-processing like spatial matching [20] can help refine the results. The results for location recognition for the StreetView dataset are presented in Table 6.4.

We first compare the BOW approach against using just the semantic similarity between images. For the top match, the BOW method performs more than 20 % better than using the semantic label descriptor. This is not surprising since different man-made structures would be considered to be the same by the semantic labeling method which retrieves scenes with similar layouts. We also observe that by considering a larger retrieval set, the difference in the recognition performance drops thereby indicating that while a correct match could be present in the retrieval set from semantic similarity, it is not necessarily the first match as appearance information has not been considered yet. When we refine the BOW output by considering the semantic rank of a retrieved image against a query, an improvement of more than 7 % (roughly 1025 images) is observed for the top retrieved image. As we increase the size of the retrieval set, the difference between the two methods starts decreasing. This emphasizes two points: semantic similarity provides a significant refinement in the set of retrieved images and the other, that this effect is more noticeable when

Table 6.4 Results for location recognition

Method	NN set size	≤10 m	≤15 m	≤20 m	≤25 m
Semantic similarity	Top-1	43.72	47.42	48.82	49.77
	Top-3	52.83	56.71	58.2	59.15
	Top-5	57.44	61.11	62.63	63.7
	Top-15	66.74	70.73	72.46	73.69
	Top-25	70.91	74.75	76.58	77.99
Bag of Words (BOW)	Top-1	60.42	67.18	69.37	70.54
	Top-3	70.32	75.23	76.9	77.73
	Top-5	74.57	78.8	80.22	80.95
	Top-15	82.06	85.12	86.26	86.96
	Top-25	85.34	87.97	89.08	89.82
BOW + Semantic similarity	Top-1	68.03	75.34	77.31	78.27
	Top-3	76.58	81.35	82.61	83.38
	Top-5	79.8	83.6	84.67	85.43
	Top-15	84.67	87.53	88.61	89.43
	Top-25	86.41	88.98	90.1	90.86

The results are for 13,448 query images from 3,362 unique locations and the reference set is composed of 36,776 images from 9,194 locations. Values under column NN set size are the size of the retrieval set considered for a query. The values in the subsequent columns indicate the percent of images correctly localized when a particular threshold is applied on the distance between a query and the geographically nearest image in its retrieval set

considering smaller sets which is useful as we desire higher precision amongst the top ranked images. A common post-processing step in BOW methods is to use a spatial verification stage to re-rank the retrieval results returned by the BOW model. In our work, we show the ability to obtain a refined the retrieval set by considering semantic neighbors for an image.

6.4 Conclusions

In this chapter, we demonstrated an approach for semantic labeling of outdoor urban scenes formulated over superpixels obtained by a bottom-up segmentation method. We showed how the attained coarse semantic labels (*building, sky, ground, trees, cars*) and their spatial layout can be used to further understanding of urban environments which are typically composed of streets and highways. This is done by introducing the *semantic label descriptor* which characterizes the spatial layout of semantic categories present in a scene. This is used to cluster locations based on their semantic similarity and induces a topology over the sequence. An implicit utility of the representation is the ability to learn semantic concepts as illustrated by the association of meta data like intersections with geographic locations. Finally, we also show how the availability of semantic information from the labeling of scenes can enhance the retrieval accuracy of appearance based approaches like the bag of words model for geo-location.

Acknowledgments Supported by the Intelligence Advanced Research Projects Activity (IARPA) via Air Force Research Laboratory, contract FA8650-12-C-7212. The U.S. Government is authorized to reproduce and distribute reprints for Governmental purposes notwithstanding any copyright annotation thereon. Disclaimer: The views and conclusions contained herein are those of the authors and should not be interpreted as necessarily representing the official policies or endorsements, either expressed or implied, of IARPA, AFRL, or the U.S. Government.

References

1. Agarwal S, Snavely N, Simon I, Seitz SM, Szeliski R (2009) Building Rome in a day. In: ICCV, pp 72–79
2. Csurka G, Dance CR, Fan L, Willamowski J, Bray C (2004) Visual categorization with bags of keypoints. In: Workshop on statistical learning in computer vision, ECCV, pp 1–22
3. Doersch C, Singh S, Gupta A, Sivic J, Efros AA (2012) What makes Paris look like Paris? ACM Trans Graph 31(4):101
4. Felzenszwalb PF, Huttenlocher DP (2004) Efficient graph-based image segmentation. Int J Comput Vis 59(2):167–181
5. Gould S, Fulton R, Koller D (2009) Decomposing a scene into geometric and semantically consistent regions. In: ICCV, pp 1–8
6. Gould S, Rodgers J, Cohen D, Elidan G, Koller D (2008) Multi-class segmentation with relative location prior. Int J Comput Vis 80(3):300–316
7. Hays J, Efros AA (2008) IM2GPS: estimating geographic information from a single image. In: CVPR, pp 1–8 (2008)
8. Hoiem D, Efros AA, Hebert M (2007) Recovering surface layout from an image. Int J Comput Vis 75(1):151–172
9. Knopp J, Sivic J, Pajdla T (2010) Avoiding confusing features in place recognition. In: ECCV, pp 748–761
10. Ladicky L, Russell C, Kohli P, Torr PHS (2009) Associative hierarchical CRFs for object class image segmentation. In: ICCV, pp 739–746
11. Ladicky L, Russell C, Kohli P, Torr PHS (2010) Graph cut based inference with co-occurrence statistics. In: ECCV (5), pp 239–253
12. Ladicky L, Sturgess P, Alahari K, Russell C, Torr PHS (2010) What, where and how many? combining object detectors and CRFs. In: ECCV (4), pp 424–437
13. Lafferty JD, McCallum A, Pereira FCN (2001) Conditional random fields: probabilistic models for segmenting and labeling sequence data. In: ICML, pp 282–289
14. Lazebnik S, Schmid C, Ponce J (2006) Beyond bags of features: spatial pyramid matching for recognizing natural scene categories. In: CVPR, pp 2169–2178
15. Leung TK, Malik J (2001) Representing and recognizing the visual appearance of materials using three-dimensional textons. Int J Comput Vis 43(1):29–44
16. Lowe DG (2004) Distinctive image features from scale-invariant keypoints. Int J Comput Vis 60(2):91–110
17. Murillo AC, Singh G, Kosecká J, Guerrero JJ (2013) Localization in urban environments using a panoramic gist descriptor. IEEE Trans Robot 29(1):146–160
18. Nistér D, Stewénius H (2006) Scalable recognition with a vocabulary tree. In: CVPR, pp 2161–2168
19. Oliva A, Torralba A (2001) Modeling the shape of the scene: a holistic representation of the spatial envelope. Int J Comput Vis 42(3):145–175

20. Philbin J, Chum O, Isard M, Sivic J, Zisserman A (2007) Object retrieval with large vocabularies and fast spatial matching. In: CVPR
21. Rabinovich A, Vedaldi A, Galleguillos C, Wiewiora E, Belongie S (2007) Objects in context. In: ICCV, pp 1–8
22. Schapire RE, Singer Y (1999) Improved boosting algorithms using confidence-rated predictions. Mach Learn 37(3):297–336
23. Schindler G, Brown M, Szeliski R (2007) City-scale location recognition. In: CVPR
24. Shotton J, Winn JM, Rother C, Criminisi A (2009) Textonboost for image understanding: multiclass object recognition and segmentation by jointly modeling texture, layout, and context. Int J Comput Vis 81(1):2–23
25. Singh G, Košecká J (2013) Visual loop closing using gist descriptors in manhattan world. In: Workshop on omnidirectional robot vision, ICRA
26. Sivic J, Zisserman A (2003) Video Google: a text retrieval approach to object matching in videos. In: ICCV, pp 1470–1477
27. Tighe J, Lazebnik S (2010) SuperParsing: scalable nonparametric image parsing with superpixels. In: ECCV (5), pp 352–365
28. Zhang H, Fang T, Chen X, Zhao Q, Quan L (2011) Partial similarity based nonparametric scene parsing in certain environment. In: CVPR, pp 2241–2248
29. Zhang H, Xiao J, Quan L (2010) Supervised label transfer for semantic segmentation of street scenes. In: ECCV (5), pp 561–574
30. Zhang W, Košecká J (2006) Image based localization in urban environments. In: 3DPVT06, pp 33–40

Chapter 7
Recognizing Landmarks in Large-Scale Social Image Collections

David J. Crandall, Yunpeng Li, Stefan Lee and Daniel P. Huttenlocher

Abstract The dramatic growth of social media websites over the last few years has created huge collections of online images and raised new challenges in organizing them effectively. One particularly intuitive way of browsing and searching images is by the geo-spatial location of where on Earth they were taken, but most online images do not have GPS metadata associated with them. We consider the problem of recognizing popular landmarks in large-scale datasets of unconstrained consumer images by formulating a classification problem involving nearly 2 million images and 500 categories. The dataset and categories are formed automatically from geotagged photos from Flickr by looking for peaks in the spatial geo-tag distribution corresponding to frequently photographed landmarks. We learn models for these landmarks with a multiclass support vector machine, using classic vector-quantized interest point descriptors as features. We also incorporate the nonvisual metadata available on modern photo-sharing sites, showing that textual tags and temporal constraints lead to significant improvements in classification rate. Finally, we apply recent breakthroughs in deep learning with Convolutional Neural Networks, finding that these models can dramatically outperform the traditional recognition approaches to this problem, and even beat human observers in some cases. (This is an expanded and updated version of an earlier conference paper [23]).

D.J. Crandall (✉) · S. Lee
Indiana University, Bloomington, IN, USA
e-mail: djcran@indiana.edu

S. Lee
e-mail: steflee@indiana.edu

Y. Li
École Polytechnique Fédérale de Lausanne, 1015 Lausanne, Switzerland
e-mail: yunpeng.li@epfl.ch

D.P. Huttenlocher
Cornell University, Ithaca, NY, USA
e-mail: dph@cs.cornell.edu

A.R. Zamir et al. (eds.), *Large-Scale Visual Geo-Localization*,
Advances in Computer Vision and Pattern Recognition,
DOI 10.1007/978-3-319-25781-5_7

7.1 Introduction

Online photo collections have grown dramatically over the last few years, with Facebook alone now hosting over 250 billion images [2]. Unfortunately, techniques for automatic photo organization and search have not kept pace, with most modern photo-sharing sites using simple techniques like keyword search based on text tags provided by users. In order to allow users to browse and search huge image collections more efficiently we need algorithms that can automatically recognize image content and organize large-scale photo collections accordingly.

A natural way of organizing photo collections is based on geo-spatial location—where on Earth an image was taken. This allows people to search for photos taken near a particular spot of interest, or to group images based on similar locations or travel itineraries. To enable this type of organization, geo-spatial coordinates or "geo-tags" can be encoded in the metadata of a photo, and Global Positioning System (GPS) receivers embedded in modern smartphones and high-end cameras can record these positions automatically when a photo is captured. However, the vast majority of online photos are not geo-tagged, and even when available, geo-tags are often incorrect due to GPS error or other noise [15].

Recognizing where a photo was taken based on its visual content is thus an important problem. Besides the potential impact on geo-localization, this is an interesting recognition problem in and of itself. Unlike many tasks like scene type recognition or tag suggestion, which are inherently subjective, place recognition is a uniquely well-posed problem; except for pathological cases like synthetic images or photos taken from space, every photo is taken at exactly one point on Earth, and so there is exactly one correct answer. Moreover, it is relatively easy to assemble large-scale training and test data for this problem by using geo-tagged images from social media sites like Flickr. This is in contrast to most other recognition problems in which producing ground truth data involves extensive manual labor, which historically has limited the size of datasets and introduced substantial bias [38].

In this chapter, we consider the problem of classifying consumer photos according to where on Earth they were taken, using millions of geo-tagged online images to produce labeled training data with no human intervention. We produce this dataset by starting with a collection of over 30 million public geo-tagged images from Flickr. We use this dataset both to define a set of category labels, as well as to assign a ground truth category to each training and test image. The key observation underlying our approach is that when many different people take photos at the same place, they are likely photographing some common area of interest. We use a mean shift clustering procedure [4] to find hotspots or peaks in the spatial distribution of geo-tagged photos, and then use large peaks to define the category labels. We then assign any photos geo-tagged within a peak to the same category label.

We call each localized hotspot of photographic activity a *landmark*. Most of our landmarks do not consist of a single prominent object; for example, many are museums, with photos of hundreds of different exhibits as well as photos containing little or no visual evidence of the landmark itself (e.g., close-ups of people's faces).

We could use visual or textual features of images to try to divide these complex landmarks into individual objects, as others have done [42], but we purposely choose not to do this; by defining the labels using only geo-tags, we ensure that the features used for testing classification algorithms (namely visual content, text tags, and timestamps) do not also bias the category labels. However, because we do not try to remove outliers or difficult images, the photographs taken at these landmarks are quite diverse (see Fig. 7.1 for some examples), meaning the labeled test datasets are noisy and challenging. Our landmark classification task is thus more similar to object category recognition than to specific object recognition. In Sect. 7.3, we discuss the details of our dataset collection approach.

Once we have assembled a large dataset of millions of images and hundreds of categories, we present and evaluate techniques for classifying the landmark at which each photo was taken. We first apply multiclass Support Vector Machines (SVMs) [5] with features based on classic bags of vector-quantized invariant feature point descriptors [8, 25]. Social photo collections also contain sources of nonvisual evidence that can be helpful for classification; for instance, social ties have been found to improve face recognition [36] and image tagging [27]. We explore incorporating the free-form text tags that Flickr users add to some photos. We also incorporate temporal evidence, using the fact that most people take series of photos over time (for instance, as they move about the tourist sites of a city). We thus analyze the *photo stream* of a given photographer, using Structured Support Vector Machines [40] to predict a sequence of category labels jointly rather than classifying a single photo at a time. Finally, inspired by the very recent success of deep learning techniques on a variety of recognition problems [20, 29, 37, 39], we apply Convolutional Neural Networks to our problem of landmark classification as an alternative to the more traditional bag-of-words models with hand-designed image features. Feature extraction, learning, and classification methods are discussed in Sect. 7.4.

In Sect. 7.5 we present a set of large-scale classification experiments involving between 10 and 500 categories and tens to hundreds of thousands of photos. We begin with the bag-of-words models of SIFT feature points, finding that the combination of image and text features performs better than either alone, and that visual features boost performance even for images that already have text tags. We also describe a small study of human accuracy on our dataset, to give a sense of the noise and difficulty of our task. We then show that using temporal context from photos taken by the same photographer nearby in time yields a significant improvement compared to using visual features alone—around 10 % points in most cases. Finally, we show that the neural nets give a further dramatic increase in performance, in some cases even beating humans, giving further evidence of the power of deep learning over traditional features on problems with large-scale datasets.

Fig. 7.1 The categories in our 10-way classification dataset, consisting of the 10 most photographed landmarks on Flickr. To illustrate the diversity and noise in our automatically generated dataset, we show five random images and five random text tags from each category (We have obscured faces to protect privacy. The landmark tagged "venice" is Piazza San Marco.)

7.2 Related Work

Visual geo-location has received increasing attention in the last few years [14, 16, 19, 22–24, 30, 31, 35, 42] driven in part by the availability of cheap training and test data in the form of geo-tagged Flickr photos. We briefly highlight some of the work most related to this chapter here, but please see Luo et al. [26] for a more comprehensive survey. The IM2GPS paper of Hays and Efros [16] estimates a latitude-longitude coordinate estimate for an image by matching against a large dataset of geo-tagged photos from Flickr, identifying nearest neighbors and producing a geo-spatial probability distribution based on the matches. Our goal is different, as we do not try to predict location directly but rather just use location to derive category labels. (For instance, in our problem formulation a misclassification with a geographically proximate category is just as bad as with one that is far away.) Moreover, the IM2GPS test set contains only 237 images that were partially selected by hand, making it difficult to generalize the results beyond that set. In contrast, we use automatically generated test sets that contain tens or hundreds of thousands of photos, providing highly reliable estimates of performance accuracy. Follow-up work by Kalogerakis et al. generalized IM2GPS to geo-localize a stream of multiple images at the same time by imposing constraints on human travel patterns [19].

Other papers have considered landmark classification tasks similar to the one we study here, although typically at a smaller scale. For example, Li et al. [22] study how to build a model of a landmark by extracting a small set of iconic views from a large set of photographs. The paper tests on three hand-chosen categories. Zheng et al. [42] have an approach similar to ours in that it finds highly photographed landmarks automatically from a large collection of geo-tagged photos. However, the test set they use is hand-selected and small—728 total images for a 124-category problem, or fewer than 6 test images per category—and their approach is based on nearest-neighbor search, which may not scale to the millions of test images we consider here. Philbin et al. [30] study building recognition in the context of how to scale bag-of-features models using random vocabulary trees and fast geometric verification, testing on a dataset of 5,000 labeled images. Crandall et al. [6] study geographic embedding and organization of photos by clustering into landmarks and also study recognition, but at a much more limited scale (classifying among landmarks of a known city).

While we approach geo-localization as a recognition problem, an alternative is to study it in the context of 3D reconstruction [35]. If a 3D model of a place is available, then new images can be geo-localized very accurately, sometimes much more accurately than GPS [7]. But 3D reconstruction is computationally expensive, and is possible only in areas having dense coverage (typically thousands of images).

We apply several widely used recognition techniques to our landmark recognition problem, based on bag-of-words models of vector-quantized, invariant feature points [8, 25]. A very large body of literature has studied these models, including how to optimize them for accuracy and speed in different contexts and tasks: see Grauman and Leibe [12] for a comprehensive overview. We also apply Convolutional

Neural Networks, which, in contrast, are arguably less well understood. Following the surprising success of deep Convolutional Neural Networks on the 2012 ImageNet recognition challenge [20], CNNs have been applied to a variety of computer vision tasks and have shown striking improvements over the state of the art [11, 21, 29, 32, 34, 37, 39, 41]. The main advantage of deep learning methods over more traditional techniques is that the image features can be learned along with the object models in a unified optimization problem, instead of using generic hand-designed features (like SIFT [25]), which are likely not optimal for most tasks. Of course, learning these features requires more data and more computational power; the resurgence of neural networks in computer vision is thanks in no small part to powerful GPUs and large annotated datasets (like the ones we have here).

7.3 Building an Internet-Scale Landmark Dataset

Social photo-sharing websites with their huge collections of publicly available, user-generated images and metadata have been a breakthrough in computer vision, giving researchers an economical way of collecting large-scale, realistic consumer imagery. However, when constructing datasets from Internet photo sources, it is critical to avoid potential biases either in selecting the images to include in the dataset, the categories to include in the classification task, or in assigning ground-truth labels to images. Biases of different types affect even the most popular vision datasets [38]. For instance, methods based on searching for photos tagged with hand-selected keywords (e.g., [16, 30]) are prone to bias, because one might inadvertently choose keywords corresponding to objects that are amenable to a particular image classification algorithm. Many researchers have also used unspecified or subjective criteria to choose which images to include in the dataset, again introducing the potential for bias toward a particular algorithm. Other object recognition datasets like PASCAL [10] and Caltech [13] have object classes that were selected by computer vision researchers, making it unclear whether these are the most important categories that should be studied. Also problematic is using the same kinds of features to produce ground-truth labels as are used by the classification algorithm [3, 33, 42]. Recent datasets like ImageNet [9] avoid many sources of bias by defining ground truth labels based on more principled approaches like semantic categories of WordNet [28], and by avoiding subconscious biases of computer vision researchers by crowd-sourcing labels with Mechanical Turk, but these approaches still require a huge amount of human labor.

We thus advocate automatic techniques for creating datasets based on properties of human activity, such as where pictures are taken, without manual intervention. To be most useful for training and testing of classifiers, the ground truth labels should be selected and produced in a way that is automatic and objective, based on sources other than the features used by the classifiers. Our approach is based on the observation that when many people take photos at the same location it is highly likely that these are photos of the same thing. We therefore define category labels by finding geo-spatial clusters of high photographic activity and assign all photos within that cluster the same label.

In particular, our dataset was formed by using Flickr's public API to retrieve metadata for over 60 million publicly accessible geo-tagged photos. We eliminated photos for which the precision of the geo-tags was worse than about a city block (precision score under 13 in the Flickr metadata). For each of the remaining 30 million photos, we considered its latitude-longitude coordinates as a point in the plane, and then performed a mean shift clustering procedure [4] on the resulting set of points to identify local peaks in the photo density distribution [6]. The radius of the disc used in mean shift allowed us to select the scale of the 'landmarks.' We used a radius of 0.001°, which corresponds to roughly 100 m at middle latitudes.[1] Since our goal is to identify locations where many *different* people took pictures, we count at most five photos from any given Flickr user toward any given peak, to prevent high-activity users from biasing the choice of categories. We currently use the top 500 such peaks as categories. After finding peaks, we rank them in decreasing order of the number of distinct photographers who have photographed the landmark. Table 7.1 shows the top 100 of these peaks, including the number of unique photographers, the geographic centroid of the cluster, and representative tags for each cluster. The tags were chosen automatically using the same technique as in [6], which looks for tags that occur very frequently on photos inside a geographic region but rarely outside of it.

We downloaded all 1.9 million photos known to our crawler that were geo-tagged within one of these 500 landmarks. For the experiments on classifying temporal photo streams, we also downloaded all images taken within 48 h of any photo taken in a landmark, bringing the total dataset to about 6.5 million photos. The images were downloaded at Flickr's medium resolution level, about 0.25 megapixels. Figure 7.1 shows random images from each of the top 10 landmarks, showing the diversity of the dataset.

7.4 Landmark Recognition

We now consider the task of image classification in the large-scale image dataset produced using the procedure described above. Since our landmark categories were selected to be nonoverlapping, these categories are mutually exclusive and thus each image has exactly one correct label. We first discuss how to classifying single images with bag-of-words models in Sect. 7.4.1, before turning to the temporal models in Sect. 7.4.2 and the deep learning-based methods in Sect. 7.4.3.

[1] Since longitude lines grow closer toward the poles, the spatial extent of our landmarks are larger at the equator than near the poles. We have not observed this to be a major problem because most population centers are near the middle latitudes, but future work could use better distance functions.

Table 7.1 The world's 100 most-photographed, landmark-sized hotspots as of 2009, according to our analysis of Flickr geo-tags, ranked by number of unique photographers

	Users	Geo-coordinate	Descriptive tags
1	4854	48.8584, 2.2943	eiffeltower, paris
2	4146	51.5080, –0.1281	trafalgarsquare, london
3	3442	51.5008, –0.1243	bigben, london
4	3424	51.5034, –0.1194	londoneye, london
5	3397	48.8531, 2.3493	cathedral, paris
6	3369	51.5080, –0.0991	tatemodern, london
7	3179	40.7485, –73.9854	empirestatebuilding, newyorkcity
8	3167	45.4340, 12.3390	venice, venezia
9	3134	41.8904, 12.4920	colosseum, rome
10	3081	48.8611, 2.3360	pyramid, paris
11	2826	40.7578, –73.9857	timessquare, newyorkcity
12	2778	40.7590, –73.9790	rockefeller, newyorkcity
13	2710	41.8828, –87.6233	cloudgate, chicago
14	2506	41.9024, 12.4574	vaticano, rome
15	2470	48.8863, 2.3430	sacrecoeur, paris
16	2439	51.5101, –0.1346	piccadillycircus, london
17	2321	51.5017, –0.1411	buckingham, london
18	2298	40.7562, –73.9871	timessquare, newyorkcity
19	2296	48.8738, 2.2950	arcdetriomphe, paris
20	2127	40.7526, –73.9774	grandcentralstation, newyorkcicity
21	2092	41.8989, 12.4768	pantheon, rome
22	2081	41.4036, 2.1742	sagradafamilia, barcelona
23	2020	51.5056, –0.0754	towerbridge, london
24	1990	38.8894, –77.0499	lincolnmemorial, washingtondc
25	1983	51.5193, –0.1270	britishmuseum, london
26	1960	52.5164, 13.3779	brandenburggate, berlin
27	1865	51.5078, –0.0762	toweroflondon, london
28	1864	45.4381, 12.3357	rialto, venezia
29	1857	40.7641, –73.9732	applestore, newyorkcity
30	1828	47.6206, –122.3490	needle, seattle
31	1828	47.6089, –122.3410	market, seattle
32	1798	51.5013, –0.1198	bigben, london
33	1789	38.8895, –77.0406	wwii, washingtondc
34	1771	50.0873, 14.4208	praga, praha
35	1767	51.5007, –0.1263	bigben, london
36	1760	48.8605, 2.3521	centrepompidou, paris
37	1743	41.9010, 12.4833	fontanaditrevi, rome
38	1707	37.7879, –122.4080	unionsquare, sanfrancisco

(continued)

Table 7.1 (continued)

	Users	Geo-coordinate	Descriptive tags
39	1688	43.7731, 11.2558	duomo, firenze
40	1688	43.7682, 11.2532	pontevecchio, firenze
41	1639	36.1124, –115.1730	paris, lasvegas
42	1629	43.7694, 11.2557	firenze, firenze
43	1611	38.8895, –77.0353	washingtonmonument, washingtondc
44	1567	41.9023, 12.4536	basilica, rome
45	1505	51.5137, –0.0984	stpaulscathedral, london
46	1462	40.7683, –73.9820	columbuscircle, newyorkcity
47	1450	41.4139, 2.1526	parcguell, barcelona
48	1433	52.5186, 13.3758	reichstag, berlin
49	1419	37.8107, –122.4110	pier39, sanfrancisco
50	1400	51.5101, –0.0986	millenniumbridge, london
51	1391	41.8991, 12.4730	piazzanavona, rome
52	1379	41.9061, 12.4826	spanishsteps, rome
53	1377	37.8026, –122.4060	coittower, sanfrancisco
54	1369	40.6894, –74.0445	libertyisland, newyorkcity
55	1362	41.8953, 12.4828	vittoriano, rome
56	1359	51.5050, –0.0790	cityhall, london
57	1349	50.8467, 4.3524	grandplace, brussel
58	1327	48.8621, 2.2885	trocadero, paris
59	1320	36.1016, –115.1740	newyorknewyork, lasvegas
60	1318	48.8656, 2.3212	placedelaconcorde, paris
61	1320	41.9024, 12.4663	castelsantangelo, rome
62	1305	52.5094, 13.3762	potsdamerplatz, berlin
63	1297	41.8892, –87.6245	architecture, chicago
64	1296	40.7613, –73.9772	museumofmodernart, newyorkcity
65	1292	50.0865, 14.4115	charlesbridge, praha
66	1270	40.7416, –73.9894	flatironbuilding, newyorkcity
67	1260	48.1372, 11.5755	marienplatz, mnchen
68	1242	40.7792, –73.9630	metropolitanmuseumofart, newyorkcity
69	1239	48.8605, 2.3379	louvre, paris
70	1229	40.7354, –73.9909	unionsquare, newyorkcity
71	1217	40.7541, –73.9838	bryantpark, newyorkcity
72	1206	37.8266, –122.4230	prison, sanfrancisco
73	1196	40.7072, –74.0110	nyse, newyorkcity
74	1193	45.4643, 9.1912	cathedral, milano
75	1159	40.4155, –3.7074	plazamayor, madrid
76	1147	51.5059, –0.1178	southbank, london
77	1141	37.8022, –122.4190	lombardstreet, sanfrancisco

(continued)

Table 7.1 (continued)

	Users	Geo-coordinate	Descriptive tags
78	1127	37.7951, –122.3950	ferrybuilding, sanfrancisco
79	1126	–33.8570, 151.2150	sydneyoperahouse, sydney
80	1104	51.4996, –0.1283	westminsterabbey, london
81	1100	51.5121, –0.1229	coventgarden, london
82	1093	37.7846, –122.4080	sanfrancisco, sanfrancisco
83	1090	41.8988, –87.6235	hancock, chicago
84	1083	52.5141, 13.3783	holocaustmemorial, berlin
85	1081	50.0862, 14.4135	charlesbridge, praha
86	1077	50.0906, 14.4003	cathedral, praha
87	1054	41.3840, 2.1762	cathedral, barcelona
88	1042	28.4189, –81.5812	castle, waltdisneyworld
89	1034	38.8898, –77.0095	capitol, washingtondc
90	1024	41.3820, 2.1719	boqueria, barcelona
91	1023	48.8638, 2.3135	pontalexandreiii, paris
92	1022	41.8928, 12.4844	forum, rome
93	1021	40.7060, –73.9968	brooklynbridge, newyorkcity
94	1011	36.6182, –121.9020	montereybayaquarium, monterey
95	1009	37.9716, 23.7264	parthenon, acropolis
96	1008	41.3953, 2.1617	casamil, barcelona
97	986	43.6423, –79.3871	cntower, toronto
98	983	52.5099, 13.3733	sonycenter, berlin
99	972	34.1018, –118.3400	hollywood, losangeles
100	969	48.8601, 2.3263	museedorsay, paris

We use these hotspots to automatically define our landmark categories. For each landmark we show the number of photographers, the latitude-longitude coordinate of the hotspot centroid, and two automatically selected tags corresponding to the most distinctive tag (i.e. most-frequent relative to the worldwide background distribution) within the landmark region and within the surrounding city-scale region

7.4.1 Single Image Classification Using Bag of Words Models

To perform image classification we adopt the bag-of-features model proposed by Csurka et al. [8], where each photo is represented by a feature vector recording occurrences of vector-quantized SIFT interest point descriptors [25]. As in that paper, we built a visual vocabulary by clustering SIFT descriptors from photos in the training set using the k-means algorithm. To prevent some images or categories from biasing the vocabulary, for the clustering process we sampled a fixed number of interest points from each image, for a total of about 500,000 descriptors. We used an efficient implementation of k-means using the approximate nearest neighbor (ANN) technique of [1] (to assign points to cluster centers during the expectation (E-step) of k-means).

The advantage of this technique over many others is that it guarantees an upper bound on the approximation error; we set the bound such that the cluster center found by ANN is within 110 % of the distance from the point to the optimal cluster center.

Once a visual vocabulary of size k had been generated, a k-dimensional feature vector was constructed for each image by using SIFT to find local interest points and assigning each interest point to the visual word with the closest descriptor. We then formed a frequency vector which counted the number of occurrences of each visual word in the image. For textual features we used a similar vector space model in which any tag used by at least three different users was a dimension in the feature space, so that the feature vector for a photo was a binary vector indicating presence or absence of each text tag. Both types of feature vectors were L2-normalized. We also studied combinations of image and textual features, in which case the image and text feature vectors were simply concatenated after normalization.

We learned a linear model that scores a given photo for each category and assigns it to the class with the highest score. More formally, let m be the number of classes and $\mathbf{x}$ be the feature vector of a photo. Then, the predicted label is

$$\hat{y} = \arg\max_{y \in \{1,\ldots,m\}} s(\mathbf{x}, y; \mathbf{w}), \tag{7.1}$$

where $\mathbf{w} = (\mathbf{w}_1^T, \ldots, \mathbf{w}_m^T)^T$ is the model and $s(\mathbf{x}, y; \mathbf{w}) = \langle \mathbf{w}_y, \mathbf{x} \rangle$ is the score for class y under the model. Note that in our settings, the photo is always assumed to belong to one of the m categories. Since this is by nature a multi-way (as opposed to binary) classification problem, we use multiclass SVMs [5] to learn the model $\mathbf{w}$, using the SVM^{multiclass} software package [18]. For a set of training examples $\{(\mathbf{x}_1, y_1), \ldots, (\mathbf{x}_N, y_N)\}$, our multiclass SVM optimizes an objective function,

$$\begin{aligned} &\min_{\mathbf{w},\xi} \frac{1}{2} \|\mathbf{w}\|^2 + C \sum_{i=1}^{N} \xi_i \\ &\text{s.t. } \forall i, \quad y \neq y_i : \langle \mathbf{w}_{y_i}, \mathbf{x}_i \rangle - \langle \mathbf{w}_y, \mathbf{x}_i \rangle \geq 1 - \xi_i, \end{aligned} \tag{7.2}$$

where C is the trade-off between training performance and margin in SVM formulations (which we simply set to $\bar{x}^{-2}$ where $\bar{x}$ is the average L2-norm of the training feature vectors). Hence, for each training example, the learned model is encouraged to give higher scores to the correct class label than to the incorrect ones. By rearranging terms it can be shown that the objective function is an upper bound on the training error.

In contrast, many previous approaches to object recognition using bag-of-parts models (such as Csurka et al. [8]) trained a set of binary SVMs (one for each category) and classified an image by comparing scores from the individual SVMs. Such approaches are problematic for n-way, forced-choice problems, however, because the scores produced by a collection of independently trained binary SVMs may not be comparable, and thus lack any performance guarantee. It is possible to alleviate this problem by using a different C value for each binary SVM [8], but this introduces additional parameters that need to be tuned, either manually or via cross validation. Here we use multiclass SVMs, because they are inherently suited for multi-category classification.

Note that while the categories in this single-photo classification problem correspond to geographic locations, there is no geographical information used during the actual learning or classification. For example, unlike IM2GPS [16], we are not concerned with pinpointing a photo on a map, but rather with classifying images into discrete categories which happen to correspond to geo-spatial positions.

7.4.2 Incorporating Temporal Information

Photos taken by the same photographer at nearly the same time are likely to be related. In the case of landmark classification, constraints on human travel mean that certain sequences of category labels are much more likely than others. To learn the patterns created by such constraints, we view temporal sequences of photos taken by the same user as a single entity and consider them jointly as a structured output.

7.4.2.1 Temporal Model for Joint Classification

We model a temporal sequence of photos as a graphical model with a chain topology, where the nodes represent photos, and edges connect nodes that are consecutive in time. The set of possible labels for each node is simply the set of m landmarks, indexed from 1 to m. The task is to label the entire sequence of photos with category labels, however we score correctness only for a single selected photo in the middle of the sequence, with the remaining photos serving as temporal context for that photo. Denote an input sequence of length n as $X = ((\mathbf{x}_1, t_1), \ldots, (\mathbf{x}_n, t_n))$, where $\mathbf{x}_v$ is a feature vector for node v (encoding evidence about the photo such as textual tags or visual information) and t_v is the corresponding timestamp. Let $Y = (y_1, \ldots, y_n)$ be a labeling of the sequence. We would like to express the scoring function $S(X, Y; \mathbf{w})$ as the inner product of some *feature map* $\Psi(X, Y)$ and the model parameters $\mathbf{w}$, so that the model can be learned efficiently using the structured SVM.

Node Features To this end, we define the feature map for a single node v under the labeling as,

$$\Psi_V(\mathbf{x}_v, y_v) = (I(y_v = 1)\mathbf{x}_v^T, \ldots, I(y_v = m)\mathbf{x}_v^T)^T, \tag{7.3}$$

where $I(\cdot)$ is an indicator function. Let $\mathbf{w}_V = (\mathbf{w}_1^T, \ldots, \mathbf{w}_m^T)$ be the corresponding model parameters with $\mathbf{w}_y$ being the weight vector for class y. Then the node score $s_V(\mathbf{x}_v, y_v; \mathbf{w}_V)$ is the inner product of the $\Psi_V(\mathbf{x}_v, y_v)$ and $\mathbf{w}_V$,

$$s_V(\mathbf{x}_v, y_v; \mathbf{w}_V) = \langle \mathbf{w}_V, \Psi_V(\mathbf{x}_v, y_v) \rangle . \tag{7.4}$$

Edge Features The feature map for an edge (u, v) under labeling Y is defined in terms of the labels y_u and y_v, the time elapsed between the two photos $\delta t = |t_u - t_v|$, and the speed required to travel from landmark y_u to landmark y_v within that time, $\text{speed}(\delta t, y_u, y_v) = \text{distance}(y_u, y_v)/\delta t$. Since the strength of the relation between two photos decreases with the elapsed time between them, we divide the full range of δt into M intervals $\Omega_1, \ldots, \Omega_M$. For δt in interval Ω_τ, we define a feature vector,

$$\psi_\tau(\delta t, y_u, y_v) = (I(y_u = y_v), I(\text{speed}(\delta t, y_u, y_v) > \lambda_\tau))^T, \tag{7.5}$$

where λ_τ is a speed threshold. This feature vector encodes whether the two consecutive photos are assigned the same label and, if not, whether the transition requires a person to travel at an unreasonably high speed (i.e., greater than λ_τ). The exact choice of time intervals and speed thresholds are not crucial. We also take into consideration the fact that some photos have invalid timestamps (e.g., a date in the twentysecond century) and define the feature vector for edges involving such photos as,

$$\psi_0(t_u, t_v, y_u, y_v) = I(y_u = y_v)(I(z = 1), I(z = 2))^T, \tag{7.6}$$

where z is 1 if exactly one of t_u or t_v is invalid and 2 if both are. Here, we no longer consider the speed, since it is not meaningful when timestamps are invalid. The complete feature map for an edge is thus,

$$\begin{aligned}\Psi_E(t_u, t_v, y_u, y_v) = (&I(\delta t \in \Omega_1)\psi_1(\delta t, y_u, y_v)^T, \ldots, I(\delta t \in \Omega_M)\psi_M(\delta t, y_u, y_v)^T, \\ &\psi_0(t_u, t_v, y_u, y_v)^T)^T\end{aligned} \tag{7.7}$$

and the edge score is,

$$s_E(t_u, t_v, y_u, y_v; \mathbf{w}_E) = \langle \mathbf{w}_E, \Psi_E(t_u, t_v, y_u, y_v)\rangle, \tag{7.8}$$

where $\mathbf{w}_E$ is the vector of edge parameters.

Overall Feature Map The total score of input sequence X under labeling Y and model $\mathbf{w} = (\mathbf{w}_V^T, \mathbf{w}_E^T)^T$ is simply the sum of individual scores over all the nodes and edges. Therefore, by defining the overall feature map as,

$$\Psi(X, Y) = (\sum_{v=1}^{n} \Psi_V(\mathbf{x}_v, y_v)^T, \sum_{v=1}^{n-1} \Psi_E(t_v, t_{v+1}, y_v, y_{v+1})^T)^T,$$

the total score becomes an inner product with $\mathbf{w}$,

$$S(X, Y; \mathbf{w}) = \langle \mathbf{w}, \Psi(X, Y)\rangle. \tag{7.9}$$

The predicted labeling for sequence X by model $\mathbf{w}$ is one that maximizes the score,

$$\hat{Y} = \underset{Y \in \mathcal{Y}_X}{\arg\max}\ S(X, Y; \mathbf{w}), \tag{7.10}$$

where $\mathscr{Y}_X = \{1, \dots, m\}^n$ is the the label space for sequence X of length n. This can be obtained efficiently using Viterbi decoding, because the graph is acyclic.

7.4.2.2 Parameter Learning

Let $((X_1, Y_1), \dots, (X_N, Y_N))$ be training examples. The model parameters $\mathbf{w}$ are learned using structured SVMs [40] by minimizing a quadratic objective function subject to a set of linear soft margin constraints,

$$\min_{\mathbf{w},\xi} \frac{1}{2} \|\mathbf{w}\|^2 + C \sum_{i=1}^{N} \xi_i \tag{7.11}$$
$$\text{s.t. } \forall i, \quad Y \in \mathscr{Y}_{X_i} : \langle \mathbf{w}, \delta\Psi_i(Y) \rangle \geq \Delta(Y_i, Y) - \xi_i,$$

where $\delta\Psi_i(Y)$ denotes $\Psi(X_i, Y_i) - \Psi(X_i, Y)$ (thus $\langle \mathbf{w}, \delta\Psi_i(Y) \rangle = S(X_i, Y_i; \mathbf{w}) - S(X_i, Y; \mathbf{w})$) and the loss function $\Delta(Y_i, Y)$ in this case is simply the number of mislabeled nodes (photos) in the sequence. It is easy to see that the structured SVM degenerates into a multiclass SVM if every example has only a single node.

The difficulty of this formulation is that the label space $\mathscr{Y}_{X_i}$ grows exponentially with the length of the sequence X_i. Structured SVMs address this problem by iteratively minimizing the objective function using a cutting-plane algorithm, which requires finding the *most violated constraint* for every training exemplar at each iteration. Since the loss function $\Delta(Y_i, Y)$ decomposes into a sum over individual nodes, the most violated constraint,

$$\hat{Y}_i = \arg\max_{Y \in \mathscr{Y}_{X_i}} S(X_i, Y; \mathbf{w}) + \Delta(Y_i, Y), \tag{7.12}$$

can be obtained efficiently via Viterbi decoding.

7.4.3 *Image Classification with Deep Learning*

Since our original work on landmark recognition [23], a number of new approaches to object recognition and image classification have been proposed. Perhaps none has been as sudden or surprising as the very recent resurgence of interest in deep Convolutional Neural Networks, due to their performance on the 2012 ImageNet visual recognition challenge [9] by Krizhevsky et al. [20]. The main advantage of these techniques seems to be the ability to learn image features and image classifiers together in one unified framework, instead of creating the image features by hand (e.g., by using SIFT) or learning them separately.

To test these emerging models on our dataset, we trained networks using Caffe [17]. We bypassed the complex engineering involved in designing a deep network by starting with the architecture proposed by Krizhevsky et al. [20], composed of five convolutional layers followed by three fully connected layers. Mechanisms for contrast normalization and max pooling occur between many of the convolutional layers. Altogether the network contains around 60 million parameters; although our dataset is sufficiently large to train this model from random initialization, we choose instead to reduce training time by following Oquab et al. [29] and others by initializing from a pretrained model. We modify the final fully connected layer to accommodate the appropriate number of categories for each task. The initial weights for these layers are randomly sampled from a zero-mean normal distribution.

Each model was trained using stochastic gradient descent with a batch size of 128 images. The models were allowed to continue until 25,000 batches had been processed with a learning rate starting at 0.001 which decayed by an order of magnitude every 2,500 batches. In practice, convergence was reached much sooner. Approximately 20 % of the training set was withheld to avoid overfitting, and the training iteration with the lowest validation error was used for evaluation on the test set.

7.5 Experiments

We now present experimental results on our dataset of nearly 2 million labeled images. We created training and test subsets by dividing the *photographers* into two evenly sized groups and then taking all photos by the first group as training images and all photos by the second group as test images. Partitioning according to user reduces the chance of "leakage" between training and testing sets, for instance due to a given photographer taking nearly identical photos that end up in both sets.

We conducted a number of classification experiments with various subsets of the landmarks. The number of photos in the dataset differs widely from category to category; in fact, the distribution of photos across landmarks follows a power-law distribution, with the most popular landmark having roughly four times as many images as the 50th most popular landmark, which in turn has about four times as many images as the 500th most popular landmark. To ease comparison across different numbers of categories, for each classification experiment we subsample so that the number of images in each class is about the same. This means that the number of images in an m-way classification task is equal to m times the number of photos in the least popular landmark, and the probability of a correct random guess is $1/m$.

7.5.1 Bag of Words Models

Table 7.2 and Fig. 7.2 present results for various classification experiments. For single image classification, we train three multiclass SVMs for the visual features, textual features, and the combination. For text features, we use the normalized counts of text tags that are used by more than two photographers. When combining image and text features, we simply concatenate the two feature vectors for each photo. We see that

Table 7.2 Classification accuracy (% of images correct) for varying categories and types of models

Categories	Random baseline	Images—BoW			Photo streams			Images—CNNs
		Visual	Text	Vis+text	Visual	Text	Vis+text	Visual
Top 10 landmarks	10.00	57.55	69.25	80.91	68.82	70.67	82.54	81.43
Landmark 200–209	10.00	51.39	79.47	86.53	60.83	79.49	87.60	72.18
Landmark 400–409	10.00	41.97	78.37	82.78	50.28	78.68	82.83	65.20
Human baseline	10.00	68.00	–	76.40	–	–	–	68.00
Top 20 landmarks	5.00	48.51	57.36	70.47	62.22	58.84	72.91	72.10
Landmark 200–219	5.00	40.48	71.13	78.34	52.59	72.10	79.59	63.74
Landmark 400–419	5.00	29.43	71.56	75.71	38.73	72.70	75.87	54.60
Top 50 landmarks	2.00	39.71	52.65	64.82	54.34	53.77	65.60	62.28
Landmark 200–249	2.00	27.45	65.62	72.63	37.22	67.26	74.09	55.87
Landmark 400–449	2.00	21.70	64.91	69.77	29.65	66.90	71.62	49.11
Top 100 landmarks	1.00	29.35	50.44	61.41	41.28	51.32	62.93	52.52
Top 200 landmarks	0.50	18.48	47.02	55.12	25.81	47.73	55.67	39.52
Top 500 landmarks	0.20	9.55	40.58	45.13	13.87	41.02	45.34	23.88

classifying individual images using the bag-of-words visual models (as described in Sect. 7.4.1) gives results that are less accurate than textual tags but nevertheless significantly better than random baseline—four to six times higher for the 10 category problems and nearly 50 times better for the 500-way classification. The combination of textual tags and visual tags performs significantly better than either alone, increasing performance by about 10 % points in most cases. This performance improvement is partially because about 15 % of photos do not have any text tags. However, even when such photos are excluded from the evaluation, adding visual features still gives a significant improvement over using text tags alone, increasing accuracy from 79.2 to 85.47 % in the top-10 category case, for example.

Table 7.2 shows classification experiments for different numbers of categories and also for categories of different rank. Of course, top-ranked landmark classes have (by definition) much more training data than lower-ranked classes, so we see significant

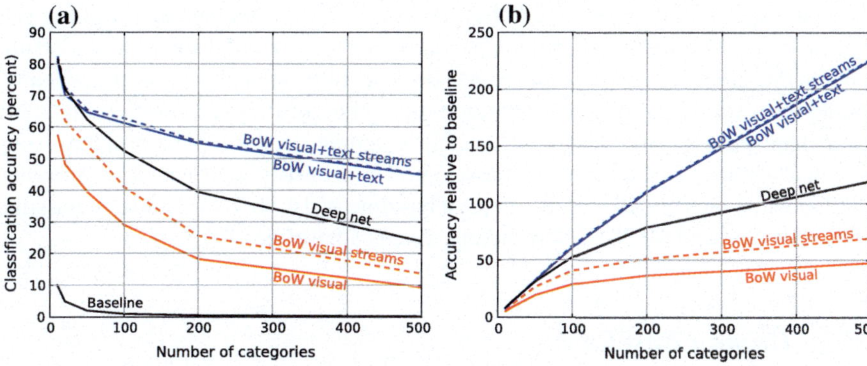

Fig. 7.2 Classification accuracy for different types of models across varying numbers of categories, measured by **a** absolute classification accuracy; and **b** ratio relative to random baseline

Table 7.3 Visual classification rates for different vocabulary sizes

# Categories	Vocabulary size				
	1,000	2,000	5,000	10,000	20,000
10	47.51	50.78	52.81	55.32	57.55
20	39.88	41.65	45.02	46.22	48.51
50	29.19	32.58	36.01	38.24	39.71
100	19.77	24.05	27.53	29.35	30.42

drops in visual classification accuracy when considering less-popular landmarks (e.g., from 57.55 % for landmarks ranked 1–10 to 41.97 % for those ranked 400–409). However, for the text features, problems involving *lower-ranked* categories are actually *easier*. This is because the top landmarks are mostly located in the same major cities, so that tags like *london* are relatively uninformative. Lower categories show much more geo-spatial variation and thus are easier for text alone.

For most of the experiments shown in Fig. 7.2, the visual vocabulary size was set to 20,000. This size was computationally prohibitive for our structured SVM learning code for the 200- and 500-class problems, so for those we used 10,000 and 5,000, respectively. We studied how the vocabulary size impacts classification performance by repeating a subset of the experiments for several different vocabulary sizes. As Table 7.3 shows, classification performance improves as the vocabulary grows, but the relative effect is more pronounced as the number of categories increases. For example, when the vocabulary size is increased from 1,000 to 20,000, the relative performance of the 10-way classifier improves by about 20 % (10.05 % points, or about one baseline) while the accuracy of the 100-way classifier increases by more than 50 % (10.65 % points, or nearly 11 baselines). Performance on the 10-way problem asymptotes by about 80,000 clusters at around 59.3 %. Unfortunately, we could not try such large numbers of clusters for the other tasks, because the learning becomes intractable.

In the experiments presented so far, we sampled from the test and training sets to produce equal numbers of photos for each category in order to make the results easier to interpret. However, our approach does not depend on this property; when we sample from the actual photo distribution our techniques still perform dramatically better than the majority class baseline. For example, in the top-10 problem using the actual photo distribution we achieve 53.58 % accuracy with visual features and 79.40 % when tags are also used, versus a baseline of 14.86 %; the 20-way classifier produces 44.78 and 69.28 % respectively, versus a baseline of 8.72 %.

7.5.2 Human Baselines

A substantial number of Flickr photos are mislabeled or inherently ambiguous—a close-up photo of a dog or a sidewalk could have been taken at almost any landmark. To try to gauge the frequency of such difficult images, we conducted a small-scale, human-subject study. We asked 20 well-traveled people each to label 50 photos taken in our top-10 landmark dataset. Textual tags were also shown for a random subset of the photos. We found that the average human classification accuracy was 68.0 % without textual tags and 76.4 % when both the image and tags were shown (with standard deviations of 11.61 and 11.91, respectively). Thus, the humans performed better than the automatic classifier when using visual features alone (68.0 % vs. 57.55 %) but about the same when both text and visual features were available (76.4 % vs. 80.91 %). However, we note that this was a small-scale study and not entirely fair: the automatic algorithm was able to review hundreds of thousands of training images before making its decisions, whereas the humans obviously could not. Nevertheless, the fact that the human baseline is not near 100 % gives some indication of the difficulty of this task.

7.5.3 Classifying Photo Streams

Table 7.2 and Fig. 7.2 also present results when temporal features are used jointly to classify photos nearby in time from the same photographer, using structured SVMs, as described in Sect. 7.4.2. For training, we include only photos in a user's photostream that are within one of the categories we are considering. For testing, however, we do not assume such knowledge (because we do not know where the photos were taken ahead of time). Hence the photo streams for testing may include photos outside the test set that do not belong to any of the categories, but only photos in the test set contribute toward evaluation. For these experiments, the maximum length of a photo stream was limited to 11, or five photos before and after a photo of interest.

The results show a significant improvement in visual bag-of-words classification when photo streams are classified jointly—nearly 12 % points for the top-10 category problem, for example. In contrast, the temporal information provides little improvement for the textual tags, suggesting that tags from contemporaneous images contain largely redundant information. In fact, the classification performance using temporal and visual features is actually slightly higher than using temporal and textual features for the top-20 and top-50 classification problems. For all of the experiments, the best performance for the bag-of-words models is achieved using the full combination of visual, textual and temporal features, which, for example, gives 82.54 % correct classification for the 10-way problem and 45.34 % for the 500-way problem—more than 220 times better than the baseline.

7.5.4 Classifying with Deep Networks

Finally, we tested our problem on what has very recently emerged as the state-of-the-art in classification technology: deep learning using Convolutional Neural Networks. Table 7.2 and Fig. 7.2 shows the results for the single image classification problem, using vision features alone. The CNNs perform startlingly well on this problem compared to the more traditional bag-of-words models. On the 10-way problem, they increase results by almost 25 % *points*, or about 2.5 times the baseline, from 57.6 to 81.4 %. In fact, CNNs with visual features significantly outperform the text features, and very narrowly beat the combined visual and text features. They also beat the photo stream classifiers for both text and visual features, despite the fact the CNNs see less information (a single image versus a stream of photos), and very slightly underperform when vision and text are both used. The CNNs also beat the humans by a large margin (81.4 % vs. 68.0 %), even when the humans saw text tags and the CNNs did not (81.4 % vs. 76.4 %).

For classification problems with more categories, CNNs outperform bag-of-words visual models by an increasing margin relative to baseline. For instance, for 50-way the CNN increases performance from 39.71 to 62.28 %, or by more than 11 baselines, whereas for 500-way the increase is 9.55–23.88 %, or over 71 baselines. However, as the number of categories grows, text features start to catch up with visual classification with CNNs, roughly matching it for the 100-way problem and significantly beating for 500-way (40.58 % vs. 23.88 %). For 500-way, the combined text and vision using bags-of-words outperform the vision-only CNNs by a factor of about 2 (45.13 % vs. 23.88 %). Overall, however, our results add to evidence that deep learning can offer significant improvements over more traditional techniques, especially on image classification problems where training sets are large.

7.5.5 Discussion

The experimental results we report here are highly precise because of the large size of our test dataset. Even the smallest of the experiments, the top-10 classification, involves about 35,000 test images. To give a sense of the variation across runs due to differences in sampling, we ran 10 trials of the top-10 classification task with different samples of photos and found the standard deviation to be about 0.15 % points. Due to computational constraints we did not run multiple trials for the experiments with large numbers of categories, but the variation is likely even less due to the larger numbers of images involved.

We showed that for the top-10 classification task, our automatic classifiers can produce accuracies that are competitive with or even better than humans, but are still far short of the 100 % performance that we might aspire to. To give a sense for the error modes of our classifiers, we show a confusion matrix for the CNNs on the 10-way task in Fig. 7.3. The four most difficult classes are all in London (Trafalgar Square, Big Ben, London Eye, Tate Modern), with a substantial degree of confusion between them (especially between Big Ben and London Eye). Classes within the same city can be confusing because it is often possible to either photograph two landmarks in the same photo, or to photograph one landmark from the other. Landmarks in the same city also show greater noise in the ground truth, since consumer GPS is only accurate to a few meters under ideal circumstances, and signal reflections in cities can make the error significantly worse.

Figure 7.4 shows a random set of photos incorrectly classified by the CNN. Several images, like (c) and (i), are close-ups of objects that have nothing to do with the landmark itself, and thus are probably nearly impossible to identify even with an optimal classifier. Other errors seem quite understandable at first glance, but could probably be fixed with better classifiers and finer-grained analysis. For instance, Fig. 7.4a is a close-up of a sign and thus very difficult to geo-localize, but a human would not have predicted Colosseum because the sign is in English. Image (e) is a crowded street scene and the classifier's prediction of Empire State Building is not unreasonable, but the presence of a double-decker bus reveals that it must be in London. Image (f) is a photo of artwork and so the classifier's prediction of the Louvre museum is understandable, although a tenacious human could have identified

Correct class \ Predicted class	Eiffel	Trafalgar	Big Ben	London Eye	Notre Dame	Tate Modern	Empire State Bldg	Piazza San Marco	Colosseum	Louvre
Eiffel	**82.3**	2.2	1.0	2.7	2.4	2.3	3.0	1.1	0.7	1.7
Trafalgar	0.7	**77.0**	2.9	2.3	1.5	4.0	2.3	4.1	1.8	3.6
Big Ben	1.2	3.7	**76.8**	8.2	2.2	3.4	1.1	1.4	0.7	1.3
London Eye	2.9	2.9	8.1	**76.3**	1.4	2.6	2.2	1.3	0.5	1.7
Notre Dame	2.0	2.9	1.7	1.6	**81.8**	1.1	1.0	2.7	2.1	3.0
Tate Modern	1.6	3.2	1.7	3.4	1.3	**81.0**	2.4	1.7	1.2	2.6
Empire State Bldg	1.6	1.9	1.2	1.5	0.6	2.3	**88.2**	0.8	0.6	1.2
Piazza San Marco	1.3	4.1	0.9	1.1	2.8	2.0	1.7	**81.1**	2.1	2.9
Colosseum	0.9	1.7	0.6	0.5	2.0	1.4	0.4	2.1	**88.7**	1.6
Louvre	1.2	2.8	0.9	1.2	2.8	4.1	1.0	3.1	1.9	**81.5**

Fig. 7.3 Confusion matrix for the Convolutional Neural Network visual classifier, in percentages. Off-diagonal cells greater than 3 % are highlighted in *yellow*, and greater than 6 % are shown in *red*

Fig. 7.4 Random images incorrectly classified by the Convolutional Neural Network using visual features, on the 10-way problem

the artwork and looked up where it is on exhibit. This small study of error modes suggests that while some images are probably impossible to geo-localize correctly, our automatic classifiers are also making errors that, at least in theory, could be fixed by better techniques with finer-grained analysis.

Regarding running times, the bag-of-words image classification on a single 2.66 GHz processor took about 2.4 s, most of which was consumed by SIFT interest point detection. Once the SIFT features were extracted, classification required only approximately 3.06 ms for 200 categories and 0.15 ms for 20 categories. SVM training times varied by the number of categories and the number of features, ranging from less than a minute on the 10-way problems to about 72 h for the 500-way structured SVM on a single CPU. We conducted our bag-of-words experiments on a small cluster of 60 nodes running the Hadoop open source map-reduce framework. The CNN image classification took approximately 4 ms per image running on a machine equipped with an NVidia Tesla K20 GPU. Starting from the pretrained ImageNet model provided a substantial speedup for training the network, with convergence ranging between about 2 h for 10 categories to 3.5 h for the 500-way problem.

7.6 Summary

We have shown how to create large labeled image datasets from geo-tagged image collections, and experimented with a set of over 30 million images of which nearly 2 million are labeled. Our experiments demonstrate that multiclass SVM classifiers using SIFT-based bag-of-word features achieve quite good classification rates for large-scale problems, with accuracy that in some cases is comparable to that of humans on the same task. We also show that using a structured SVM to classify the stream of photos taken by a photographer, rather than classifying individual photos, yields dramatic improvement in the classification rate. Such temporal context is just one kind of potential contextual information provided by photo-sharing sites. When these image-based classification results are combined with text features from tagging, the accuracy can be hundreds of times the random guessing baseline. Finally, recent advances in deep learning have pushed the state of the art significantly, demonstrating dramatic improvements over the bags-of-words classification techniques. Together these results demonstrate the power of large labeled datasets and the potential for classification of Internet-scale image collections.

Acknowledgments The authors thank Alex Seewald and Dennis Chen for configuring and testing software for the deep learning experiments during a Research Experiences for Undergraduates program funded by the National Science Foundation (through CAREER grant IIS-1253549). The research was supported in part by the NSF through grants BCS-0537606, IIS-0705774, IIS-0713185, and IIS-1253549, by the Intelligence Advanced Research Projects Activity (IARPA) via Air Force Research Laboratory, contract FA8650-12-C-7212, and by an equipment donation from NVidia Corporation. This research used the high-performance computing resources of Indiana University which are supported in part by NSF (grants ACI-0910812 and CNS-0521433), and by the Lilly Endowment, Inc., through its support for the Indiana University Pervasive Technology Institute, and in part by the Indiana METACyt Initiative. It also used the resources of the Cornell University Center for Advanced Computing, which receives funding from Cornell, New York State, NSF, and other agencies, foundations, and corporations. The U.S. government is authorized to reproduce and distribute reprints for overnmental purposes notwithstanding any copyright annotation thereon. Disclaimer: The views and conclusions contained herein are those of the authors and should not be interpreted as necessarily representing the official policies or endorsements, either expressed or implied, of IARPA, AFRL, NSF, or the U.S. government.

References

1. Arya S, Mount DM (1993) Approximate nearest neighbor queries in fixed dimensions. In: ACM-SIAM symposium on discrete algorithms
2. Bort J (2013) Facebook stores 240 billion photos and adds 350 million more a day. In: Business insider
3. Collins B, Deng J, Li K, Fei-Fei L (2008) Towards scalable dataset construction: an active learning approach. In: European conference on computer vision
4. Comaniciu D, Meer P (2002) Mean shift: a robust approach toward feature space analysis. IEEE transactions on pattern analysis and machine intelligence
5. Crammer K, Singer Y (2001) On the algorithmic implementation of multiclass kernel-based vector machines. J Mach Learn Res
6. Crandall D, Backstrom L, Huttenlocher D, Kleinberg J (2009) Mapping the world's photos. In: International world wide web conference
7. Crandall D, Owens A, Snavely N, Huttenlocher D (2013) SfM with MRFs: discrete-continuous optimization for large-scale structure from motion. IEEE transactions on pattern analysis and machine intelligence 35(12)
8. Csurka G, Dance C, Fan L, Willamowski J, Bray C (2004) Visual categorization with bags of keypoints. In: ECCV workshop on statistical learning in computer vision

9. Deng J, Dong W, Socher R, Li L, Li K, Fei-Fei L (2009) ImageNet: a large-scale hierarchical image database. In: IEEE conference on computer vision and pattern recognition
10. Everingham M, Van Gool L, Williams CKI, Winn J, Zisserman A (2008) The PASCAL VOC. http://www.pascal-network.org/challenges/VOC/voc2008/workshop/
11. Girshick R, Donahue J, Darrell T, Malik J (2013) Rich feature hierarchies for accurate object detection and semantic segmentation. arXiv preprint arXiv:1311.2524
12. Grauman K, Leibe B (2011) Visual object recognition. Morgan & Claypool Publishers
13. Griffin G, Holub A, Perona P (2007) Caltech-256 object category dataset. Tech rep, California Institute of Technology
14. Hao Q, Cai R, Li Z, Zhang L, Pang Y, Wu F (2012) 3d visual phrases for landmark recognition. In: IEEE conference on computer vision and pattern recognition
15. Hauff C (2013) A study on the accuracy of Flickr's geotag data. In: International ACM SIGIR conference
16. Hays J, Efros AA (2008) IM2GPS: estimating geographic information from a single image. In: IEEE conference on computer vision and pattern recognition
17. Jia Y (2013) Caffe: an open source convolutional architecture for fast feature embedding. http://caffe.berkeleyvision.org/
18. Joachims T (1999) Making large-scale SVM learning practical. In: Schölkopf B, Burges C, Smola A (eds) Advances in kernel methods—support vector learning. MIT Press
19. Kalogerakis E, Vesselova O, Hays J, Efros A, Hertzmann A (2009) Image sequence geolocation with human travel priors. In: IEEE international conference on computer vision
20. Krizhevsky A, Sutskever I, Hinton G (2012) ImageNet classification with deep convolutional neural networks. In: Advances in neural information processing systems
21. Lee S, Zhang H, Crandall D (2015) Predicting geo-informative attributes in large-scale image collections using convolutional neural networks. In: IEEE winter conference on applications of computer vision
22. Li X, Wu C, Zach C, Lazebnik S, Frahm J (2008) Modeling and recognition of landmark image collections using iconic scene graphs. In: European conference on computer vision
23. Li Y, Crandall D, Huttenlocher D (2009) Landmark classification in large-scale image collections. In: IEEE international conference on computer vision
24. Li Y, Snavely N, Huttenlocher D, Fua P (2012) Worldwide pose estimation using 3d point clouds. In: European conference on computer vision
25. Lowe DG (2004) Distinctive image features from scale-invariant keypoints. Int J Comput Vis 60(2):91–110
26. Luo J, Joshi D, Yu J, Gallagher A (2011) Geotagging in multimedia and computer vision—a survey. Multimedia Tools Appl 51(1):187–211
27. McAuley JJ, Leskovec J (2012) Image labeling on a network: using social-network metadata for image classification. In: European conference on computer vision
28. Miller G (1995) WordNet: a lexical database for English. Commun ACM 38(11):39–41
29. Oquab M, Bottou L, Laptev I, Sivic J (2014) Learning and transferring mid-level image representations using convolutional neural networks. In: IEEE conference on computer vision and pattern recognition
30. Philbin J, Chum O, Isard M, Sivic J, Zisserman A (2007) Object retrieval with large vocabularies and fast spatial matching. In: IEEE conference on computer vision and pattern recognition
31. Raguram R, Tighe J, Frahm JM (2012) Improved geometric verification for large scale landmark image collections. In: British machine vision conference
32. Razavian AS, Azizpour H, Sullivan J, Carlsson S (2014) CNN features off-the-shelf: an astounding baseline for recognition. arXiv preprint arXiv:1403.6382
33. Schroff F, Criminisi A, Zisserman A (2007) Harvesting image databases from the web. In: IEEE international conference on computer vision
34. Sermanet P, Eigen D, Zhang X, Mathieu M, Fergus R, LeCun Y (2013) Overfeat: Integrated recognition, localization and detection using convolutional networks. CoRR. http://arxiv.org/abs/1312.6229

35. Snavely N, Seitz SM, Szeliski R (2008) Modeling the world from internet photo collections. Int J Comput Vis 80(2)
36. Stone Z, Zickler T, Darrell T (2008) Autotagging facebook: social network context improves photo annotation. In: 1st IEEE workshop on internet vision
37. Taigman Y, Yang M, Ranzato M, Wolf L (2013) DeepFace: closing the gap to human-level performance in face verification. In: IEEE conference on computer vision and pattern recognition
38. Torralba A, Efros A (2011) Unbiased look at dataset bias. In: IEEE conference on computer vision and pattern recognition
39. Toshev A, Szegedy C (2013) DeepPose: human pose estimation via deep neural networks. arXiv preprint arXiv:1312.4659
40. Tsochantaridis I, Hofmann T, Joachims T, Altun Y (2004) Support vector machine learning for interdependent and structured output spaces. In: International conference on machine learning
41. Zeiler MD, Fergus R (2014) Visualizing and understanding convolutional networks. In: European conference on computer vision
42. Zheng Y, Zhao M, Song Y, Adam H, Buddemeier U, Bissacco A, Brucher F, Chua T, Neven H (2009) Tour the world: building a web-scale landmark recognition engine. In: IEEE conference on computer vision and pattern recognition

Part III
Geometric Matching Based Geo-localization

Geometric alignment of query image to geo-referenced 3D and elevation models

Data-driven and semantic geo-localization methods (Part I and II of the book) provide a fast but rather coarse estimation of the location. In contrast, the techniques that perform geometric matching of a query against a 3D model yield a more precise estimation (often 6 DOF that includes both rotation and translation) at the expense of more computation and needing more elaborate reference data. In this part, we introduce several methods that perform geo-localization through geometric matching against reference 3D models, namely DEMs and structure-from-motion reconstructed 3D models, and discuss the challenges each pose.

Chapter 8
Worldwide Pose Estimation Using 3D Point Clouds

Yunpeng Li, Noah Snavely, Daniel P. Huttenlocher and Pascal Fua

Abstract This chapter addresses the problem of determining where a photo was taken by estimating a full 6-DOF-plus-intrincs camera pose with respect to a large geo-registered 3D point cloud, bringing together research on image localization, landmark recognition, and 3D pose estimation. Our method scales to datasets with hundreds of thousands of images and tens of millions of 3D points through the use of two new techniques: a co-occurrence prior for RANSAC and bidirectional matching of image features with 3D points. We evaluate our method on several large datasets, and show state-of-the-art results on landmark recognition as well as the ability to locate cameras to within meters, requiring only seconds per query.

8.1 Introduction

Localizing precisely where a photo or video was taken is a key problem in computer vision with a broad range of applications, including consumer photography ("where did I take these photos again?"), augmented reality [30], photo editing [13], and autonomous navigation [19]. Information about camera location can also aid in more general scene understanding tasks [9, 15]. With the rapid growth of online photo sharing sites and the creation of more structured image collections such as Google's Street View, increasingly any new photo can in principle be localized with respect to this growing set of existing imagery.

Y. Li · P. Fua
EPFL, Lausanne, Switzerland
e-mail: yuli@cs.cornell.edu

P. Fua
e-mail: pascal.fua@epfl.ch

N. Snavely (✉) · D.P. Huttenlocher
Cornell University, Ithaca, NY, USA
e-mail: snavely@cs.cornell.edu

D.P. Huttenlocher
e-mail: dph@cs.cornell.edu

A.R. Zamir et al. (eds.), *Large-Scale Visual Geo-Localization*,
Advances in Computer Vision and Pattern Recognition,
DOI 10.1007/978-3-319-25781-5_8

In this chapter, we approach the image localization problem as that of *worldwide pose estimation*: given an image, automatically determine a camera matrix (position, orientation, and camera intrinsics) in a geo-referenced coordinate system. As such, we focus on images with completely unknown pose (i.e., with no GPS). In other words, we seek to extend the traditional pose estimation problem, applied in robotics and other domains, to accurate geo-registration at world scale—or at least as much of the world as we can index. Our focus on precise camera geometry is in contrast to most of prior work on image localization that has taken an image retrieval approach [4, 25], where an image is localized by finding images that match it closely without recovering explicit camera pose. This limits the applicability of such methods in areas such as augmented reality where precise pose is important. Moreover, if we can establish the precise pose for an image, we then instantly have strong priors for determining what parts of an image might be sky (since we know where the horizon must be) or even what parts are roads or buildings (since the image is now automatically registered with a map). Our ultimate goal is to automatically establish exact camera pose for as many images on the Web as possible, and to leverage such priors to understand images at world scale.

Our method directly establishes correspondence between 2D features in an image and 3D points in a very large point cloud covering many places around the world, then computes a camera pose consistent with these feature matches. This approach follows recent work on direct 2D-to-3D registration [17, 24], but at a dramatically larger scale—we use a 3D point cloud created by running structure from motion (SfM) on over 2 million images, resulting in over 800,000 reconstructed images and more than 70 million 3D points, covering hundreds of distinct places around the globe. This dataset, illustrated in Fig. 8.1, is drawn from three individual datasets: a landmarks dataset created from over 200,000 geo-tagged high-resolution Flickr photos of world's top 1,000 landmarks, the recent San Francisco dataset with over a million images covering downtown San Francisco [4], and a smaller dataset from a university campus with accurate ground truth query image locations [8].

While this model only sparsely covers the Earth's surface, it is "worldwide" in the sense that it includes many distinct places around the globe, and is of a scale more than an order of a magnitude beyond what has been attempted by previous 2D-to-3D pose estimation systems (e.g., [17, 24]). At this scale, we found that noise in the feature matching process—due to repeated features in the world and the difficulty of nearest neighbor matching at scale—necessitated new techniques. Our main contribution is a scalable method for accurately recovering 3D camera pose from a single photograph taken at an unknown location , going well beyond the rough identification of position achieved by today's large-scale image localization methods. Our 2D-to-3D matching approach to image localization is advantageous compared with image retrieval approaches because the pose estimate provides a powerful geometric constraint for validating a hypothesized location of an image, thereby improving recall and precision. Even more critically, we can exploit powerful priors over sets of 3D points, such as their co-visibility relations, to address both scalability and accuracy. We show state-of-the-art results compared with other localization methods,

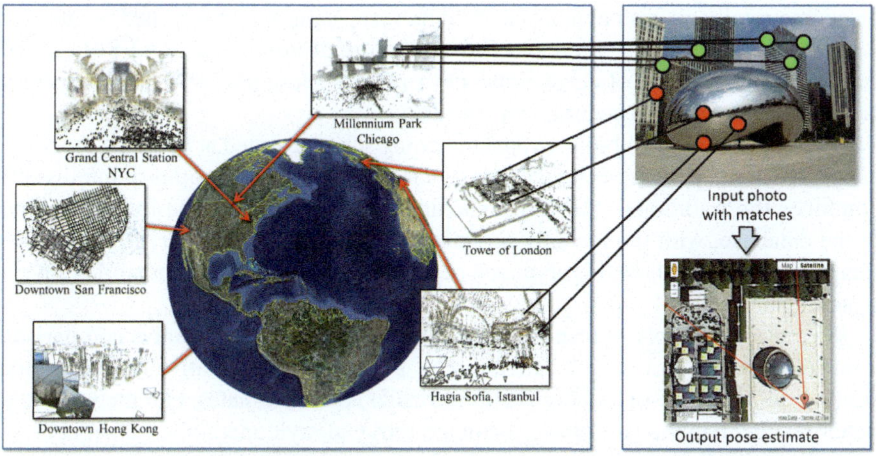

Fig. 8.1 *A worldwide point cloud database.* In order to compute the pose of a query image, we match it to a database of geo-referenced structure from motion point clouds assembled from photos of places around the world. Our database (*left*) includes a street view image database of downtown San Francisco and Flickr photos of hundreds of landmarks spanning the globe; a few selected point cloud reconstructions are shown here. We seek to compute the geo-referenced pose of new query images, such as the photo of Chicago on the *right*, by matching to this worldwide point cloud. Direct feature matching is very noisy, producing many incorrect matches (shown as *red* features). Hence, we propose robust new techniques for the pose estimation problem

and require only a few seconds per query, even when searching our entire worldwide database.

A central technical challenge is that of finding good correspondences to image features in a massive database of 3D points. We start with a standard approach of using approximate nearest neighbors to match SIFT [20] features between an image and a set of database features, then use a hypothesize-and-test framework to find a camera pose and a set of inlier correspondences consistent with that pose. However, we find that with such large 3D models the retrieved correspondences often contain so many incorrect matches that standard matching and RANSAC techniques have difficulty finding the correct pose. We propose two new techniques to address this issue. The first is the use of statistical information about the co-occurrence of 3D model points in images to yield an improved RANSAC scheme, and the second is a bidirectional matching algorithm between 3D model points and image features.

Our first contribution is based on the observation that 3D points produced by SfM methods often have strong co-occurrence relationships; some visual features in the world frequently appear together (e.g., two features seen at night in a particular place), while others rarely appear in the same image (e.g., a daytime and nighttime feature). We find such statistical co-occurrences by analyzing the large numbers of images in our 3D SfM models, then use these co-occurrences to define a new sampling prior for RANSAC. This prior samples minimal subsets of matched points that are likely to co-occur, and we find that such matches are much more likely to be geometrically consistent than standard uniform sampling of matched points. This sampling technique can often succeed with a small number of RANSAC rounds even when the fraction of correct matches (inliers) is less than 1 %. This robustness is critical for speed and accuracy in our task.

Second, we present a bidirectional matching scheme aimed at boosting the recovery of true correspondences between image features and model points. It intelligently combines the traditional "forward matching" from features in the image to points in the database, with the recently proposed "inverse matching" [17] from points to image features. We show this approach performs better than either forward or inverse matching alone.

We present a variety of results of our method, including quantitative comparisons with recent work on image localization [4, 17, 24] and qualitative results demonstrating full 6-degree-of-freedom (plus intrinsics) pose estimates. Our method yields better results than the bag-of-words image-retrieval-style method of Chen et al. [4] when both use only image features. Further, we achieve nearly the same accuracy with image features alone as their approach does when additionally provided with an approximate geo-tag for each query image. We evaluate localization accuracy on a smaller dataset with precise geo-tags, and show examples of the recovered field of view superimposed on satellite photos for both indoor and outdoor images.[1]

8.2 Related Work

Our task of worldwide pose estimation is related to several areas of recent interest in computer vision.

Landmark recognition and localization. The problem of "where was this photo taken?" can be answered in several ways. Some techniques approach the problem as that of classification into one of a predefined set of places (e.g., "Eiffel Tower," "Arc de Triomphe")—i.e., the "landmark recognition/classification" problem [16, 35]. Other methods create a database of localized imagery and formulate the problem as one of image retrieval, after which the query image can be associated with the location of the retrieved images. For instance, in their `im2gps` work, Hays and Efros seek to characterize the location of arbitrary images (e.g., of forests and deserts) with a rough probability distribution over the surface of Earth, but with coarse confidences on the order of hundreds of kilometers [9]. In follow-up work, human travel priors are used to improve performance for sequences of images [11], but the resulting locations are still coarse. Others seek to localize urban images more precisely by matching to databases of street-side imagery [4, 12, 14, 25, 33, 34], often using bag-of-words retrieval techniques [22, 26]. Our work differs from these retrieval-based methods in that we seek not just a rough camera position (or distribution over positions), but

[1] An earlier version of this work appeared at the European Conference on Computer Vision, 2012 [18].

instead a full camera matrix, including position, orientation, and focal length. To that end, we match to a geo-registered 3D point cloud and find pose with respect to these points. Other work in image retrieval also uses co-occurrence information, but in a different way from what we do. Chum et al. use co-occurrence of visual words to improve matching [6] by identifying confusing combinations of visual words, while we use co-occurrence to guide sampling of good matches within a RANSAC framework.

Localization from point clouds. More similar to our approach are methods that leverage results of SfM techniques. Irschara et al. [10] use SfM reconstructions to generate a set of "virtual" images that cover a scene, then index these as documents using BoW methods. Direct 2D-to-3D approaches have recently been used to establish correspondence between a query image and a reconstructed 3D model, bypassing an intermediate image retrieval step [17, 24]. While "inverse matching" from 3D points to image features [17] can sometimes find correct matches very quickly through search prioritization, results with this method becomes more difficult on the very large models we consider here. Similarly, the large scale will also pose a severe challenge to the method of Sattler et al. [24] as the matches become more noisy; this system already needs to perform RANSAC for up to a minute to ensure good results on much smaller datasets. In contrast, our method, aided by co-occurrence sampling and bidirectional search techniques, is able to handle much larger scales while requiring only a few seconds per query image. Finally, our co-occurrence sampling method is related to the view clustering approach of Lim et al. [19], but uses much more detailed statistical information.

8.3 Efficient Pose Estimation

Our method takes as input a database of geo-registered 3D points $\mathcal{P}$ resulting from structure from motion on a set of database images d. We are also given a bipartite graph G specifying, for each 3D point, the database images it appears in, i.e., a point $p \in \mathcal{P}$ is connected to an image $J \in d$ if p was detected and matched in image J. For each 3D point p we denote the set of images in which p appears (i.e., its neighbors in G) as A_p. Finally, one or more SIFT [20] descriptors is associated with each point p, derived from the set of descriptors in the images A_p that correspond to p; in our case we use either the centroid of these descriptors or the full set of descriptors. To simplify the discussion we initially assume one SIFT descriptor per 3D point.

For a query image I (with unknown location), we seek to compute the pose of the camera in a geo-referenced coordinate system. To do so, we first extract a set of SIFT feature locations and descriptors $\mathcal{Q}$ from I. To estimate the camera pose of I, a straightforward approach is to find a set of correspondences, or matches, between the 2D image features $\mathcal{Q}$ and 3D points $\mathcal{P}$ (e.g., using approximate nearest neighbor search). The process yields a set of matches $\mathcal{M}$, where each match $(q, p) \in \mathcal{M}$ links an image feature $q \in \mathcal{Q}$ to a 3D point $p \in \mathcal{P}$. Because these matches are corrupted by outliers, a pose is typically computed from $\mathcal{M}$ using robust techniques such as RANSAC coupled with a minimal pose solver (e.g., the 3-point algorithm for pose with known focal length). To reduce the number of false matches, nearest neighbor methods often employ a *ratio test* that requires the distance to the nearest neighbor to be at most some fraction of the distance to the second nearest neighbor.

As the number of points in the database grows larger, several problems with this approach begin to appear. First, it becomes harder to find true nearest neighbors due to the approximate nature of high-dimensional search. Moreover, the nearest neighbor might very well be an incorrect match (even if a true match exists in the database) due to similar-looking visual features in different parts of the world. Even if the closest match is correct, there may still be many other similar points, such that the distances to the two nearest neighbors have similar values. Hence, in order to get good recall of correspondence, the ratio test threshold must be set ever higher, resulting in poor precision (i.e., many outlier matches). Given such noisy correspondence, RANSAC methods will need to run for many rounds to find a consistent pose, and may fail outright. To address this problem, we introduce two techniques that yield much more efficient and reliable pose estimates from noisy correspondences: a co-occurrence-based sampling prior for speeding up RANSAC and a bidirectional matching scheme to improve the set of putative matches.

8.3.1 Sampling with Co-occurrence Prior

As a brief review, RANSAC operates by selecting samples from $\mathcal{M}$ that are minimal subsets of matches for fitting hypothesis models (in our case, pose estimates) and then evaluating each hypothesis by counting the number of inliers. The basic version of RANSAC forms samples by selecting each match in $\mathcal{M}$ uniformly at random. There is a history of approaches that operate by biasing the sampling process towards better subsets. These include guided-MLESAC [31], which estimates the inlier probability of each match based on cues such as proximity of matched features; PROSAC [5], which samples based on a matching quality measure; and GroupSAC [21], which selects samples using cues such as image segmentation. In our approach, we use *image co-occurrence* statistics of 3D points in the database images (encoded in the bipartite graph G) to form high-quality samples. This leads to a powerful sampling scheme: choosing subsets of matched 3D points that we believe are likely to co-occur in new query images, based on prior knowledge from the SfM results. In other words, if we denote with $\mathcal{P}_{\mathcal{M}}$ the subset of 3D points involved in the set of feature matches $\mathcal{M}$, then we want to sample with higher probability subsets of $\mathcal{P}_{\mathcal{M}}$ that co-occur frequently in the database, hence biasing the sampling towards more probable subsets. Unlike previous work, which tends to use simple evidence from the query image, our setting allows for a much more powerful prior due to the fact that we have multiple (for some datasets, hundreds) of images viewing each 3D point, and can hence leverage statistics not available in other domains. This sampling scheme enables our method to easily handle inlier rates as low as 1 %, which is essential as

Fig. 8.2 *Examples of frequently co-occurring points as seen in query images.* Notice that such points are not always close to each other, in either 3D space or the 2D images

we use a permissive ratio test to ensure high enough recall of true matches. Figure 8.2 shows some examples of frequently co-occurring points; note that these points are not always nearby in the image or 3D space. Instead, these points tend to be grouped according to other properties of scenes, features, and images, such as what views of a scene are common (i.e., "canonical" views often photographed), what points are most repeatable, and what features tend to fire under a given illumination condition (e.g., night vs. day or sunny vs. cloudy). These kinds of correlations are much easier to derive when starting from SfM models produced by large image collections.

Given a set of putative matches $\mathcal{M}$, and a minimal number of matches K we need to sample to fully constrain the camera pose, the goal in each round of RANSAC is to select such a subset of matched points,[2] $\{p_1, \ldots, p_K\} \subseteq \mathcal{P}_{\mathcal{M}}$, proportional to an estimated probability that they jointly correspond to a valid pose, i.e.,

$$\Pr_{\text{select}}(p_1, \ldots, p_K) \propto \Pr_{\text{valid}}(p_1, \ldots, p_K). \tag{8.1}$$

As a proxy for this measure, we define the likelihood to be proportional to their empirical co-occurrence frequency in the database, taking the view that if a set of putative points were often seen together before, then they are likely to be good matches if seen together in a new image. Specifically, we define:

$$\Pr_{\text{select}}(p_1, \ldots, p_K) \propto \left|A_{p_1} \cap \cdots \cap A_{p_K}\right|, \tag{8.2}$$

i.e., the number of database images in which *all* the K points are visible. If all of the image sets $A_{p_1}, \ldots, A_{p_K}$ are identical and have large cardinality, then $\Pr_{\text{select}}$ is high; if any two are disjoint, then $\Pr_{\text{select}}$ is 0.

As it is quite expensive to compute and store such joint probabilities for K larger than 1 or 2 (in our case, 3 or 4), we instead opt to draw the points for a given sample one at a time in a greedy fashion, where the i-th point is selected by marginalizing over all possible future choices:

$$\Pr_{\text{select}}(p_i | p_1, \ldots, p_{i-1}) \propto \sum_{p_{i+1}, \ldots, p_K} \left|A_{p_1} \cap \cdots \cap A_{p_K}\right|. \tag{8.3}$$

[2]Here we assume that each point p is matched to at most one feature in $\mathcal{Q}$, and hence appears at most once in $\mathcal{M}$. We find that this is almost always the case in practice.

In practice, the summation over future selections $(p_{i+1}, \ldots, p_K)$ can still be slow. To avoid this expensive forward search, we approximate it using simply the co-occurrence frequency of the first i points, i.e.,

$$\tilde{\Pr}_{\text{select}}(p_i | p_1, \ldots, p_{i-1}) \propto \left| A_{p_1} \cap \cdots \cap A_{p_i} \right|. \tag{8.4}$$

Given precomputed image sets A_p, this quantity can be evaluated efficiently at runtime using fast set intersection.[3]

We also tried defining $\Pr_{\text{select}}$ using other measures, such as the Jaccard index and the cosine similarity between $A_{p_1} \cap \cdots \cap A_{p_{i-1}}$ and A_{p_i}, but found that using simple co-occurrence frequency performed just as well as these more sophisticated alternatives.

8.3.2 Bidirectional Matching

The RANSAC approach described above assumes a set of putative matches; we now return to the problem of computing such a set in the first place. Matching an image feature to a 3D point amounts to retrieving the feature's nearest neighbor in the 128-D SIFT space, among the set of points $\mathcal{P}$ in the 3D model (using approximate nearest neighbor techniques such as [2]), subject to a ratio test. Conversely, one could also match in the other direction, from 3D points to features, by finding for each point in $\mathcal{P}$ its nearest neighbor among image features $\mathcal{Q}$, subject to the same kind of ratio test. We call the first scheme (image feature to point) *forward matching* and the second (point to feature) *inverse matching*. Again, we begin by assuming there is a single SIFT descriptor associated with each point.

We employ a new bidirectional matching scheme combining forward and inverse matching. A key observation is that visually similar points are more common in our 3D models than they are in a query image, simply because our models tend to have many more points (millions) than an image has features (thousands). A prominent point visible in a query image sometimes cannot be retrieved during forward matching, because it is confused with other points with similar appearance. However it is often much easier to find the correct match for such a point in the query image, where the corresponding feature is more likely to be unique. Hence inverse matching can help recover what forward matching has missed. On the other hand, inverse matching alone is inadequate for large models, even with prioritization [17], due to the much higher proportion of irrelevant points for any given query image and hence the increased difficulty in selecting relevant ones to match. This suggests a two-step approach:

---

[3] While our method requires that subsets of three or four points often be co-visible in the database images, this turns out to be a very mild assumption given the further constraints we use to determine correct poses, described below.

1. **Forward matching**. Find a set of primary matches using the conventional forward matching scheme, and designate as *preferred matches* a subset of them with low distance ratios (and hence relatively higher confidence);
2. **Inverse matching**. Augment the set of primary matches by performing a prioritized inverse matching [17], starting from the preferred matches as the initial seed model points to search for in the images. The final pose estimation is carried out on the augmented set of matches.

This inverse matching step is prioritized in that it considers the points in the model to match to the image in decreasing order of co-occurrence frequency with the initial seed points. This ordering scheme is implemented as a priority queue on model points using co-occurrence frequency as the priority; these priorities are updated occasionally as more good matches are found (see [17] for more details).

We apply these two steps in a cascade: we attempt pose estimation as soon as the primary matches are found and skip the second step if we already have enough inliers to successfully estimate the pose.

As mentioned above, a 3D point can have multiple descriptors since it is associated with features from multiple database images. Hence we can choose to either compute and store a single average descriptor for each point (as in [17, 24]) or keep all the individual descriptors; we evaluate both options in our experiments. In the latter case, we relax the ratio test so that, besides meeting the ratio threshold, a match is also accepted if both the nearest neighbor and the second nearest neighbor (of the query feature) are descriptors of the same 3D point. This is necessary to avoid "self confusion," since descriptors for the same point are expected to be similar. While this represents a less strict test, we found that it works well in practice. The same relaxation also applies to the selection of preferred matches. For inverse matching, we always use average descriptors.

8.4 Evaluation Datasets

To provide a quantitative evaluation of the localization performance of our method, we have tested on three datasets, both separately and combined into a single point cloud; some are from the literature to facilitate benchmarking. Table 8.1 summarizes each dataset. The sizes of the Dubrovnik and Rome datasets used in [17] are included for comparison; our combined model is about two orders of magnitude larger than the Rome dataset.

Landmarks. The first dataset consists of a large set of geo-tagged photos (i.e., photos with latitude and longitude) of famous places downloaded from Flickr. We first created a list of geo-tagged Flickr photos from the world's top 1,000 landmarks derived via clustering on geo-tags by Crandall et al. [7]. We ran SfM on each of these 1,000 individual collections to create a set of point cloud models [1]. These models are in an arbitrary coordinate system, and are related to geographic coordinates by an unknown similarity transform. To solve for this transform, we first estimated the

Table 8.1 Statistics of the datasets used for evaluation, including the sizes of the reconstructed 3D models and the number of test images

	Images in 3D model	Points in 3D model	Test images
Landmarks	205,162	38,190,865	10,000
SF-0	610,773	30,342,328	803
SF-1	790,409	75,410,077	803
Quad	4,830	2,022,026	348
Dubrovnik [17]	6,044	1,975,263	800
Rome [17]	15,179	4,067,119	1,000

SF-1 refers to the **San Francisco** dataset with image histogram equalization and upright SIFT features [4], while SF-0 is the one without. Note that SF-0 and SF-1 are derived from the same image set

gravity vector (i.e., the "up vector") of each model by finding the direction in the SfM model most orthogonal to the greatest number of camera x-vectors [29], and then refining this vector by detecting lines in the images and estimating a vertical vanishing point [32]. We use this gravity vector to transform the model to be "upright," and finally geo-registered the reconstructed 3D model using the image geo-tags, so that its coordinates can be mapped to actual locations on the globe. Since the geo-tags are quite noisy, we used RANSAC to estimate the required 2D translation, 1D rotation, and scale. This sometimes produced inaccurate results, which could be alleviated in the future by more robust SfM and geo-registration methods [28, 32]. Finally, we took the union of these SfM models to form a single, geo-referenced point cloud. Some of the individual models are illustrated in Fig. 8.1.

For evaluation, we created a set of test images by removing a random subset of 10,000 images from the reconstruction. This involves removing them from the image database and their contribution to the SIFT descriptors of points, and deleting any 3D points that are no longer visible in at least two images. Withholding the test images slightly reduces the database size, yielding the sizes shown in Table 8.1. Each test image has a known landmark ID, which can be compared with the ID inferred from an estimated camera pose for evaluation. This ID information is somewhat noisy due to overlapping landmarks, but can provide an upper bound on the false registration rate for the dataset. Since the test images come from the original reconstructions, it should be possible to achieve a 100 % recall rate.

San Francisco. We also use the recently published San Francisco dataset [4], which contains 640×480 resolution perspective images cropped from omnidirectional panoramas. Two types of images are provided: about 1M perspective central images ("PCIs"), and 638 K perspective frontal images ("PFIs") of rectified facades. Each database image, as well as each of 803 separate test images taken by cameraphones (not used in reconstruction), comes with a building ID, which can be used to evaluate the performance of image retrieval or, in our case, pose estimation. We reconstructed our 3D model using only the PCIs (as the PFIs have non-standard imaging geometry). We reconstructed two SfM models, one (SF-0) using the raw PCIs (to be consistent with the other datasets), and one (SF-1) using upright SIFT features extracted from histogram-equalized versions of the database images (as recommended in [4]). The model was geo-registered using provided geo-tags. We ignore images that were not reconstructed by SfM.

Quad. The first two datasets only provide coarse ground truth for locations, in the form of landmark/building identifiers. Although geo-tags exist for the test images, they typically have errors in the tens (or hundreds) of meters, and are thus too imprecise for fine evaluation of positional accuracy. We therefore also use the Quad dataset from Crandall et al. [8], which comes with a database of images of the Arts Quad at Cornell University as well as a separate set of test images with accurate, sub-meter error geo-tags. We ran SfM on the database images, and use the accurately geo-tagged photos to test localization error.

8.5 Experiments and Results

To recap: to register a query image, we estimate its camera pose by extracting SIFT features $\mathcal{Q}$, finding potential matches $\mathcal{M}$ with the model points $\mathcal{P}$ through nearest neighbor search plus a ratio test, and running co-occurrence RANSAC to compute the pose, followed by bidirectional matching if this initially fails. For the minimal pose solver, we use the 3-point algorithm if the focal length is approximately known, e.g., from EXIF data, or the 4-point algorithm [3] if the focal length is unknown and needs to be estimated along with the extrinsics. Finally a local bundle adjustment is used to refine the pose. We accept the pose if it has at least 12 inlier matches, as in [17, 24].

Precision and recall of registration. We first test the effectiveness of exploiting point co-occurrence statistics in RANSAC using *registration rate*, i.e., the percentage of query images that are registered to the 3D model (i.e., that pass the 12-inlier threshold). We later estimate the likelihood of false registrations.

Figure 8.3 (left) shows the registration rates on the **Landmarks** dataset. For RANSAC with co-occurrence prior, we always use 0.9 as the ratio test threshold. For regular RANSAC *without* co-occurrence we experimented with three thresholds ($r = 0.7, 0.8, 0.9$), the best of which at 10,000 RANSAC rounds ($r = 0.8$) has performance roughly equal to running just 10 rounds *with* co-occurrence. These results demonstrate the advantage of using co-occurrence to guide sampling, especially when the number of rounds is small. For this experiment, we used average SIFT descriptors for points (cf. Sect. 8.3). When using *all* of its associated descriptors for each point, we observed the same trend as in Fig. 8.3. The overhead incurred by co-occurrence sampling is only a small fraction of the total RANSAC time; thus its impact on overall speed is almost negligible.

We also assess the performance gain from bidirectional matching, the results of which are shown in Fig. 8.3 (right). Experiments were performed with average descriptors as well as with all feature descriptors for the points. The results show

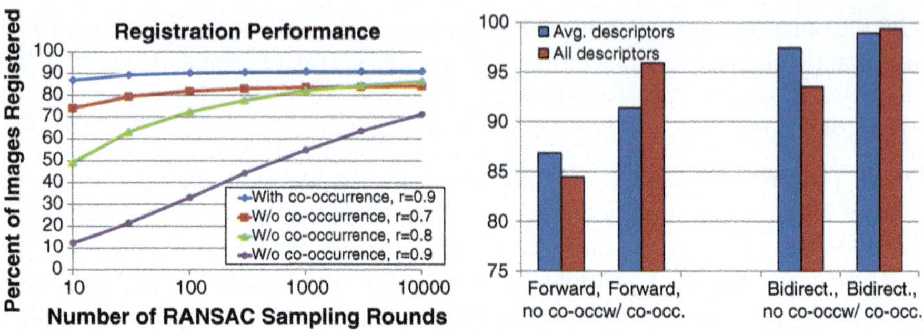

Fig. 8.3 *Registration rates on* **Landmarks**. *Left* comparison of results with and without co-occurrence prior for sampling under forward matching. At 10,000 rounds, RANSAC starts to approach the running time of the ANN-based matching. *Right* comparison of forward versus bidirectional matching. We used 10,000 RANSAC rounds and selected 0.8 as the ratio threshold for the experiments without using co-occurrence (0.9 otherwise). For comparison, applying the systems of [17, 24] on this same dataset yielded registration rates of 33.09 and 16.20 % respectively

that bidirectional matching significantly boosts the registration rate, whether or not co-occurrence based RANSAC is used. Similarly, the use of co-occurrence is also always beneficial, with or without bidirectional matching, and the advantage is more pronounced when all feature descriptors are used, as this produces more matches but also more outliers. Since co-occurrence together with bidirectional matching produced the highest performance, we use this combination for the remaining experiments.

To estimate the precision of registration, namely the fraction of query images *correctly* registered, and equivalently the false registration rate, we consider the 1000-way landmark classification problem. The inferred landmark ID is simply taken to be the one with the most points registered with the image. The classification rate among the registered images is 98.1 % when using average point descriptors and 97.9 % when using all descriptors. However, we found that this does not mean that the remaining 2 % of images are all false registrations, since some of our landmarks visually overlap and thus the classification objective is not always unambiguous. To better estimate the false registration rate, we tested our method with a set of 1468 "negative images" that are photos of other landmarks geographically distant from the top 1000 in our dataset. Of these, 10 images were registered (both for average/all descriptors), which corresponds to a false registration rate of 0.68 %. Figure 8.4 (left) shows an example false registration. Indeed, the false registrations are almost always due to identical-looking signs and logos.

While it is difficult to quantitatively evaluate the accuracy of the full camera poses on this dataset, we visualize a few recovered camera poses for the test set in Fig. 8.5; many of the poses are surprisingly visually accurate. Later, we describe the use of the **Quad** dataset to quantitatively evaluate localization error.

We also test our method on the recent **San Francisco** dataset of Chen et al. [4], and compare with their state-of-the-art system for large-scale location recognition (based

Fig. 8.4 *Examples of false registrations.* Side-by-side: query image and its closest image in the database by the number of common 3D points. *Left* from **Landmarks**. The US flag appears both on the space shuttle and in the Grand Central Railway Station. *Right* from **San Francisco**. The piers have nearly identical appearance

Fig. 8.5 *Estimated poses for the* **Landmarks** *dataset.* A few images that were successfully registered, along with their estimate pose overlaid on a map. Each map has been annotated to indicate the estimated position, orientation, and field of view of the photo, and the image itself is drawn with a *red line* showing the horizon estimated from the pose, as well as axes showing the estimated up (*blue*), north (*green*), and east (*red*) directions (Best viewed on screen when enlarged.)

on a bag-of-visual-word-style retrieval algorithm [23]). This is a much more challenging benchmark: the database images have different characteristics (panorama crops) from the test images (cell phone photos), and both have considerably lower resolution on average than those in the Landmarks dataset. Moreover, unlike Landmarks, there is no guarantee that every test image is recognizable given the database images. As in [4] we evaluate our method using the recall rate, which corresponds to the percentage of correctly registered query images. We consider a registration correct if the query image is registered to points of the correct building ID according to the ground truth annotation. The results are summarized in Table 8.2. Using the same images and features, our method outperforms that of [4] by a large margin even when the latter uses the extra GPS information. Although a maximum recall rate of 65 % for SF-1 was reported in [4], achieving this requires the additional use of the PFIs (on top of GPS) specific to this data set. Again, our method produces not just a nearby landmark or building ID, but also a definitive camera pose, including its location and orientation, as illustrated in Fig. 8.6. This pose information could be used for further tasks, such as annotating the image.

All the recall rates for our method correspond to false registration rates between 4.1 and 5.3 %, which are comparable to the 95 % precision used in [4]. As before,

Table 8.2 Percentage of query images correctly localized ("recall rate") for the **San Francisco** dataset

Our method (no GPS)	SF-0	SF-1	Chen et al. [4]	SF-0	SF-1
Avg. descriptors	50.2	58.0	No GPS	20	41
All descriptors	54.2	62.5	With GPS	–	49

For our method, we report results with (SF-1) and without (SF-0) histogram equalization and upright SIFT. We also experimented with both using average descriptors and keeping all descriptors. For [4] we cite the recall rates (if provided) for the variants that use the same kind of perspective images (PCIs) as we do

Fig. 8.6 *Estimated poses for selected* **San Francisco** *query images.* The annotations are the same as in Fig. 8.5. Our method produces reasonable poses for most of these benchmark images

most false registrations are due to logos and signs, though a few are due to highly similar buildings (as in Fig. 8.4, right). Sometimes correct registrations are judged as incorrect because of missing building IDs in the ground truth, which leads to an underestimate of both recall and precision.

Localization error. In order to evaluate the accuracy of estimated camera positions, we tested our method on the **Quad** dataset, which has accurately geo-tagged query images. This is also a challenging dataset, because of the differences in season between the database images and the query images. Our method succeeded in registering 68.4 % of the query images using average descriptors and 73.0 % using all desciptors. The localization error has a mean of 5.5 m and a median of 1.6 m, with about 90 % for images having errors of under 10 m and 95 % under 20 m. Hence despite relatively larger errors in database image geo-tags used to geo-register the 3D model, our method was able to achieve good localization accuracy comparable to that of consumer GPS.

Scalability. To further study the scalability of our method, we merged the reconstructed 3D models for our three datasets into a single large one by concatenating them together; compared with the individual datasets, the merged set has many more things that could potentially confuse a query image. For this test we used the San Francisco model without histogram equalization or upright SIFT, so that all three models are reconstructed using the same type of features and hence are more potent distractors of each other. The combined model contains over 800K images and 70M points. We run the same registration experiment for each of the three sets of query images on the combined model, and compare the results with those from running on the individual models. Table 8.3 shows the registration performance on the com-

Table 8.3 Recall rates (percent) on the combined model compared with those on individual models

Model\Query images	**Landmarks**	**San Francisco**	**Quad**
Individual	98.95	50.2	68.4
Combined	98.90	47.7	61.2

Combining the models has essentially no impact on the false registration rates. Average point descriptors are used in these experiments

bined model for each test set under the same criteria as on the individual models.[4] The performance gap is negligible for Landmarks and small (around 2 %) for San Francisco. While the gap is somewhat larger for the Quad images (about 7 %), this is likely due to the fact that the Quad model is far smaller than the other two, with fewer than 5K images and just over 2M points. Hence placing it into the combined model corresponds to more than an order of magnitude increase in the amount of irrelevant information. In this context, the decrease in registration rate for the Quad query images can be considered quite modest. Furthermore our method maintains essentially the same level of localization accuracy (mean = 4.9 m, median = 1.9 m) when given the combined model. This shows the scalability of our method and its robustness to irrelevant information.

For completeness, we also tested our method on the **Dubrovnik** and **Rome** datasets from [17]. We achieved a registration rate of 100 % on Dubrovnik and 99.7 % on Rome, using a single average descriptor per point, which compares favorably to results reported in [17, 24] (94.1/92.4 % and 98.0/97.7 %, respectively). We also verified by swapping the two sets that no Dubrovnik images were falsely registered to Rome and vice versa.

Our system takes a few (one to five) seconds per query image of medium resolution (1–2 megapixels), excluding the time to extract the SIFT keys, when running single-threaded on a Intel Xeon 2.67 GHz CPU. The variation in running time across query images depends on the number of features extracted, as well as whether or not the second inverse matching stage is invoked (so failed registrations are in general more expensive than successful ones). While not real-time, this is quite fast considering the size of the database, and could easily be parallelized.

Discussion. Most of the false registrations by our method involve some sort of signs or logos, which tend to be feature-rich and are identical at different places. This suggests that false registrations can be largely reduced if we can learn to recognize these types of objects, or take into account contextual information. Our method can require a significant amount of memory for storing descriptors, particularly when a point is assigned all of its corresponding SIFT features. In the future, this could be improved through the use of more compact descriptors [27], or by intelligently compressing the database.

[4]Registered Quad images are counted as correct if they were registered to the Quad part of the combined model, which they all did.

8.6 Conclusion

We presented a method for camera pose estimation at a worldwide scale; for the level of accuracy in pose we aim for, this is to our knowledge the largest such system that exists. Our method leverages reconstructed 3D point cloud models aided by two new techniques: co-occurrence based RANSAC and bidirectional matching, which greatly improve its reliability and efficiency. We evaluated our method on several large datasets and show state-of-the-art results. Moreover, comparable performance is maintained when we combine these datasets into an even greater one, further demonstrating the effectiveness and scalability of our method.

References

1. Agarwal S, Snavely N, Simon I, Seitz SM, Szeliski R (2009) Building rome in a day. In: ICCV
2. Arya S, Mount DM (1993) Approximate nearest neighbor queries in fixed dimensions. In: ACM-SIAM symposium on discrete algorithms
3. Bujnak M, Kukelova Z, Pajdla T (2008) A general solution to the p4p problem for camera with unknown focal length. In: CVPR
4. Chen DM, Baatz G, Köser K, Tsai SS, Vedantham R, Pylvänäinen T, Roimela K, Chen X, Bach J, Pollefeys M, Girod B, Grzeszczuk R (2011) City-scale landmark identification on mobile devices. In: CVPR
5. Chum O, Matas J (2005) Matching with prosac—progressive sample consensus. In: CVPR, pp 220–226
6. Chum O, Matas J (2010) Unsupervised discovery of co-occurrence in sparse high dimensional data. In: CVPR
7. Crandall D, Backstrom L, Huttenlocher D, Kleinberg J (2009) Mapping the world's photos. In: WWW
8. Crandall D, Owens A, Snavely N, Huttenlocher D (2011) Discrete-continuous optimization for large-scale structure from motion. In: CVPR
9. Hays J, Efros AA (2008) IM2GPS: estimating geographic information from a single image. In: CVPR
10. Irschara A, Zach C, Frahm JM, Bischof H (2009) From structure-from-motion point clouds to fast location recognition. In: CVPR
11. Kalogerakis E, Vesselova O, Hays J, Efros AA, Hertzmann A (2009) Image sequence geolocation with human travel priors. In: ICCV
12. Knopp J, Sivic J, Pajdla T (2010) Avoiding confusing features in place recognition. In: ECCV
13. Kopf J, Neubert B, Chen B, Cohen MF, Cohen-Or D, Deussen O, Uyttendaele M, Lischinski D (2008) Deep photo: model-based photograph enhancement and viewing. SIGGRAPH Asia Conf Proc 27(5), 116:1–116:10
14. Kroepfl M, Wexler Y, Ofek E (2010) Efficiently locating photographs in many panoramas. In: GIS
15. Lalonde JF, Efros AA, Narasimhan SG (2011) Estimating the natural illumination conditions from a single outdoor image. IJCV
16. Li Y, Crandall DJ, Huttenlocher DP (2009) Landmark classification in large-scale image collections. In: ICCV
17. Li Y, Snavely N, Huttenlocher DP (2010) Location recognition using prioritized feature matching. In: ECCV
18. Li Y, Snavely N, Huttenlocher DP, Fua P (2012) Worldwide pose estimation using 3d point clouds. In: ECCV

19. Lim H, Sinha SN, Cohen MF, Uyttendaele M (2012) Real-time image-based 6-dof localization in large-scale environments. In: CVPR
20. Lowe DG (2004) Distinctive image features from scale-invariant keypoints. IJCV 60(2):91–110
21. Ni K, Jin H, Dellaert F (2009) Groupsac: efficient consensus in the presence of groupings. In: ICCV
22. Nistér D, Stewénius H (2006) Scalable recognition with a vocabulary tree. In: CVPR, pp 2118–2125
23. Philbin J, Chum O, Isard M, Sivic J, Zisserman A (2007) Object retrieval with large vocabularies and fast spatial matching. In: CVPR
24. Sattler T, Leibe B, Kobbelt L (2011) Fast image-based localization using direct 2D-to-3D matching. In: ICCV
25. Schindler G, Brown M, Szeliski R (2007) City-scale location recognition. In: CVPR
26. Sivic J, Zisserman A (2003) Video google: a text retrieval approach to object matching in videos. In: ICCV, pp 1470–1477. http://www.robots.ox.ac.uk/vgg
27. Strecha C, Bronstein AM, Bronstein MM, Fua P (2010) LDAHash: improved matching with smaller descriptors. In: EPFL-REPORT-152487
28. Strecha C, Pylvanainen T, Fua P (2010) Dynamic and scalable large scale image reconstruction. In: CVPR
29. Szeliski R (2006) Image alignment and stitching: a tutorial. Found Trends Comput Graph Comput Vis 2(1)
30. Takacs G, Xiong Y, Grzeszczuk R, Chandrasekhar V, chao Chen W, Pulli K, Gelfand N, Bismpigiannis T, Girod B (2008) Outdoors augmented reality on mobile phone using loxel-based visual feature organization. In: Proceedings of the multimedia information retrieval
31. Tordoff B, Murray DW (2002) Guided sampling and consensus for motion estimation. In: ECCV
32. Wang CP, Wilson K, Snavely N (2013) Accurate georegistration of point clouds using geographic data. In: Proceedings of international conference on 3D vision
33. Zamir AR, Shah M (2010) Accurate image localization based on google maps street view. In: ECCV
34. Zhang W, Kosecka J (2006) Image based localization in urban environments. In: International symposium on 3D data processing, visualization and transmission
35. Zheng YT, Zhao M, Song Y, Adam H, Buddemeier U, Bissacco A, Brucher F, Chua TS, Neven H (2009) Tour the world: building a web-scale landmark recognition engine. In: CVPR

Chapter 9
Exploiting Spatial and Co-visibility Relations for Image-Based Localization

Torsten Sattler, Bastian Leibe and Leif Kobbelt

Abstract Image-based localization techniques aim to estimate the position and orientation from which a given query images was taken with respect to a 3D model of the scene. Recent advances in Structure-from-Motion, which allow us to reconstruct large scenes in little time, create a need for image-based localization approaches that handle large-scale models consisting of millions of 3D points both efficiently and effectively in order to localize as many query images as possible in as little time as possible. While multiple efficient localization methods based on prioritized feature matching have been proposed recently, they lack the effectiveness of slower approaches. In this chapter, we show that we can increase the effectiveness of approaches based on prioritized 2D-to-3D matching at little to no additional run-time costs by exploiting both spatial and co-visibility relations between the 3D points in the model. The resulting localization framework incorporates both 2D-to-3D and 3D-to-2D matching and achieves state-of-the-art efficiency and effectiveness.

9.1 Introduction

Nowadays, devices used for localization and navigation, e.g., smart phones and navigation devices for cars, rely on sensor data such as GPS and digital compasses. While sufficient for many scenarios, these localization methods do not achieve the accuracy required for advanced applications such as navigation for autonomous vehicles [14], Structure-from-Motion (SfM) [1, 10, 29], or location-based applications where the user is more interested in what she actually observers rather than her position.

T. Sattler (✉)
Department of Computer Science, ETH Zürich, Zürich, Switzerland
e-mail: torsten.sattler@inf.ethz.ch

B. Leibe · L. Kobbelt
RWTH Aachen University, Aachen, Germany
e-mail: leibe@vision.rwth-aachen.de

L. Kobbelt
e-mail: kobbelt@cs.rwth-aachen.de

A.R. Zamir et al. (eds.), *Large-Scale Visual Geo-Localization*,
Advances in Computer Vision and Pattern Recognition,
DOI 10.1007/978-3-319-25781-5_9

Fortunately, both problems can be solved by using images for localization: Cameras can be used to determine the position and orientation of the user with high precision [11] while being readily available and significantly cheaper than specialized sensors such as laser scanners. The high accuracy of the estimated camera poses, i.e., the positions and orientations from which images were taken, in turn enables us to display virtual content, e.g., information about the buildings that are visible in a photo that the user took, by using Augmented Reality techniques [4].

Traditionally, the image-based localization problem of determining the camera pose for a given query image has been cast as an image retrieval problem [28, 31, 32]. Given a set of geo-tagged database images, retrieval-based methods first find a subset of photos that depict the same location as the query image and then use the geo-tags of the retrieved database photos to estimate the pose of the query image [32]. The precision of the resulting pose directly depends on the accuracy of the geo-tags of the database images, i.e., obtaining geo-tags through GPS is not sufficient. In contrast, 3D structure-based localization approaches offer the desired pose accuracy by relating 2D image measurements with points in a 3D model of the scene. These resulting 2D–3D matches are then used to estimate the position and orientation from which a query photo was taken. Recent advances in SfM [1, 10] allow us to reconstruct large environments in a few hours or days, creating the need for structure-based localization methods that work on a truly large scale.

In this chapter, we analyze how to exploit spatial and co-visibility relations between 3D points in the scene models to improve the effectiveness and efficiency of prioritized image-based localization approaches. Spatial cues are used to identify scene points likely to be visible in the query image, which are subsequently matched against the features found in the query photo. This back-matching scheme enables us to exploit the complimentary advantages and disadvantages of the two search directions: When matching features against the 3D model, there might be multiple points with similar appearance and it is thus hard to find the correct match. By matching points that are likely visible against the query image, we can recover such correspondences at little additional costs. At the same time, matching only points likely to be visible avoids the problem of finding multiple conflicting correspondences from comparing multiple points, distributed all over the model, with similar descriptors against the query image. The proposed scheme finds additional correspondences that enable us to improve the localization effectiveness, i.e., to localize more query images. Co-visibility information are then used to avoid unnecessary matching operations and to identify wrong matches before attempting camera pose estimation. The resulting large-scale localization approach is both efficient, i.e., offers fast localization times, and effective. This chapter is based on our previous publication [25],[1] but compares the proposed approach with methods based on similar ideas that were published at the same time [5, 15] and shows that the proposed approach offers the best combination of effectiveness and efficiency. To provide an introduction and

[1] ©Springer-Verlag Berlin Heidelberg 2012.

pointers for readers not familiar with image-based localization, we also provide a very detailed discussion of directly related work.

The rest of this chapter is organized as follows. Sect. 9.2 introduces the image-based localization problem and discusses related work. Sect. 9.3 analyzes the complimentary advantages and disadvantages of 2D-to-3D and 3D-to-2D matching. Section 9.4 describes our *Active Search* approach that exploits spatial cues to integrate 3D-to-2D search into a prioritized 2D-to-3D matching framework [24]. Section 9.5 shows how to incorporate co-visibility information to increase the localization efficiency of our method. Finally, Sect. 9.6 provides an experimental evaluation of the resulting approach.

9.2 The Image-Based Localization Problem

As outlined in the introduction, we define the image-based localization problem as the problem of computing the camera pose, i.e., the position and orientation from which a given query image was taken, relative to a 3D model of the scene. We solve this problem by establishing 2D–3D matches between 2D measurements in the query image and 3D points in a scene model. These 2D–3D matches are then used to estimate the camera pose by using a n-point-pose solver [11], i.e., an algorithm that estimates the camera pose from n 2D–3D matches. In order to robustly handle wrong matches, we apply the solver inside a RANSAC loop [9].

In the case that the scene has been reconstructed using SfM, each 3D point in the model has been triangulated from two or more local image features, e.g., SIFT [18], found in the set of database images used for the reconstruction. Consequently, we can associate each 3D point with the descriptors of its corresponding image features to obtain a representation that contains both the 3D structure and the local appearance of the scene. Using this representation, we can establish 2D–3D matches by extracting local features in the query image and comparing their descriptors with the descriptors associated with the 3D points. We can thus reduce the image-based localization problem to a descriptor matching problem, which we can solve using nearest neighbor search. In order to handle large-scale scenes, we need to perform nearest neighbor search as efficiently as possible. Due to the high dimensionality of the descriptors typically used and the resulting curse of dimensionality, we use approximate nearest neighbor search. However, there is a trade-off between matching efficiency and localization effectiveness and we want to make sure that we find enough correct matches to facilitate camera pose estimation for as many query images as possible. At the same time, we need to ensure that we do not find too many wrong matches since RANSAC's run-time increases exponentially in the percentage of wrong matches [9]. Thus, it is common to use Lowe's ratio test [18] to reject ambiguous matches in order to avoid finding too many wrong matches. Given the two nearest neighboring descriptors $\mathbf{d}_1$, $\mathbf{d}_2$ for a query $\mathbf{d}_q$, the test accepts a match between $\mathbf{d}_q$ and $\mathbf{d}_1$ only if the nearest neighbor is sufficiently closer to $\mathbf{d}_q$ than the second nearest neighbor, i.e., if

$$\|\mathbf{d}_q - \mathbf{d}_1\|_2 < \tau \cdot \|\mathbf{d}_q - \mathbf{d}_2\|_2 \tag{9.1}$$

holds. The constant τ is usually chosen from the range [0.6, 0.9].

A conceptually simple image-based localization approach represents each 3D point by the mean of all its descriptors. These mean descriptors are then used to construct a kd-tree [20] that is used to accelerate nearest neighbor search for the features detected in the query image [24]. While this 2D-to-3D matching approach has been shown to achieve state-of-the-art effectiveness on a standard benchmark dataset, it is not very efficient as the nearest neighbor search takes multiple seconds for models containing a few million 3D points [24]. In order to make nearest neighbor search more efficient, Li et al. proposed a prioritized 3D-to-2D matching scheme based on the co-visibility relations between points in the SfM model [16]. Two reconstructed points are deemed co-visible if they can be seen together in at least one database image. If a 3D-to-2D match is found for a 3D point p, the priorities of all points co-visible with p are increased. The initial priority of each point is proportional to the number of database images it is visible in and Li et al. use a set of seed points, distributed all over the model and selected by solving a set cover problem, to allow their approach to quickly find the region of the model that is depicted in the query image. While being significantly faster than the tree-based approach, this Point-to-Feature (P2F) matching scheme is also significantly less effective since 3D-to-2D matching is inherently less reliable than 2D-to-3D search as outlined in Sect. 9.1. Consequently, Sattler et al. propose a prioritized 2D-to-3D matching framework based on using a fixed quantization of the descriptor space in the form of a visual vocabulary [24]. In an offline step, they assign each 3D point descriptor to its closest visual word from a vocabulary of 100k cluster centers using approximate nearest neighbor search. Given a query image, they first assign each feature to its closest word and then use exhaustive search through all point descriptors assigned to the same word to find the two nearest neighbors required by the ratio test. The number of 3D point descriptors assigned to the same word thus defines the search costs for each query feature and Sattler et al. consider the features in ascending order of costs, stopping the search once a fixed number of 2D-to-3D matches is found. While the resulting Vocabulary-based Prioritized Search (VPS) approach is both more efficient and effective than P2F, it does not quite reach the effectiveness of the simple tree-based approach since it is affected by quantization artifacts, i.e., it cannot find matches between features and points if their descriptors are assigned to different visual words. In Sect. 9.4, we will show how we can recover such lost matches through 3D-to-2D search by exploiting spatial relations between the 3D points in the model to quickly determine candidate points for 3D-to-2D matching. While computationally efficient, our resulting *Active Search* step induces additional run-time costs. However, we show in Sect. 9.5 that we can avoid many unnecessary computations during both 3D-to-2D matching and RANSAC-based pose estimation by also considering the co-visibility relation between the 3D points. The resulting approach achieves state-of-the-art localization efficiency and effectiveness, even outperforming the tree-based search approach.

Simultaneous to our original publication [25], two other approaches have been proposed that follow ideas similar to the ones presented in this chapter. Choudhary and Narayanan use a prioritized 2D-to-3D matching approach similar to VPS to quickly find high-quality matches that are then used to seed 3D-to-2D search [5]. They use the co-visibility relations between the points contained in the 3D model to obtain a probabilistic formulation for 3D-to-2D search, which they use to prioritize matching. Compared to the approach presented in this chapter, this Visibility Probability (Vis. Prob.) scheme achieves a lower effectiveness and a lower localization accuracy at run-times similar to our method while it is only feasible to run their scheme on a subset of all model points. The method proposed in [5], similar to the other approaches discussed above, aims at solving the localization problem as efficiently as possible. In contrast, Li et al. sacrifice efficiency for effectiveness [15]. They build on the tree-based approach discussed above and increase the ratio test threshold to ensure that they miss only few correct matches. Since this also increases the false positive matching ratio significantly, they adapt RANSAC's sampling scheme to randomly select n matches based on the probability of their 3D points being visible together. If the pose cannot be estimated from the resulting 2D-to-3D matches alone, Li et al. use P2F, seeded at the previously found correspondences, to recover additional matches. The resulting method is significantly slower than the approach proposed in this chapter while achieving a slightly higher localization effectiveness.[2] We notice that the methods from [5, 15] first perform 2D-to-3D matching before applying 3D-to-2D search. In contrast, the framework proposed in this chapter is able to automatically select the preferable search direction. Similar to the RANSAC sampling scheme of Li et al. applying the RANSAC pre-filtering step discussed in Sect. 9.5 also enables us to accelerated RANSAC-based pose estimation by avoiding to generate samples unlikely to yield correct poses.

By adapting RANSAC's sampling scheme, Li et al. are able to focus on generating samples that are likely to yield a correct pose and can thus handle a higher outlier ratio than classic RANSAC approaches. Svarm et al. take this concept one step further for the case that the focal length of the query camera, the gravity direction, and the approximate height from which the query image was taken are known [30]. They propose a deterministic outlier filter for the resulting 2D registration problem, which allows them to reliably handle inlier ratios of 1 % or less.

All related methods discussed above are based on directly comparing the descriptors of the features found in the query image and the 3D point descriptors. Consequently, they need to keep the point descriptors in memory at all times or suffer from slow hard disk accesses when loading the descriptors on demand. The high dimensionality of the commonly used SIFT descriptors thus induces a high memory consumption, which theoretically limits the scalability of these approaches. As a

[2]Chapter 11 discusses the method from Li et al. in more detail.

result, multiple solutions to decrease the memory consumption have been proposed recently [3, 8, 16, 26]. Another solution to this problem is to apply localization approaches based on image retrieval [2, 12, 27]. These approaches first retrieve a set of database images showing the same part of the model as the query photo and then establish 2D–3D correspondences by matching the features found in the query image against the 3D points visible in the retrieved images. In theory, such approaches are more scalable than the direct matching strategies discussed above since they do not require the actual point descriptors during retrieval. However, in practice it is safe to assume that we have a rough prior on where a query image was taken, e.g., from cell phone tower ids. Thus, it is sufficient to match the query image against a restricted model instead of the whole world, which can be done by distributing image-based localization approaches over multiple machines based on the region depicted in the 3D model. In this chapter, we thus focus on direct matching methods and refer the interested reader to [23] for an overview over retrieval-based approaches.

Ideally, we would like to run image-based localization directly on mobile devices such as smart phones and tablets in real-time. Yet, all methods described above rely on powerful, but slow to compute and memory consuming SIFT features, rendering them unsuitable for mobile applications. In order to obtain a real-time capable approach, Lim et al. use simple image patches to track 3D points once they have been matched and only perform matching with more complex descriptors if the number of tracked points falls below a given threshold [17]. While their approach requires a copy of the 3D model to be stored on the device, Middelberg et al. propose a server-client approach that requires only constant memory on the mobile device [19]. They use keyframe-based Parallel Tracking and Mapping [13] to track the camera pose on the device in real-time. The keyframes are then send to a server running the method proposed in this chapter, which returns a global pose for the keyframes. Incorporating this global information into the local map, they are able to prevent drift, resulting in a truly scalable mobile localization approach.

9.3 Choosing the Right Matching Direction

In order to estimate the camera pose, we need to establish 2D–3D matches between the 2D features extracted from the query image and the 3D points in the scene model. As outlined above, this is done using nearest neighbor search and Lowe's ratio test. Using direct matching, we can choose whether we match the descriptors of the 2D features against the point descriptors (2D-to-3D matching) or match the point descriptors against the query image (3D-to-2D matching). In the following, we analyze the advantages and disadvantages of both matching directions.

From a technical point of view, there is little difference between matching in either direction: Given a query feature f with descriptor $\mathbf{d}_f$, 2D-to-3D matching finds the two nearest neighboring point descriptors $\mathbf{d}_{p_1}$, $\mathbf{d}_{p_2}$ belonging to 3D points p_1 and p_2 using approximate nearest neighbor search. Since each 3D point is associated with two or more image descriptors, we define the second nearest neighbor $\mathbf{d}_{p_2}$ as the point descriptor with minimal distance to $\mathbf{d}_f$ among all descriptors belonging to points other than p_1. We accept a 2D-to-3D match between f and p_1 if Lowe's ratio test is passed. If multiple matches passing the ratio test are found for a single 3D point, we only keep the match with the smallest absolute descriptor distance.

Let $D(p)$ be the set of descriptors associated with a 3D point p and let

$$\text{dist}(p, f) = \text{min}_{\mathbf{d} \in D(p)} \|\mathbf{d} - \mathbf{d}_f\|_2 \tag{9.2}$$

be the minimal descriptor distance between the descriptor of a feature f and the descriptors from $D(p)$. During 3D-to-2D search, we again use approximate nearest neighbor search to find the two 2D features f_1, f_2 with minimal distances dist(p, f_1), dist(p, f_2), again enforcing that $f_1 \neq f_2$, and accept a 3D-to-2D match between p and f_1 if Lowe's ratio test is fulfilled. If multiple matches are found for a single 2D feature, we again only keep the match with the smallest descriptor distance.

While technically similar, both matching directions have complimentary advantages and disadvantages, caused by the fact that the behavior of the ratio test depends on the density of the descriptor space: A query descriptor needs to be much closer to its nearest neighbor in a densely populated descriptor space than in a sparser space in order to pass the ratio test. Applying the ratio test in a sparser space thus leads to a higher false positive matching rate than applying the test in a denser space. Typically, there are orders of magnitude more 3D points in large scene models than there are features in a query image. Thus, the matches found by 2D-to-3D matching are inherently more reliable than the ones found by pure 3D-to-2D search.

In addition, globally ambiguous structures present a problem for 3D-to-2D matching. Imagine a set of 3D points with similar or identical descriptors, representing a structural element that repeats over the model.[3] If matching one of these points against the query image yields a 3D-to-2D correspondence, then it is rather likely that all points from the set yield matches. Since the second nearest neighbor found by 3D-to-2D search is a 2D feature from the query image, 3D-to-2D matching is not able to detect this problem and thus might accept a wrong match that has the closest descriptor distance to a feature from the image but is not geometrically consistent. In contrast, 2D-to-3D matching approaches are able to reject any match found with one of the points in the set as too ambiguous. However, this implies that the ratio test might also reject correct matches as too ambiguous. When considering larger and larger models, the chance that two points from different parts of the model have similar local appearance increases and 2D-to-3D matching will find fewer and fewer matches due to the ratio test.

Due to their complimentary advantages and disadvantages, it is clear that combining both search directions should yield better localization results than using pure 2D-to-3D or 3D-to-2D matching. Since they are inherently more reliable than 3D-to-2D correspondences, we should first find a set of 2D-to-3D matches. These corre-

---

[3]Notice that in the case of SfM models, globally similar points cannot be observed together in a single database image since such locally ambiguous structures are removed by applying the ratio test during the pairwise image matching phase of SfM.

spondences essentially represent hypotheses about which parts of the 3D model are visible in the query image and we can thus use 3D-to-2D matching to find additional matches in these regions, enabling us to recover matches lost to 2D-to-3D search due to the behavior of the ratio test. In the next section, we will thus detail how to select candidate points for 3D-to-2D search based on spatial proximity in 3D and discuss different strategies for combining both search directions.

9.4 Active Search: Exploiting Spatial Relations

As has been shown in [16, 24], prioritized matching is an important tool for efficient image-based localization. While one main focus of this chapter is to combine both search directions in order to increase the localization effectiveness, we are also interested in an approach that is as efficient as possible. Based on the fact that prioritized 2D-to-3D search is both more efficient and effective than prioritized 3D-to-2D matching [24], we thus decided to integrate 3D-to-2D search into our Vocabulary-based Prioritized Search (VPS) framework [24]. In the following, we briefly review VPS before explaining how to integrate 3D-to-2D search into this method.

As illustrated in Fig. 9.1, VPS first assigns each feature in a query image to its closest visual word from a vocabulary containing 100k words. Since VPS only considers 3D point descriptors assigned to the same word when searching for the nearest neighbors for a query feature, the number of point descriptors assigned to its word defines the search costs for the feature. Storing the number of assigned point descriptors, we can determine the search costs for a feature without actually performing the search. VPS sorts the features in ascending order of search costs and then performs nearest neighbor search for each feature, starting with the one with the lowest costs. Once 100 matches passing the ratio test with threshold $\tau = 0.7$ (cf. Eq. 9.1) have been found, the correspondence search is terminated. The found matches are then used to estimate the camera pose inside a RANSAC-loop. An image is considered as

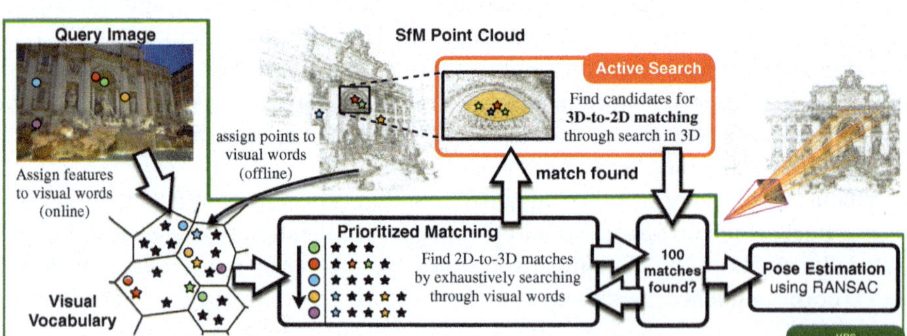

Fig. 9.1 This chapter shows how to integrate 3D-to-2D search into the Vocabulary-based Prioritized Search (VPS) approach from [24] through *Active Search*

successfully localized if the best pose found by RANSAC has at least 12 inliers. Notice that considering only the point descriptors assigned to the same word as the query feature for nearest neighbor search naturally introduces quantization artifacts. Thus, VPS will miss certain matches when a feature and its corresponding point descriptors are assigned to different visual words. As a consequence, VPS has a lower effectiveness than the tree-based search approach [24]. Fortunately, we can also use 3D-to-2D search to recover such matches lost to quantization at the cost of little computational overhead. In Sect. 9.4.1, we introduce our *Active Search* step designed to recover correspondences lost due to quantization or the ratio test. Sect. 9.4.2 then explains different strategies to combine 2D-to-3D and 3D-to-2D matching into a single prioritization scheme. Finally, Sect. 9.4.3 proves that Active Search is more efficient than the conventional strategy of searching through multiple words in order to reduce quantization artifacts [22].

9.4.1 The Active Search Mechanism

Each 2D-to-3D match found for a 3D point p represents a hypothesis about which part of the 3D model is visible in the query image. If the match is correct, we can expect that other points from the same part of the model can also be observed. Choudhary and Narayana exploit this fact and try to match 3D points co-visible with p against the image [5]. They use a probabilistic prioritization scheme for 3D-to-2D search based on the co-visibility relations between the 3D points and are able to show that it is sufficient to find only few additional 3D-to-2D matches to successfully localize the query image. The drawback of their method is that they need to explicitly store the probabilities that points can be observed together, resulting in a large memory overhead. As a consequence, they only use a subset of all 3D points in the model to keep the memory consumption feasible. In contrast, we propose to exploit the fact that co-visibility can to some degree be inferred from spatial proximity between two 3D points: Finding a correct match for a point p implies that the structure surrounding p is visible in the query image. Consequently, we can expect to find some points among the N_{3D} nearest neighbors of p in 3D that will yield 3D-to-2D correspondences when matched against the image. These spatial nearest neighbors can be found efficiently using spatial subdivision schemes such as kd-trees and storing these search structures requires only a small memory overhead.

Figure 9.2 illustrates our *Active Search* mechanism[4]: After finding a 2D-to-3D match using VPS, we search for the N_{3D} nearest neighbors in 3D for the matching point. Each of these neighbors is a potential candidate for 3D-to-2D matching that we will eventually match against the query image. In order to accelerate 3D-to-2D search, we exploit the search structure already used by VPS to avoid the construction of an additional data structure: We build a vocabulary tree [21] on top of the fine vocabulary used by VPS to define the search costs and to limit the search space for

[4]Source code is available at http://www.graphics.rwth-aachen.de/localization.

each feature in the query image. By construction, the nodes on a level l in the tree define a quantization of the descriptor space, i.e., each level of the tree corresponds to a visual vocabulary. We use the smaller vocabularies defined by the lower levels in the tree to perform 3D-to-2D matching, which automatically leads to reduced quantization artifacts. Notice that we obtain the assignments of feature and point descriptors to the lower levels in the vocabulary tree for free since the assignments to the leaf nodes uniquely define a path through the tree and thus the corresponding words on the lower levels. We build a vocabulary tree with branching factor 10 and choose the level in the tree that is used for 3D-to-2D matching depending on the number of features found in the query image to limit the search costs for each point. We use level two (corresponding to a vocabulary containing 100 words) if at most 5k features are found in the query image and level three (1k words) otherwise. As with VPS, we stop the search once 100 2D–3D matches have been found.

We use the *integer mean per visual word* representation for 3D point descriptors from [24]. Let $W(p)$ be the set of visual words for the vocabulary W containing descriptors from $D(p)$ and let $w(\mathbf{d})$ be the word to which descriptor $\mathbf{d}$ is assigned. The *integer mean per visual word* strategy represents a point as a set of mean descriptors

$$D_W(p) = \bigcup_{w \in W(p)} \{\text{mean}(\{\mathbf{d} \mid \mathbf{d} \in D(p) \wedge w(\mathbf{d}) = w\})\} \tag{9.3}$$

such that only a single descriptor is stored for p for each word from $W(p)$. We use the common quantization of SIFT descriptors that represents each descriptor entry using an integer value from [0, 255] and round the entries of the mean descriptors to the nearest integer values. Furthermore, let $w_l(\mathbf{d})$ be the word on level l in the tree belonging to descriptor $\mathbf{d}$. To perform 3D-to-2D search for a point p, we match each descriptor $\mathbf{d} \in D_W(p)$ against all feature descriptors assigned to $w_l(\mathbf{d})$ and search

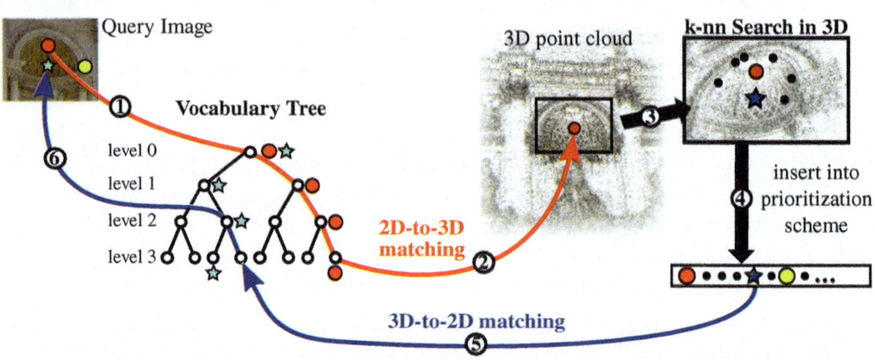

Fig. 9.2 The *Active Search* mechanism: After finding a 2D-to-3D match, we identify spatially close points surrounding the matching point, which are eventually matched against the images to find 3D-to-2D matches. Matching the points against the image on a lower level of the vocabulary tree used during 2D-to-3D search enables us to recover matches lost to quantization artifacts. Reproduced with permission from [25], ©Springer-Verlag Berlin Heidelberg

for the two nearest neighbors minimizing the distance to the descriptors from $D_W(p)$ (cf. Eq. 9.2). We then apply Lowe's ratio test with $\tau = 0.6$, where the lower threshold compared to 2D-to-3D search reflects that we trust 2D-to-3D matches more than 3D-to-2D correspondences. This is also reflected by our choice that we always prefer 2D-to-3D over 3D-to-2D matches, i.e., we keep the 2D-to-3D correspondence (f, p) even if it has a higher descriptor distance than some 3D-to-2D match (p, f').

Notice that similar to the 2D-to-3D search performed by VPS, we model 3D-to-2D search as finding the nearest neighboring descriptors assigned to a set of visual words. As for 2D-to-3D matching, we can pre-compute the number of descriptor comparisons required to match a point p against the image before performing the actual search: Let $F(w_l)$ be the set of query features assigned to word w_l on level l in the tree. The search costs for a point p are thus given by

$$\text{cost}(p) = \sum_{\mathbf{d} \in D_W(p)} |F(w_l(\mathbf{d}))| \ . \tag{9.4}$$

9.4.2 Prioritization Strategies

In the previous section, we have shown how to obtain candidates for 3D-to-2D matching. In order to take full advantage of prioritized search, we have to combine both search directions into a single prioritization scheme. Figure 9.3 illustrates the three combinations that are possible: Once we find a 2D-to-3D match that passes the ratio test and identify its N_{3D} nearest neighbors in 3D, we can choose to first match all candidate points against the image and only continue 2D-to-3D search if less than 100 matches have been found. The number of features found in a query image does not increase with the size of the 3D models. For larger models, there thus will be significantly more 3D points assigned to each word in the fine vocabulary than there are features assigned to words in the coarser vocabulary defined by level l in

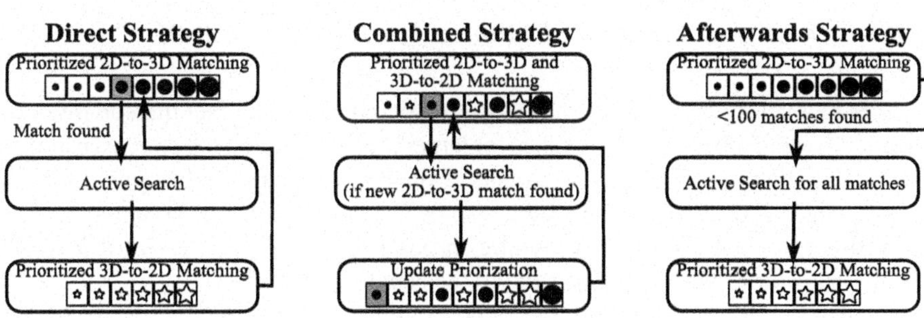

Fig. 9.3 There are three possible combinations to merge 2D-to-3D and 3D-to-2D search candidates into a single prioritized matching scheme. The size of the *circles* is proportional to the search costs of the 2D features and 3D points they represent

the tree. This *direct* prioritization strategy, used by [5], thus has the advantage that it firsts considers the cheaper points before continuing 2D-to-3D search. However, this strategy has the distinct disadvantage of finding many matches in a small region of the query image, resulting in unstable configurations for camera pose estimation and thus a lower localization accuracy.

A second possibility, used in [15], is to only perform 3D-to-2D search if not enough correspondences could be found through 2D-to-3D matching alone. Compared to the *direct* strategy, this *afterwards* approach usually leads to a much more uniform distribution of feature matches in the query image, leading to a higher localization accuracy. Besides potentially higher overall search costs for larger models, this strategy might also cause problems when considering smaller 3D models since the latter induce sparser descriptor spaces for 2D-to-3D matching. As a result, 2D-to-3D search will find more wrong matches and we might actually terminate the correspondence search before enough correct matches have been found to facilitate camera pose estimation. The *direct* strategy can recover the additional correspondences needed for pose estimation through 3D-to-2D search, but the *afterwards* strategy will never attempt 3D-to-2D search in this case, resulting in a lower localization effectiveness. Notice that this problem does not affect [15] as they attempt to find as many matches as possible and thus do not use prioritized search.

Using the same prioritization function, the number of required descriptor comparisons, to prioritize features and 3D points enables a third strategy that jointly prioritizes both matching directions. This *combined* strategy uses a single prioritization queue containing both features and points, sorted in ascending order of search costs. Consequently, this strategy will always pick the search direction that is cheaper to evaluate, which intuitively corresponds to our goal of obtaining an efficient image-based localization approach. For larger models, this strategy will degenerate towards the *direct* prioritization scheme. Yet, we will show in Sect. 9.6 that for the models considered in this chapter, the *combined* strategy offers an accuracy similar to the *afterwards* scheme while achieving an effectiveness similar to the *direct* strategy.

9.4.3 The Computational Complexity of Active Search

Besides recovering matches lost to 2D-to-3D search due to the fact that the ratio test rejects more and more correct matches for larger models, recovering correspondences lost to VPS due to quantization artifacts is one of our main motivations for using *Active Search*. This second goal could also be achieved using soft assignments [22] for 2D-to-3D search, i.e., by searching through the $c > 1$ closest word for each feature instead of restricting matching to the same word.

In general, $|P|/|W|$ 3D points are contained in each word, where $|P|$ is the number of 3D points and $|W|$ is the number of visual words in the fine vocabulary used by VPS. Assuming a constant number of words, the increase in search costs caused by soft assignments is thus linear in the number of points in the model.

In contrast to soft assignments, *Active Search* attempts to match at most $100 \cdot N_{3D}$ points against the image since the correspondence search is terminated after finding 100 matches. At the same time, the search costs for each query features are not altered. In the worst case, each 3D point is matched against all $|F|$ features in the query image. In practice, it is reasonable to assume that $|F|$ is bounded by some constant. *Active Search* thus only adds a constant number of descriptor comparisons since N_{3D} is also a constant. Since both $|F|$ and the number of candidate points for 3D-to-2D search are bounded by some constant, updating the common prioritization queue can also be done in constant time. As a result, finding the N_{3D} nearest neighbors in 3D asymptotically becomes the dominant cost for *Active Search*. Using spatial subdivision schemes, nearest neighbor search in 3D can be done in time $O(N_{3D} \cdot \log |P|)$, i.e., the overall computational complexity of *Active Search* is $O(\log |P|)$. Thus, *Active Search* is a more efficient approach for recovering matches lost due to quantization than the classical strategy of using soft assignments.

9.5 Exploiting Co-visibility Information

In Sect. 9.4, we have exploited the fact that some of the 3D points spatially close to a matching point p should also be visible in the query image to efficiently obtain a set of candidate points for 3D-to-2D search. However, spatial proximity alone does not guarantee that we can find matches for the candidate points. Consider two spatially close points p_1 and p_2 observed in database images I_1 and I_2, respectively. If the viewpoints from which I_1 and I_2 were taken differ too much, then it is unlikely that both p_1 and p_2 have a matching feature in a query image due to the limited robustness to viewpoint changes of both feature detectors and descriptors. As a result, a large fraction of candidate points for 3D-to-2D search might never yield 3D-to-2D matches, resulting in many unnecessary descriptor comparisons. Fortunately, we can detect such points before performing matching by exploiting the co-visibility information readily provided by the SfM process. In the example above, we may observe that there is no database image seeing both p_1 and p_2. In the case that we have found a 2D-to-3D match for p_1, this information allows us to decide not to consider p_2 for 3D-to-2D matching as it has no chance of yielding a match. In the following, we formalize this filtering step as an operation on the so-called *Visibility Graph* [16]. We then show that this *Visibility Graph* can also be used to detect and remove wrong matches before applying RANSAC-based camera pose estimation. Finally, we argue that the *Visibility Graph* only approximates the true co-visibility relations and introduce a clustering scheme designed to yield a better approximation.

The Visibility Graph The database images and 3D points in a SfM model define the bipartite *Visibility Graph* $G = (P_G \cup C_G, E)$ (cf. Fig. 9.4a). Each vertex in P_G corresponds to a 3D point while each node in C_G corresponds to a database image. The graph contains an edge $\{p, c\}$ between a point node $p \in P_G$ and a camera node $c \in C_G$ iff the corresponding database image observes the 3D point associated with

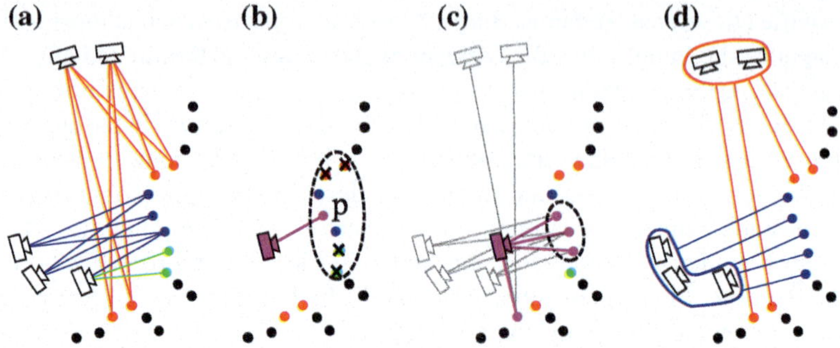

Fig. 9.4 **a** The *Visibility Graph* defined by the points and cameras in a SfM model. **b** Our *point filter* detects and removes candidate points that are not co-visible with the point p that triggered *Active Search* (*pink*). **c** Our *RANSAC pre-filter* only keeps matches falling into the largest connected component induced by all matches. **d** We cluster cameras to obtain a better approximation to the true co-visibility relationship. Reproduced with permission from [25], ©Springer-Verlag Berlin Heidelberg

the node p. Two points with vertices p_1, p_2 are co-visible if there exists a camera vertex c such that both edges $\{p_1, c\}$ and $\{p_2, c\}$ are contained in E.

Filtering candidates for 3D-to-2D search Once we have found the N_{3D} spatial nearest neighbors surrounding a matching point corresponding to the vertex $p \in P_G$, we check for each neighboring point whether it is co-visible with p. Given the *Visibility Graph*, this can be done efficiently by testing whether both point vertices are connected to a common camera node. All neighbors that are not co-visible with p are eliminated and not used for 3D-to-2D search (cf. Fig. 9.4b).

A RANSAC pre-filter Consider the set of matches $M = \{\{f, p\}\}$ found by applying VPS together with *Active Search*. Given perfect co-visibility information, there should be at least one database image observing all correctly matching 3D points. In contrast, it is reasonable to expect that wrongly matching points are scattered more or less randomly over the model.[5] Thus, we could easily detect and remove wrong matches by identifying the camera node c connected to most matching points and removing all matches whose 3D points are not connected to c in G. However, due to the approximate nature of the matching process applied for SfM, there might not be a single database image observing all correctly matching 3D points. We thus detect the largest connected component in the subgraph $G(M) = (P_G(M) \cup C_G(M), E(M))$ induced by the matches, where $P_G(M) = \{p \in P_G \mid \exists f \text{ s.t. } \{f, p\} \in M\}$, $E(M) = \{\{p, c\} \in E \mid p \in P_G(M)\}$, and $C_G(M) = \{c \in C_G \mid \exists p \text{ s.t. } \{p, c\} \in E(M)\}$. All matches whose 3D point is not contained in this component are removed before applying RANSAC (cf. Fig. 9.4c).

[5] Remember that matches found for globally repetitive structures are rejected as too ambiguous during 2D-to-3D search and thus also do not trigger *Active Search*.

Clustering cameras Obviously, the set of database images represents a discrete approximation to the space of viewing conditions under which the scene can be observed. Thus, the *Visibility Graph* defined by a SfM reconstruction only approximates the true co-visibility relations between the 3D points in the model. The two filtering steps thus might be too aggressive and filter out valid candidate points or correct matches, potentially decreasing the number of images that can be localized. In order to avoid this problem, we propose to obtain a new *Visibility Graph* by clustering cameras taken from similar viewpoints. For each database image c, we find the k closest cameras surrounding c in 3D. Let $\text{sim}(c)$ be the subset of these closest cameras whose viewing direction differs from c's by at most 60°. We obtain an intermediate graph G_{sim} by replacing each camera node c by a new node $\text{sim}(c)$ and connecting $\text{sim}(c)$ to all point nodes that were connected to a camera $c' \in \text{sim}(c)$ in G (cf. Fig. 9.4d). Finally, we select a minimal subset of camera nodes from G_{sim} such that each original camera vertex is contained in one of the selected nodes. This Set Cover problem is solved in a greedy fashion, iteratively selecting the camera node from G_{sim} that contains the largest number of nodes $c \in C_G$ not yet contained in any previously selected camera node [16]. We then use the resulting new *Visibility Graph* G'_{sim} when applying the two filtering steps defined above.

9.6 Experimental Evaluation

In this section, we evaluate the impact of the different parameters of our method on its localization efficiency and effectiveness and compare it to the current state-of-the-art. In the following, we first discuss the evaluation procedure used in this chapter and the used datasets. We then analyze the parameters of the proposed *Active Search* method and the impact of both visibility filtering steps before comparing the performance of the resulting approach with existing methods.

Experimental setup and datasets We evaluate the performance of the proposed *Active Search* method on three datasets that form a standard benchmark for image-based localization methods [5, 15, 16, 24, 25]: Both the Dubrovnik and Rome models were reconstructed from photos downloaded by Flickr [16]. The Dubrovnik model consists of a single connected component depicting the old city of Dubrovnik, Croatia, while the Rome model contains multiple landmark buildings, each represented by a connected component. Pictures taken at regular intervals using a single calibrated camera were used to reconstruct three landmarks in Vienna, Austria, which are contained in the Vienna model [12]. Query images for the Dubrovnik and Rome models were obtained by removing cameras from the original reconstructions. The points visible in at least two remaining database images then form the models used for our experiments. The scale of the Dubrovnik reconstruction is known, allowing us to use the original camera positions as ground truth for measuring the localization accuracy of our approach. The query images for the Vienna dataset were downloaded

Table 9.1 Details on the datasets used for experimental evaluation

	# cameras	# 3D points	# descriptors	# query images	Avrg. # feat. per query image
Dubrovnik	6044	1,886,884	9,606,317	800	9,678.14
Rome	15,179	4,067,119	21,515,110	1000	7,279.91
Vienna	1324	1,123,028	4,854,056	266	9,707.29

from Panoramio. Each query image has a maximum dimensionality of 1600×1600 pixels. Details on the datasets can be found in Table 9.1.

We use a fine vocabulary containing 100k words [24] and use a modified version of the FLANN library [20], where we removed back-tracking to accelerate the tree traversal, to build a vocabulary tree with branching factor 10 on top of this vocabulary. After finding 100 matches, we terminate the correspondence search and estimate the camera pose by applying the 6-point DLT solver [11], which estimates the full projection matrix for an uncalibrated camera from six 2D–3D matches, inside a LO-RANSAC-loop [7] that uses the SPRT test [6] for early model rejection. As in [16, 24], we consider a query image as successfully localized or registered if the best pose estimated by RANSAC has at least 12 inliers.

All experiments were performed on a single thread of an Intel i7-920 CPU with 2.79 GHz. We repeated each experiment 10 times to account for RANSAC's random nature and report the mean registration and rejection times (both excluding feature extraction) and the mean number of localized images.

The impact of the prioritization strategies We start evaluating *Active Search* by analyzing the impact of the three prioritization strategies on both localization effectiveness and efficiency. For this experiment, we only rely on spatial relations and do not perform any filtering based on co-visibility information. Figure 9.5 compares *Active Search* with the three prioritization strategies and different values for N_{3D}, the number of spatial nearest neighbors, with VPS and the kd-tree approach from [24]. For VPS, we report both the performance using all point descriptors and using the *integer mean per word* representation that is also used by our method. As evident by the number of localized images, incorporating *Active Search* into VPS enables us to achieve a localization effectiveness similar to or even better than the tree-based approach, independently of the value of N_{3D} and the chosen prioritization strategy. Over all datasets, the *direct* prioritization strategy, which first performs 3D-to-2D matching for all candidates before subsuming 2D-to-3D search, achieves the best effectiveness while the *afterwards* strategy, which only performs 3D-to-2D search if less than 100 2D-to-3D correspondences could be found, localizes the least number of images. However, as predicted in Sect. 9.4.2, the *direct* prioritization strategy also causes the highest localization error due to its tendency to find many matches in small regions in the query image (cf. Table 9.2). We notice that the *combined* strategy, which always prefers the cheaper search direction, offers a good compromise between the

(a) (b) (c)

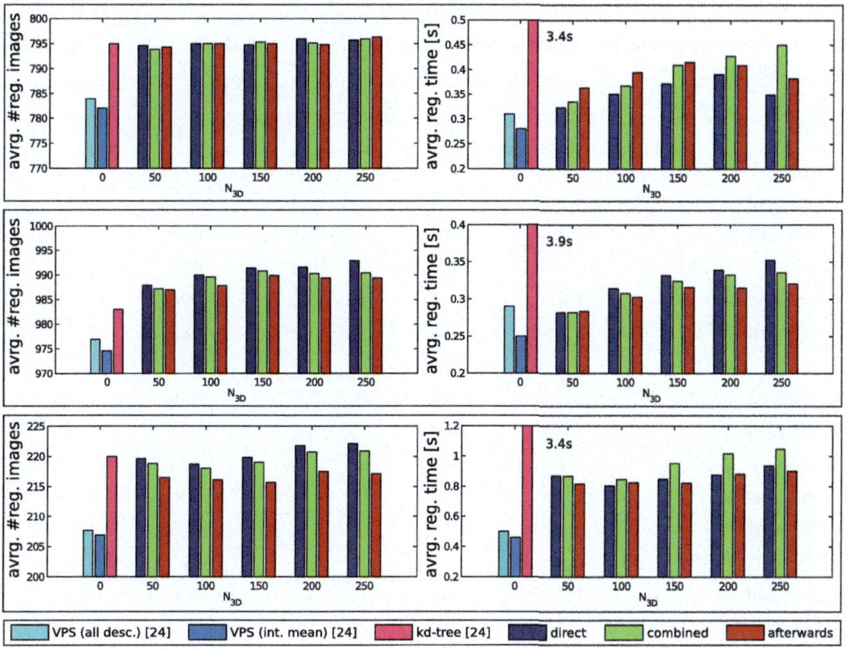

Fig. 9.5 The effect of the three prioritization strategies on (*left column*) the mean number of localized images and (*right column*) the mean localization times for different values of N_{3D}. *Active Search* closes the gap in effectiveness between VPS and tree-based search by recovering matches previously lost to VPS due to quantization. **a** Dubrovnik. **b** Rome. **c** Vienna. Reproduced with permission from [25], ©Springer-Verlag Berlin Heidelberg.

direct and *afterwards* schemes both in terms of effectiveness and pose accuracy. We thus decide to use the *combined* strategy for all further experiments.

Notice that all three prioritization strategies are able to register more images than the kd-tree approach on the Rome dataset, which is the largest of the three models used for experimental evaluation. As detailed in Sect. 9.3, larger models induce a denser descriptor space, forcing the ratio test used for 2D-to-3D matching to reject more and more correct matches as too ambiguous. Figure 9.5 thus shows that *Active Search* is able to recover such matches by employing 3D-to-2D search.

Visibility Filtering As evident from the right column of Fig. 9.5, applying *Active Search* introduces additional computations that slow down the localization times compared to our original VPS approach. The filtering steps introduced in Sect. 9.5 were designed to counter this decrease in efficiency by identifying and avoiding unnecessary computations. The left column of Fig. 9.6 details the impact of both filtering steps on localization times and the number of localized images. As can be seen, both filters are able to increase the localization efficiency of *Active Search* by reducing its run-time. The point filter that removes candidates for 3D-to-2D matching that are not co-visible with the matching point that triggered *Active Search* has the

Table 9.2 Localization errors for the three strategies measured on the Dubrovnik dataset as the distance between the mean camera position from 10 repetitions of the experiment and the ground truth position

Method/ Strategy	N_{3D}	Median [m]	Quartiles [m]		#images with error	
			1st	3rd	<18.3 m	>400 m
Direct	50	*1.9*	*0.5*	*7.6*	*674*	12
Combined		1.3	0.4	5.3	707	10
Afterwards		1.3	0.4	6.1	693	15
Direct	100	*2.0*	*0.6*	*10.4*	*664*	12
Combined		1.3	0.4	6.1	694	8
Afterwards		1.3	0.4	5.2	688	16
Direct	150	*2.2*	*0.6*	*9.8*	*647*	*16*
Combined		1.3	0.4	5.8	696	13
Afterwards		1.4	0.4	5.2	697	14
Direct	200	*2.4*	*0.7*	*11.0*	*655*	*13*
Combined		1.3	0.4	5.6	700	9
Afterwards		1.4	0.4	5.6	693	10
Direct	250	*2.6*	*0.7*	*11.5*	*642*	*17*
Combined		1.3	0.4	5.6	700	12
Afterwards		1.3	0.4	5.5	694	13

smallest impact on the Rome dataset. The reason for this behavior is that this dataset consists of individual landmarks seen from many perspectives. Thus, most spatial neighboring points are actually co-visible and the filter only removes few candidate points. The RANSAC pre-filter that detects and removes wrong matches has the largest impact on the Vienna datasets due to the higher false positive matching rate observed for this smallest of the three datasets. Based on the results from Fig. 9.6, we use both filters and choose $N_{3D} = 200$, which maintains most of the effectiveness achieved without filtering while resulting in localization times similar to or faster than VPS.

As can be seen from the left column of Fig. 9.6, applying the filters also decreases the localization effectiveness. This decrease is caused by the fact that the *Visibility Graph* defined by each SfM model only approximates the true co-visibility relations between the 3D points. In Sect. 9.5, we therefore introduced a camera clustering scheme designed to better approximate the true co-visibility relations. The right column of Fig. 9.6 details the impact of clustering cameras on both the localization effectiveness (top part of each plot) and efficiency (bottom part of each plot) when varying the number k of close cameras used to define the clusters. As can be seen, using camera clusters instead of the original database images ($k = 0$) enables us to localize a similar amount of query images compared to not using any visibility filters while increasing the registration times only slightly. This is an interesting result when comparing our RANSAC pre-filter with the adapted RANSAC sampling

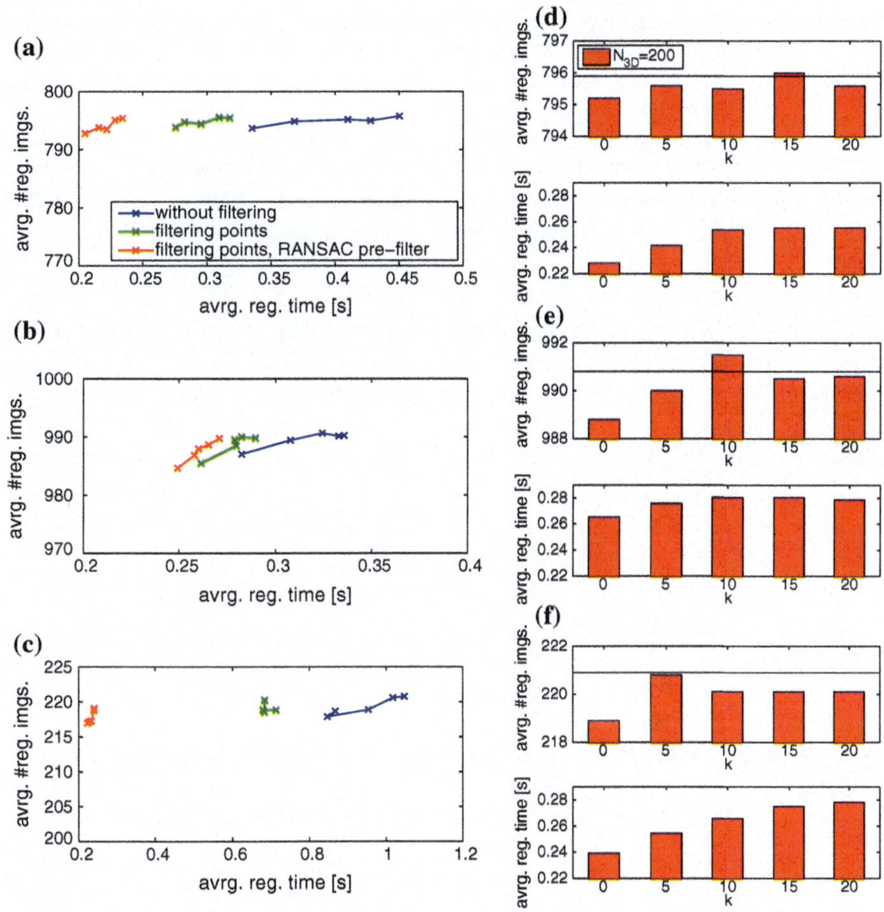

Fig. 9.6 (*Left column*) The filtering steps proposed in Sect. 9.5 increase the efficiency of *Active Search* at the cost of effectiveness. The curves were created by setting N_{3D} to values from {50, 100, 150, 200, 250}, where $N_{3D} = 250$ corresponds to the rightmost points. (*Right column*) Clustering cameras increases the effectiveness (*top*) compared to using the original database image ($k = 0$). It even allows us to localize a similar number of query photos compared to not using visibility filtering (*black line*). Clustering cameras also slightly increases the registration times (*bottom*). **a**, **d** Dubrovnik. **b**, **e** Rome. **c**, **f** Vienna. Reproduced with permission from [25], ©Springer-Verlag Berlin Heidelberg

Table 9.3 *Active Search* achieves a localization effectiveness similar to state-of-the-art methods that are orders of magnitude slower while offering a better effectiveness than methods with similar run-times

Method	Dubrovnik			Rome			Vienna		
	# reg. images	Registr. time [s]	Reject. time [s]	# reg. images	Registr. time [s]	Reject. time [s]	# reg. images	Registr. time [s]	Reject. time [s]
Active search	795.5	**0.25**	0.56	991.5	0.28	2.14	**220.1**	**0.27**	**0.52**
WPE [15]	**800**	"Few seconds"		**997**	"Few seconds"		–	–	–
[30]	798	5.06		–	–	–	–	–	–
kd-tree [24]	795	3.4	14.45	983	3.97	6.27	**220**	3.44	2.72
Vis. prob. [5]	788	**0.25**	**0.51**	977	0.27	**0.61**	**219**	0.40	**0.49**
VPS (int. mean) [24]	782.0	0.28	1.70	974.6	**0.25**	1.66	206.9	0.46	2.43
P2F [16]	753	0.73	2.70	921	0.91	2.93	204	0.55	1.96
P2F+F2P [16]	753	0.70	3.96	924	0.87	4.67	205	0.54	3.62

Table 9.4 Localization accuracy on the Dubrovnik dataset

Method	# localized images	Median error [m]	Quartile errors [m]		#images with error	
			1st	3rd	<18.3 m	>400 m
Active search	795.5	1.4	**0.4**	5.3	704	9
[30]	**798**	**0.56**	–	–	**771**	**3**
Vis. prob. [5]	788	3.1	0.88	11.83		
VPS (int. mean) [24]	782.0	1.3	0.5	**5.1**	675	13
P2F [16]	753	9.3	7.5	13.4	655	–

method from [15]. The latter selects the 2D–3D matches required to estimate a camera pose based on co-visibility information and evaluates the estimated pose on all matches, while our approach uses a smaller set of matches for both pose estimation and evaluation. At the same time, our RANSAC pre-filter is both easier to implement than the modified sampling scheme from [15] and allows us to adapt the number of required RANSAC iterations based on the detected inlier ratio, which is not possible with the method from [15].

As can be expected, the optimal choice of k is dataset dependent, but setting $k = 10$ offers a good compromise between effectiveness and efficiency on all test sets. Thus, we finalize our choice of parameters for *Active Search* by using the *combined* strategy with $N_{3D} = 200$ and $k = 10$.

Comparison with State-of-the-Art We compare our method against the current state-of-the-art for localization approaches based on direct matching. Specifically, we compare *Active Search* against the Worldwide Pose Estimation (WPE) [15] and Visibility Probability (Vis. Prob.) [5] approaches that are also based on combining 2D-to-3D and 3D-to-2D search and exploiting co-visibility information, as well as the method from [30] that performs exhaustive geometric verification. P2F+F2P denotes the prioritized Point-to-Feature (P2F) matching method that applies 2D-to-3D matching if P2F fails [16]. A more detailed description of these related approaches can be found in Sect. 9.2. As can be seen from Table 9.3, *Active Search* nearly achieves the same localization effectiveness as methods that are one order of magnitude or more slower. Over all 10 repetitions of the experiment, we found that only 3 and 7 query images cannot be localized for the Dubrovnik and Rome datasets, respectively. Thus, *Active Search* has a similar effectiveness as WPE, which was designed to maximize the effectiveness of image-based localization by recovering as many correct matches as possible through the use of a higher ratio test threshold while our approach was designed to offer a good compromise between effectiveness and efficiency. *Active Search* achieves essentially the same run-times as VPS (with the exception of the Vienna dataset, where *Active Search* is significantly faster), even though it requires additional effort for 3D-to-2D search. Compared to Vis. Prob., *Active Search* has a similar efficiency, which is interesting as the latter only uses about ∼3 % of all 3D points in the original models while our method uses the full point clouds. *Active Search* offers a better localization effectiveness than Vis. Prob., which demonstrates that exploiting simple spatial relations to select candidates for 3D-to-2D search is sufficient and it is not necessary to use the more complex method based on co-visibility probabilities proposed in [5].

Table 9.4 compares the localization accuracy obtained with *Active Search* on the Dubrovnik dataset with other methods. As can be seen, *Active Search* achieves a much higher accuracy than P2F and Vis. Prob., which mainly rely on 3D-to-2D matches. Vis. Prob. essentially uses the *direct* prioritization strategy, resulting in unstable configurations for pose estimation where many matches are found in a small region of the query image. Svarm et al. [30] do not rely on RANSAC but use a deterministic approach that achieves a higher positional accuracy. Thus, we could also apply their method on the matches found by *Active Search*.

9.7 Conclusion

In this chapter, we have demonstrated that we can exploit spatial and co-visibility relations between 3D points in large-scale SfM models to increase the effectiveness of image-based localization approaches based on direct matching. We have argued that in order to localize as many query images as possible, we need to combine both 2D-to-3D and 3D-to-2D search, where 3D-to-2D matching should be triggered by 2D-to-3D correspondences since the latter matches are inherently more reliable. We have shown that we can use nearest neighbor search in 3D to efficiently obtain candidate points for 3D-to-2D search, which only induces a small memory overhead to store the data structures required for nearest neighbor search in 3D. Since spatial proximity does not necessarily imply that we can actually find matches for the candidate points, we have shown how to exploit co-visibility information to filter out candidate points unlikely to yield correspondences as well as removing wrong matches before applying RANSAC-based camera pose estimation. In addition, we have demonstrated that clustering cameras based on spatial proximity and similar viewing directions enables us to better deal with the approximate nature of the co-visibility information extracted from SfM reconstructions. We have experimentally shown the importance of exploiting both spatial and co-visibility relations for image-based localization. Integrating the proposed methods into an approach using prioritized 2D-to-3D and 3D-to-2D matching results in an image-based localization method that achieves state-of-the-art localization effectiveness and efficiency while other state-of-the-art methods only succeed in one of these two categories.

References

1. Agarwal S, Snavely N, Simon I, Seitz S, Szeliski R (2009) Building Rome in a day. In: International conference on computer vision
2. Cao S, Snavely N (2013) Graph-based discriminative learning for location recognition. In: IEEE conference on computer vision and pattern recognition
3. Cao S, Snavely N (2014) Minimal scene descriptions from structure from motion models. In: IEEE conference on computer vision and pattern recognition
4. Castle RO, Klein G, Murray DW (2008) Video-rate localization in multiple maps for wearable augmented reality. In: International symposium on wearable computers
5. Choudhary S, Narayanan PJ (2012) Visibility probability structure from SfM datasets and applications. In: European conference on computer vision
6. Chum O, Matas J (2008) Optimal randomized RANSAC. IEEE Trans Pattern Anal Mach Intell 30(8):1472–1482
7. Chum O, Matas J, Obdržálek S (2004) Enhancing RANSAC by generalized model optimization. In: Asian conference on computer vision
8. Donoser M, Schmalstieg D (2014) Discriminative feature-to-point matching in image-based locallization. In: IEEE conference on computer vision and pattern recognition
9. Fischler M, Bolles R (1981) Random sample consensus: a paradigm for model fitting with applications to image analysis and automated cartography. Commun ACM 24(6):381–395
10. Frahm JM, Fite-Georgel P, Gallup D, Johnson T, Raguram R, Wu C, Jen YH, Dunn E, Clipp B, Lazebnik S, Pollefeys M (2010) Building Rome on a cloudless day. In: ECCV
11. Hartley RI, Zisserman A (2004) Multiple view geometry in computer vision, 2nd edn. Cambridge University Press
12. Irschara A, Zach C, Frahm JM, Bischof H (2009) From structure-from-motion point clouds to fast location recognition. In: IEEE conference on computer vision and pattern recognition
13. Klein G, Murray D (2007) Parallel tracking and mapping for small AR workspaces. In: International symposium on mixed and augmented reality
14. Lee GH, Fraundorfer F, Pollefeys M (2013) Structureless pose-graph loop-closure with a multi-camera system on a self-driving car. In: Intelligent robots and systems
15. Li Y, Snavely N, Huttenlocher D, Fua P (2012) Worldwide pose estimation using 3D point clouds. In: European conference on computer vision
16. Li Y, Snavely N, Huttenlocher DP (2010) Location recognition using prioritized feature matching. In: European conference on computer vision
17. Lim H, Sinha SN, Cohen MF, Uyttendaele M (2012) Real-time image-based 6-DOF localization in large-scale environments. In: IEEE conference on computer vision and pattern recognition
18. Lowe D (2004) Distinctive image features from scale-invariant keypoints. Int J Comput Vision 60(2):91–110
19. Middelberg S, Sattler T, Untzelmann O, Kobbelt L (2014) Scalable 6-DOF localization on mobile devices. In: European conference on computer vision, pp 461–468

20. Muja M, Lowe DG (2009) Fast approximate nearest neighbors with automatic algorithm configuration. In: International conference on computer vision theory and applications
21. Nister D, Stewenius H (2006) Scalable recognition with a vocabulary tree. In: IEEE conference on computer vision and pattern recognition
22. Philbin J, Chum O, Isard M, Sivic J, Zisserman A (2008) Lost in quantization: Improving particular object retrieval in large scale image databases. In: IEEE conference on computer vision and pattern recognition (2008)
23. Sattler T (2013) Efficient & effective image-based localization. Ph.D. thesis, RWTH Aachen University, Aachen, Germany
24. Sattler T, Leibe B, Kobbelt L (2011) Fast image-based localization using direct 2D-to-3D matching. In: International conference on computer vision
25. Sattler T, Leibe, B, Kobbelt L (2012) Improving image-based localization by active correspondence search. In: European conference on computer vision
26. Sattler T, Leibe B, Kobbelt L (2012) Towards fast image-based localization on a city-scale. In: Outdoor and large-scale real-world scene analysis, Lecture Notes in Computer Science, vol 7474. Springer, Berlin, pp 191–211
27. Sattler T, Weyand T, Leibe B, Kobbelt L (2012) Image retrieval for image-based localization revisited. In: British machine vision conference
28. Schindler G, Brown M, Szeliski R (2007) City-scale location recognition. In: IEEE Conference on computer vision and pattern recognition
29. Snavely N, Seitz SM, Szeliski R (2006) Photo tourism: exploring photo collections in 3D. In: ACM SIGGRAPH
30. Svarm L, Enqvist O, Oskarsson M, Kahl F (2014) Accurate localization and pose estimation for large 3D models. In: IEEE conference on computer vision and pattern recognition (2014)
31. Zamir AR, Shah M (2010) Accurate image localization based on google maps street view. In: Daniilidis K, Maragos P, Paragios P (eds) ECCV 2010, Part IV. LNCS, vol 6314. Springer, Heidelberg, pp 255–268
32. Zhang W, Kosecka J (2006) Image based localization in urban environments. In: International symposium on 3D data processing, visualization, and transmission

Chapter 10
3D Point Cloud Reduction Using Mixed-Integer Quadratic Programming

Hyun Soo Park, Yu Wang, Eriko Nurvitadhi, James C. Hoe, Yaser Sheikh and Mei Chen

Abstract Large-scale 3D image localization requires computationally expensive matching between 2D feature points in the query image and a 3D point cloud. In this chapter, we present a method to accelerate the matching process and reduce the memory footprint by analyzing the view statistics of points in a training corpus. Given a training image set that is representative of common views of a scene, our approach identifies a compact subset of the 3D point cloud for efficient localization, while achieving comparable localization performance to using the full 3D point cloud. We demonstrate that the problem can be precisely formulated as a mixed-integer quadratic program and present a point-wise descriptor calibration process to improve matching. We show that our algorithm outperforms the state-of-the-art greedy algorithm on standard datasets, on measures of both point-cloud compression and localization accuracy.

H.S. Park (✉) · Y. Wang (✉) · J.C. Hoe · Y. Sheikh
Carnegie Mellon University, Pittsburgh, PA, USA
e-mail: hyunsoop@cs.cmu.edu

Y. Wang
e-mail: yuw@andrew.cmu.edu

J.C. Hoe
e-mail: jhoe@ece.cmu.edu

Y. Sheikh
e-mail: yaser@cs.cmu.edu

E. Nurvitadhi · M. Chen
Intel Corporation, Hillsboro, OR, USA
e-mail: eriko.nurvitadhi@intel.com

M. Chen
e-mail: mei.chen@intel.com

A.R. Zamir et al. (eds.), *Large-Scale Visual Geo-Localization*,
Advances in Computer Vision and Pattern Recognition,
DOI 10.1007/978-3-319-25781-5_10

10.1 Introduction

Advances in structure from motion techniques have made it possible to construct 3D point cloud models at "city-scale" with millions of feature points in a matter of hours [1]. To localize a query image, we need to find correspondences between the 2D local features in the image and points in the 3D point cloud model. This correspondence search is equivalent to a nearest neighbor search when a descriptor (such as SIFT [25]) is associated with each 3D point. Although effective for laboratory datasets, the memory footprint and computational requirements of matching becomes prohibitive as larger and larger descriptor sets are considered.

Large-scale 3D point clouds offer significant opportunities for reduction. First, physical constraints, such as roads/walkways and the average height of typical photographers, produce considerable structure in the positions and orientations that are likely to occur in a space. These correlations are captured in the 3D point cloud and can be used to identify "low-value" points that are unlikely to be used for future localization. Second, the spatial distribution of 3D point clouds reflects the texture statistics of the environment as well as the view statistics of the photographers. Since we are interested only in ensuring that future cameras can be accurately and efficiently localized, the point cloud can be culled to better reflect view statistics independently. Finally, during 3D reconstruction, certain points may simply not be accurately reconstructed and are therefore unlikely to help for future localizations. In this chapter, we present a method to reduce the 3D point cloud explicitly based on the view statistics of a training corpus. It determines a compact subset of the full 3D point database that delivers comparable camera registration performance as using the full set. Given an image that is representative of view-sampling patterns in a given scene, our algorithm seeks a compact subset of the points such that at least b points in the subset are visible from a query image—a constraint inherited directly from typical perspective-n-point algorithms.

This problem is related to the maximum coverage problem, or the K-cover problem. The objective of the maximum coverage problem is to find K 3D points that maximize the sum of the number of 2D–3D correspondences for all training images; whereas for our problem, the cost of the subset, e.g., the number of 3D points in the subset, is minimized while keeping the number of correspondences in each image larger than a threshold (the required minimal number of correspondences to localize the image accurately). Karp [14] noted that the maximum coverage problem is NP-complete, so is our problem. Li et al. [24] and Irschara et al. [11] presented greedy techniques that approximately solve the problem. However, these greedy approaches are suboptimal and it is difficult to characterize or analyze their precise behavior. Thus, even small design modifications of the objective function are difficult to incorporate into the algorithm, and may necessitate the design of an entirely new greedy procedure. Our approach formulates this problem as a mixed-integer linear/quadratic programming problem. This allows us to obtain an optimal 3D point subset (using branch-and-bound) to deliver better data compression rate and camera registration performance than greedy approaches. This also allows flexibility to users to design and modify the objective function based on the application.

Contributions: We present an algorithm for 3D point cloud reduction that outperforms a state-of-the-art approach on standard datasets. We formulate the problem as a mixed-integer program that is closely integrated with camera pose estimation algorithms (e.g., the number of inliers parameter in a RANSAC pose estimator). This formulation allows explicit characterization of occurrence and co-occurence constraints, facilitating future algorithm development. We also introduce a generative measure to find a correct match based on Mahalanobis distance, which allows us to obviate the need for finding the second nearest neighbor. This measure provides much tighter threshold directly learned from a database and results in higher success rate for RANSAC-based matching.

10.2 Related Work

Image localization is one of the core problems in computer vision and robotics. There is a significant body of the literature related to matching a set of images against a large non-coincidental image repository for localization. We discuss existing localization methods in this section.

Image localization techniques often incorporate other sensory data that show a strong correlation with images. Cozman and Krotkov [5] introduced localization of an image taken from unknown territory using temporal changes in sun altitudes. Jacobs et al. [12] incorporate weather data reported by satellite imagery to localize widely distributed cameras. They find matches between weather conditions on images over a year and the expected weather changes indicated by satellite imagery. As GPS becomes a viable solution for localization in many applications, GPS-tagged images can help to localize images that do not have the tags. Zhang and Kosecka [37] built a GPS-tagged image repository in urban environments and find correspondences between a query image and the database based on SIFT features [25]. Hays and Efros [10] leveraged GPS-tagged internet images to estimate a probability distribution over the earth and Kalogerakis et al. [13] extended the work to disambiguate locations of the images without distinct landmarks. They applied a travel prior in the form of temporal information in the images. Baatz et al. [2] estimated image location based on a 3D elevation model of mountain terrains and evaluated their method on the scale of a country (Switzerland).

Pure image based localization has also been studied extensively. Torralba et al. [35] used global context to recognize a scene category using a hidden Markov model framework. Se et al. [31] applied a RANSAC framework for global camera registration. Robertson and Cipolla [29] estimated image positions relative to a set of rectified views of facades registered onto a city map. Zamir and Shah [36] leveraged the Google Street View images that provide accurate geo-location. Cummins and Newman [6] showed a visual SLAM system that reliably estimates camera poses. Structure from motion has also been employed for large-scale image localization.

Snavely et al. [33] exploited structure from motion to browse a photo collection from the exact location where it was taken. They used hundreds of images to register in 3D. Agarwal et al. [1] presented a parallelizable system that can reconstruct hundreds of thousands of images (city scale) within two days. Frahm et al. [8] showed larger scale reconstruction (millions of images) that can be executed on a single PC.

As the database becomes larger, the matching process between a query image and the database becomes computationally expensive. A number of algorithms have been introduced to accelerate the matching process [15–19]. A vocabulary tree proposed by Nistér and Stewénius [27] has been widely adopted in image localization. Havlena et al. [9] indexed images using visual vocabularies to measure the similarity between images and construct a graph that prioritizes image matching. Irschara et al. [11] synthesized views by exploiting the relationship between images and the point cloud and indexed synthesized view points based on coverage of projected 3D points. Tree building and search method based on N-best paths were proposed by Schindler et al. [30], while a vector quantization method was adopted by Baatz et al. [3]. Chen et al. [4] applied the visual vocabulary tree to localize images from various mobile devices. A graph representation of an image set can reduce the search space significantly. Simon et al. [32] presented a method to find a minimal set of images that can represent whole image sets via 3D reconstruction. Snavely et al. [34] employs skeletal graph models to identify a compact subset among a highly redundant image collection. Li et al. [22] represented a collection of images with an iconic image graph that enabled them to search on a tree structure. Li and Kosecka [21] showed a method to select discriminative features that are optimized for location recognition. Ni et al. [26] adapted compact image epitomes for query image localization.

Our image localization approach exploits a 3D point cloud reconstructed by a collection of images. Our database representation includes 3D points and corresponding feature vectors. We find a compact subset of the 3D point cloud such that a query image can find at least b matches. This approach is related to the method by Li et al. [24]. They proposed a prioritization scheme to avoid having to match every point in the model against the query image. They show that using a reduced set of points with the highest priority is better than using all 3D points, as it permits registration of more images while reducing the time needed for registration. Since this method does not constrain the set of possible views, it outperforms the algorithm by Irschara et al. [11] in terms of the number of images that can be registered. However, while inverse matching from 3D points to image features can find correct matches quickly through search prioritization, it has difficulty on larger models. In their most recent work by Li et al. [23], the method of "inverse-matching" is used in conjunction with the traditional 2D-to-3D "forward matching" to create a bidirectional matching scheme for enhancing the similarities of correspondences. Our method produces the optimal solution given the objective function, whereas existing methods suffer from sub-optimality. Also our formulation allows us to design and modify the objective function depending on applications.

10.3 Method

A camera pose in 3D (translation and orientation) can be estimated by finding correspondences between 3D points from a database and 2D points from the image captured by the camera. A state-of-the-art perspective-n-point algorithm allows us to recover the pose parameters minimally from four 2D–3D point correspondences [20] given the camera intrinsic parameters. Finding the correspondences is computationally prohibitive for real-time image localization when the number of 3D points in the database is large [1, 8, 23]. In Sect. 10.3.1, we introduce an algorithm to intelligently reduce the number of 3D points in the database based on training images. Using this reduced database, we present a novel criterion for finding 2D–3D correspondences in Sect. 10.3.2.

10.3.1 Database Reduction

Given 3D points in the database, we seek a compact subset of the points such that at least b points from the subset are visible in each query image. This problem is closely related to the maximum coverage problem, or K-cover problem. In this section, we start from the well known maximum coverage problem and derive our objective function in the form of a mixed-integer quadratic programming problem.

Let $\mathscr{S} = \{\mathbf{p}_i\}_{i=1,\dots,N}$ be a set of all 3D points, $\mathbf{p}$, where N is the number of points and let $\mathscr{S}_{\text{in}} \subset \mathscr{S}$ be a subset of points where $\mathbf{p}_i$ is discarded if $\mathbf{p}_i \notin \mathscr{S}_{\text{in}}$. We want to find a $\mathscr{S}_{\text{in}}$ that satisfies a user-defined criterion.

The maximum coverage problem is to find K points that maximize the number of correspondences, i.e.,

$$\begin{aligned} \underset{\mathbf{x}}{\text{maximize}} \quad & \mathbf{1}^{\mathsf{T}}\mathbf{A}\mathbf{x} \\ \text{subject to} \quad & \begin{matrix} \mathbf{1}^{\mathsf{T}}\mathbf{x} \le K \\ \mathbf{x} \in \{0, 1\}^N \end{matrix}, \end{aligned} \tag{10.1}$$

where $\mathbf{x}$ is a binary vector whose ith element is one if the $\mathbf{p}_i$ is kept or zero otherwise and $\mathbf{1}$ is a vector whose elements are all one, i.e., if $\mathbf{p}_i \in \mathscr{S}_{\text{in}}$, then $\mathbf{x}_i = 1$. $\mathbf{A}$ is an F by N visibility matrix where F is the number of images, i.e., if the jth point is visible from the ith image, $\mathbf{A}_{ij} = 1$, otherwise $\mathbf{A}_{ij} = 0$. Note that $\mathbf{Ax}$ is a vector, whose ith element is the number of correspondences for the ith image. Equation (10.1) does not explicitly encode the fact that there must be at least b number of correspondences to estimate the image pose parameters in 3D. By incorporating such a constraint, the problem can be formulated as:

$$\begin{aligned} \underset{\mathbf{x}}{\text{minimize}} \quad & \mathbf{1}^{\mathsf{T}}\mathbf{x} \\ \text{subject to} \quad & \begin{matrix} \mathbf{A}\mathbf{x} \ge b\mathbf{1} \\ \mathbf{x} \in \{0, 1\}^N \end{matrix}, \end{aligned} \tag{10.2}$$

where b is the minimum number of 2D–3D correspondences to be kept for each image. Equation (10.2) directly minimizes the number of elements in $\mathscr{S}_{\text{in}}$ while maintaining the number of correspondences larger than b. Therefore, the smallest subset of 3D points can be obtained.

When prior knowledge about a 3D point is available, a different weight on each point can produce the desired solution (Eq. (10.2) has the same weight on all points). For instance, a point that is frequently matched to the training images is more favorable because it is more likely to be matched to a query image. This prior knowledge can be explicitly applied by weighting on each point individually:

$$\begin{aligned} \underset{\mathbf{x}}{\text{minimize}} \quad & \mathbf{q}^{\mathsf{T}}\mathbf{x} \\ \text{subject to} \quad & \begin{array}{l} \mathbf{Ax} \geq b\mathbf{1} \\ \mathbf{x} \in \{0, 1\}^N \end{array}, \end{aligned} \tag{10.3}$$

where $\mathbf{q}$ is a weight vector. Binary relations between two correspondences can also be considered. For example, a spatially widely distributed 3D point configuration is favorable because it can enhance the accuracy of image localization; the uncertainty of pose estimation becomes higher when the distance between 3D points becomes smaller. By encouraging high 2D/3D Euclidean distance between two correspondences or penalizing co-occurence in the same image, a desirable solution that produces highly accurate image localization can be achieved. This relation can be encoded in the form of a binary quadratic programming problem as follows:

$$\begin{aligned} \underset{\mathbf{x}}{\text{minimize}} \quad & \frac{1}{2}\mathbf{x}^{\mathsf{T}}\mathbf{Q}\mathbf{x} + \mathbf{q}^{\mathsf{T}}\mathbf{x} \\ \text{subject to} \quad & \begin{array}{l} \mathbf{Ax} \geq b\mathbf{1} \\ \mathbf{x} \in \{0, 1\}^N \end{array}, \end{aligned} \tag{10.4}$$

where $\mathbf{Q}$ is a symmetric matrix. $\mathbf{Q}_{ij}$ accounts for a weight on the binary relation between the ith and jth points. Figure 10.1a shows that the quadratic term can contribute towards accurate and robust image localization. This is also observed in the image itself; the solution from quadratic programming produces more widely distributed correspondences.

The inequality constraint $\mathbf{Ax} \geq b\mathbf{1}$ in Eqs. (10.2)–(10.4) is a hard constraint, i.e., if the total number of correspondences of the ith image is less than b, there is no feasible solution. For some training images, the number of correspondences do not need to be greater than b because the images are not informative. The hard constraint can be relaxed by introducing a slack variable as follows:

$$\begin{aligned} \underset{\mathbf{x},\boldsymbol{\xi}}{\text{minimize}} \quad & \frac{1}{2}\mathbf{x}^{\mathsf{T}}\mathbf{Q}\mathbf{x} + \mathbf{q}^{\mathsf{T}}\mathbf{x} + \lambda\mathbf{1}^{\mathsf{T}}\boldsymbol{\xi} \\ \text{subject to} \quad & \begin{array}{l} \mathbf{Ax} \geq b\mathbf{1} - \boldsymbol{\xi} \\ \mathbf{x} \in \{0, 1\}^N \\ \boldsymbol{\xi} \in \{\mathbb{N}_+\}^F \end{array}, \end{aligned} \tag{10.5}$$

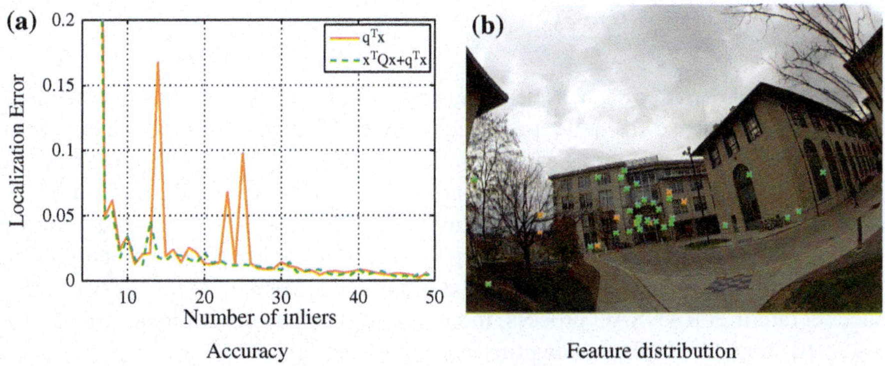

Fig. 10.1 A quadratic term in Eq. (10.4) allows us to design the binary relation between two points. We penalize co-occurring feature points for accurate camera pose localization. **a** The quadratic term in Eq. (10.4) contributes to produce a more stable/accurate localization. **b** *Green points* are points that are selected by solving Eq. (10.4) while *orange points* are selected by solving Eq. (10.3). By penalizing the co-occurence, the quadratic term encourages widely distributed correspondence configuration. The variance of 2D point distribution is 303.65 and 174.24 for *green* and *orange points*, respectively

where $\boldsymbol{\xi}$ is a semi-positive integer vector that allows small violation of the inequality constraint and λ determines the hardness of the inequality constraint. When λ goes to infinite, the constraint becomes hard, i.e., the same as Eq. (10.4), and when λ approaches zero, the constraint becomes soft. This slack variable also prevents the solution from overfitting to the training images. As shown in Fig. 10.2a, λ controls the shape of graph. $\lambda = 0.1$ well generalizes the shape of the testing graph shown in Fig. 10.4a. $\mathbf{Q}$, $\mathbf{q}$, and λ are user designable parameters based on applications.

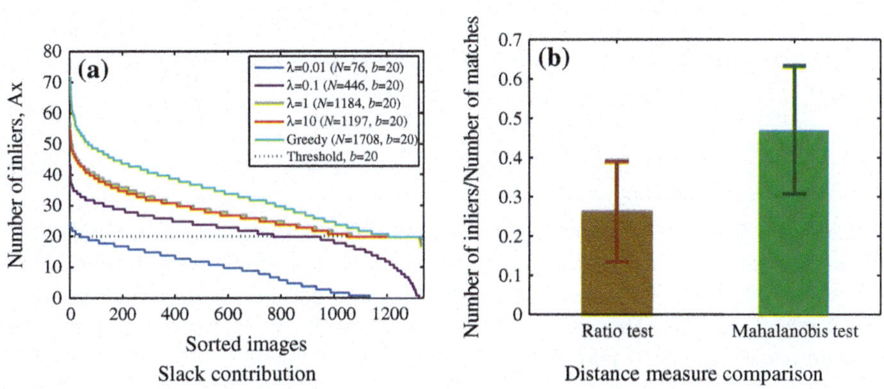

Fig. 10.2 **a** Slack variable in Eq. (10.5) allows small violation of the inequality constraint. This prevents the solution to overfit the training data. When $\lambda = 0.1$, it predicts the underlying distribution of test images well. Compare the shape of the testing set graph shown in Fig. 10.4a, c. **b** Mahalanobis distance measure rejects more false positive matches, i.e., fewer noisy matches. As a result, the RANSAC success rate becomes higher

10.3.2 Matching Process

The number of correspondences per image decreases as the size of database becomes smaller. The database size reduction accelerates the matching process whereas finding correct 2D–3D matches becomes more challenging. The number of inliers is significantly lower than the number of outliers ($<5\,\%$). We intend the number of correspondences to be similar to b when solving Eq. (10.5), where b is a small number compared to the number of all feature points in a query image. Lower probability of choosing the inlier requires many iterations for RANSAC [7]. To have a 95 % success rate in a RANSAC process, the required number of iterations is more than 4.8×10^5 for 5 % of inliers using the four point method [20].

A previous method for finding correct matches relies on a discriminative measure, i.e., the ratio test [25] to reject false positive matches. If the ratio between distances of the nearest neighbor and the second nearest neighbor is smaller than a threshold (0.7 is popularly used), the 2D and 3D points are considered to be a correct match. This ratio test ensures that the match is distinctive. However, the ratio criterion produces many false positives for our case where correct matches are extremely sparse among all detected feature points in the query image. To reduce the number of false positive matches, Li et al. [24] used a bidirectional matching method with different ratio tests, and Li et al. [23] applied a co-occurence prior distribution for false positive rejection. Unlike previous methods, we utilize a generative measure; Mahalanobis distance, by learning a distribution of descriptors directly from the training images. This approach can provide a more strict criterion for false positive detection.

A 3D point associates with at least two similar descriptors because two 2D points are minimally required to be triangulated to estimate the 3D point. In many cases, a point corresponds to more than two descriptors from the training images. From many descriptors associated with a single 3D point, we learn the covariance of the descriptors, $\mathbf{C}_\mathbf{y} = \mathbf{Y}\mathbf{Y}^\mathsf{T}$, where $\mathbf{Y} = \left[\mathbf{y}_1 - \bar{\mathbf{y}} \cdots \mathbf{y}_m - \bar{\mathbf{y}}\right]$, $\mathbf{y}_i$ is a descriptor vector, $\bar{\mathbf{y}}$ is the mean descriptor vector, and m is the number of the descriptors associated with the 3D point. The covariance enables us to define a Mahalanobis distance between 3D and 2D descriptors to measure the similarity, i.e.,

$$d(\bar{\mathbf{y}}, \mathbf{z}) = \sqrt{(\mathbf{z} - \bar{\mathbf{y}})^\mathsf{T}\, \mathbf{C}_\mathbf{y}^{-1}\, (\mathbf{z} - \bar{\mathbf{y}})}, \tag{10.6}$$

where $\mathbf{z}$ is a query descriptor. The main benefit of this distance is that we can estimate the probability of the match given the query descriptor. Also the distance is a normalized measure, therefore a single threshold can be applied to all points to reject false positives. This generative measure is a much tighter criterion than the discriminative measure in previous works while generalizing well across all matches as long as the query image is drawn from the distribution of the training images. Figure 10.2b shows the effectiveness of our distance measure. The ratio between the number of inliers and matches is higher than the ratio test, which results in higher success rate for RANSAC given the same number of iterations (Fig. 10.3).

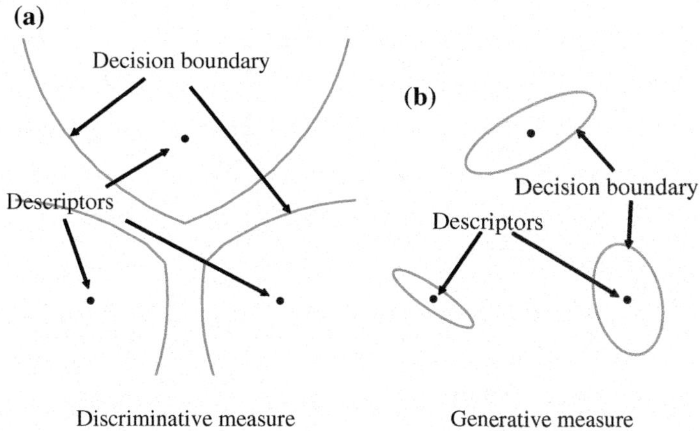

Fig. 10.3 We present a novel method to reject false positive matches by learning a distribution of descriptors from the training images. **a** Previous methods have used a discriminative measure, the ratio test. If the ratio between distances of the nearest and the second nearest neighbor is less than 0.7, it is considered to be a match. **b** Instead, we learn the covariance of descriptors. This measure results in a more strict criterion, which enables us to reject false positive matches. We show the decision boundary of matching

In practice, we look for bidirectionally consistent matches: we match from a query image to the 3D database and vice versa, and keep only the consistent matches. From these consistent matches, we apply Mahalanobis distance test based on Eq. (10.6), i.e., if $d(\bar{\mathbf{y}}, \mathbf{z}) > t$, we reject the match. t is a threshold. For the covariance matrix, we use $\widetilde{\mathbf{C}}_\mathrm{y} = \mathbf{C}_\mathrm{y} + \mathbf{I}$ to avoid ill-conditioned inverse operation. Instead of representing a distribution of descriptors with a D by D matrix where D is the feature dimension, only diagonal elements in $\widetilde{\mathbf{C}}_\mathrm{y}$ is used (off-diagonal terms are extremely sparse) for compute and memory efficiency. We use the variance of each element of the descriptors by assuming that there is no correlation across elements in a descriptor.

10.4 Result

We test our algorithm on four real datasets: two from our own data collection and two from standard benchmark datasets provided by Li et al. [24]. All quantitative evaluation is compared to the baseline greedy algorithm presented by Li et al. [24]. For our data collection, we reconstruct 3D points from videos that exhaustively scan the space of interest. We use a standard structure from motion algorithm to reconstruct the 3D scene. We ask people to wear head-mounted cameras and move freely in the space. We use one of the videos as the test set and the rest of the videos as the training set. Two sequences, outdoor and indoor, are used for our evaluation. For the standard datasets, we used Dubrovnik and Rome data reconstructed by internet photos. We use known camera intrinsic parameters available from the meta data and estimate the camera pose in 3D using the efficient perspective-n-point algorithm [20]. While the minimal

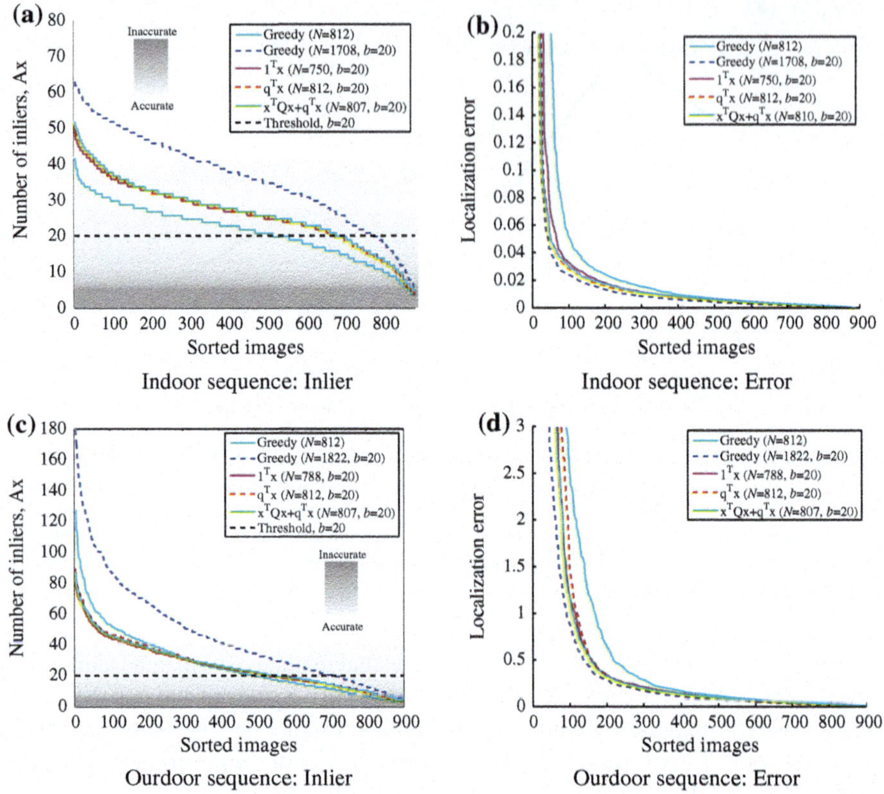

Fig. 10.4 **a** and **c** Our method outperforms a greedy solution which has comparable size, particularly when the number of inliers is between 10 and 20 where camera localization becomes inaccurate. The *gray* gradient at the *bottom* of the graph shows the accuracy changes as the number of inliers increase. To get the same threshold, b, with our method, the greedy algorithm selects more than twice as many points than our approach. **b** and **d** Image localization accuracy is measured. Our method consistently performs better than comparable size of the greedy solution. Note that the quadratic term produces the highest accuracy among our solutions

number of correspondences for image localization is four given intrinsic parameters, accurate and robust estimation typically requires more correspondences. Only when the number of inliers from RANSAC is higher than 10, the image is considered to be registered. The element of the unary weight, $\mathbf{q}$, is set to $\mathbf{q}_i = q_{\max} - q_i$, where q_i is the number of times the ith point is visible and $q_{\max} = \max\{q_i\}_{i=1,\ldots,N}$. For the binary weight, $\mathbf{Q}$, we penalize the co-occurence of two points by linearly increasing the $\mathbf{Q}_{ij}$ whenever the ith and jth points are observed from the same image. To solve the mixed-integer quadratic programming problem, we use the commercial optimization software Gurobi.[1]

[1] http://www.gurobi.com.

10.4.1 Our Datasets: Outdoor and Indoor

We set the threshold, b, from the inequality constraint in Eq. (10.4) to 20 to ensure that the number of correspondences is greater than 10 (registration threshold). Figure 10.4a shows the number of inliers after RANSAC (images are sorted in descending order based on the number of inliers). Our algorithm outperforms the greedy algorithm, particularly around 10–20 inliers. This is an important range for image localization because the accuracy of the localization degrades significantly within the range (the grey gradient encodes the accuracy of localization in Fig. 10.4a). Figure 10.4b shows that the solution obtained by our method is more accurate than the greedy method. The same observation also applies to the outdoor sequence as

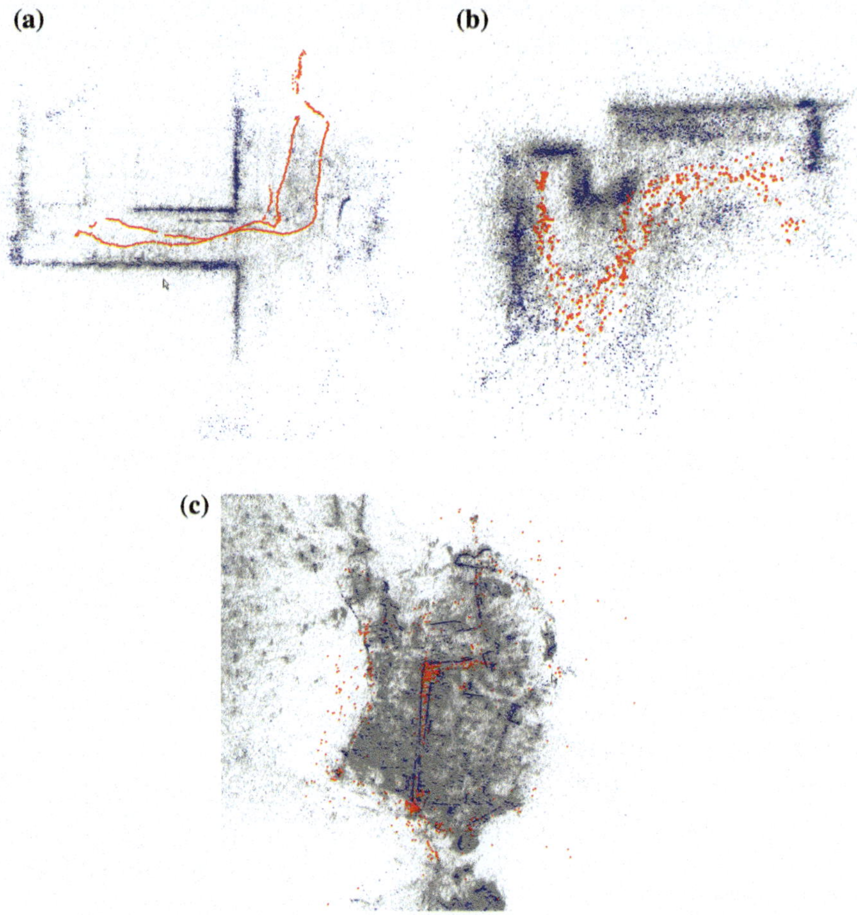

Fig. 10.5 Our method significantly reduces the number of points. *Gray points* are original 3D points, *blue points* are the reduced subset, and *red points* are reconstructed image locations. **a** Outdoor sequence. **b** Indoor sequence. **c** Dubrovnik

shown in Fig. 10.4c, d. In both sequences, our method achieves over 90 % image registration rate. Figure 10.5a, b show the qualitative results of our method.

10.4.2 Standard Datasets: Dubrovnik and Rome

We apply our algorithm to two standard datasets provided by Li et al. [24]. While these datasets contain many images (6000 for Dubrovnik and 16000 for Rome), the number of image samples per area is lower than our datasets, i.e., covering area per image is fairly large. To account for sparse training images, we raise b to 80. Due to memory constraints, we only employ binary linear programming, i.e., Eqs. (10.2) and (10.3). Figure 10.6a shows our method can register about 80 % of query images and consistently outperforms the greedy solution. Our localization error is also lower

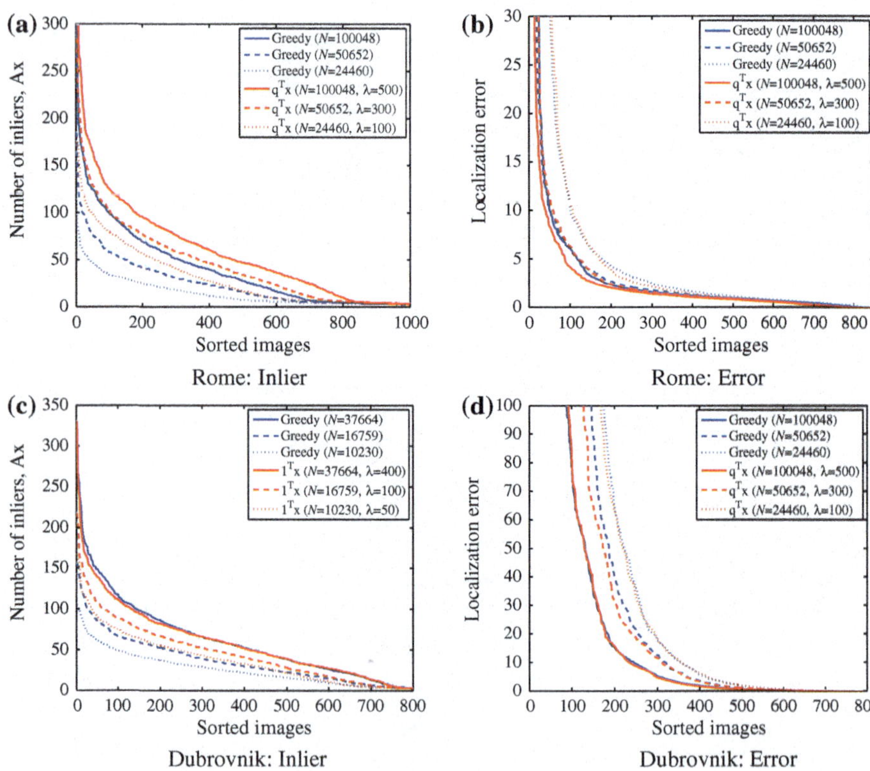

Fig. 10.6 **a** and **c** We reduce the 3D points by varying slack parameter, λ. As λ decreases, the reduction rate significantly increases. The difference between the greedy method and ours is more emphasized. For all experiments, our method consistently registers more images than the greedy method. **b** and **d** our method is more accurate

Table 10.1 Registration performance for Rome data (885 query images)

	N	$\mathbf{1}^\mathsf{T}\mathbf{x}$	Greedy		N	$\mathbf{q}^\mathsf{T}\mathbf{x}$	Greedy
$\lambda = 1$	94708	786	662	$\lambda = 500$	100048	806	676
$\lambda = 0.5$	57263	730	540	$\lambda = 300$	50652	722	603
$\lambda = 0.1$	20157	584	416	$\lambda = 100$	24460	620	466

Table 10.2 Registration performance for Dubrovnik data (780 query images)

	N	$\mathbf{1}^\mathsf{T}\mathbf{x}$	Greedy		N	$\mathbf{q}^\mathsf{T}\mathbf{x}$	Greedy
$\lambda = 0.5$	46820	711	716	$\lambda = 400$	37664	711	704
$\lambda = 0.25$	17436	602	604	$\lambda = 100$	16759	649	637
$\lambda = 0.125$	10502	560	536	$\lambda = 50$	10230	598	576

than the greedy algorithm as shown in Fig. 10.6b. Our experiments on Dubrovnik data produce similar observations as shown in Fig. 10.6c, d. Detailed comparison is listed in Tables 10.1 and 10.2. Figure 10.5c shows camera registration (red points) using the reduced set of 3D point cloud (blue points).

10.5 Discussion

In this chapter, we present an algorithm to find a compact subset of 3D points based on view-statistics for efficient image localization. This compact subset of 3D points speeds up the matching process and results in a comparable image localization rate. We formulate the problem as a mixed-integer quadratic programming problem. This formulation allows a user to design the objective function depending on applications unlike existing greedy methods. We demonstrate that our solution outperforms a related greedy method on standard datasets. We also introduce a method to calibrate descriptors associated with a 3D point by learning a distribution of the descriptors directly from the training images. This generative measure provides higher success rate on a RANSAC based matching.

Our formulation of the database reduction problem allows a user to design the objective function based on the application. This enables users to manipulate the desired output easily and to incorporate domain knowledge. One interesting future direction might be to learn $\mathbf{Q}$ and $\mathbf{q}$ for different target camera motions and different types of databases. For example, camera motion in a vehicle is different from that of a first-person camera. The reduced point set should reflect and benefit from the prior knowledge about camera motion. Also if the database contains a subset of GPS-tagged images, weights on points corresponding to the tagged images should be different. The optimal weight can be learned from different datasets.

We show that the data reduction problem is inherently a mixed integer quadratic programming problem. It is a non-convex optimization that requires high capacity of memory and heavy computation where the global optimum cannot be guaranteed. For our outdoor dataset (30,000 points), it took 30 min on an Intel i7 CPU@2.5 GHz with 16 GB memory to solve the mixed integer quadratic programming problem using a branch-and-bound method. Such memory requirement and computation are apparent limitations of our method. However, efficient optimization algorithms have solved large-scale problems via spectral or semidefinite relaxation [28]. Rapid advances in computing power and memory capacity will also enable us to handle the problem at large scale.

References

1. Agarwal S, Snavely N, Simon I, Seitz S, Szeliski R (2009) Building Rome in a Day. In: ICCV
2. Baatz G, Saurer O, Köser K, Pollefeys M (2012) Large scale visual geo-localization of images in mountainous terrain. In: ECCV
3. Baatz G, Köser K, Chen D, Grzeszczuk R, Pollefeys M (2012) Handling urban location recognition as a 2D homothetic problem. In: ECCV
4. Chen D, Baatz G, Köser K, Tsai SS, Vedantham R, Pylvänäinen T, Roimela K, Chen X, Bach J, Pollefeys M, Girod B, Grzeszczuk R (2011) City-scale landmark identification on mobile devices. In: CVPR
5. Cozman F, Krotkov E (1995) Robot localization using a computer vision sextant. In: ICRA
6. Cummins M, Newman P (2008) FAB-MAP: probabilistic localization and mapping in the space of appearance. In: IJRR
7. Fischler MA, Bolles RC (1981) FAB-MAP: random sample consensus: a paradigm for model fitting with applications to image analysis and automated cartography. In: Communications of the ACM
8. Frahm J-M, Georgel P, Gallup D, Johnson T, Raguram R, Wu C, Jen Y-H, Dunn E, Clipp B, Lazebnik S, Pollefeys M (2010) Building Rome on a cloudless day. In: ECCV
9. Havlena M, Torii A, Pajdla T (2010) Efficient structure from motion by graph optimization. In: ECCV
10. Hays J, Efros AA (2008) IM2GPS: estimating geographic information from a single image. In: CVPR
11. Irschara A, Zach C, Frahm J-M, Bischof H (2009) From structure-from-motion point clouds to fast location recognition. In: CVPR
12. Jacobs N, Satkin S, Roman N, Speyer R, Pless R (2007) Geolocating static cameras. In: ICCV
13. Kalogerakis E, Vesselova O, Hays J, Efros AA, Hertzmann A (2009) Image sequence geolocation with human travel priors. In: ICCV
14. Karp RM (1972) Reducibility among combinatorial problems. In: Complexity of computer computations
15. Sattler T, Leibe B, Kobbelt L (2011) Fast image-based localization using direct 2D-to-3D matching. In: ICCV
16. Choudhary S, Narayanan PJ (2012) Visibility probability structure from SfM datasets and applications. In: ECCV
17. Sattler T, Leibe B, Kobbelt L (2012) Improving image-based localization by active correspondence search. In: ECCV
18. Donoser M, Schmalstieg D (2014) Discriminative feature-to-point matching in image-based localization. In: CVPR

19. Cao S, Snavely N (2013) Graph-based discriminative learning for location recognition. In: CVPR
20. Lepetit V, Moreno-Noguer F, Fua P (2009) EPnP: an accurate O(n) solution to the PnP problem. In: IJCV
21. Li F, Kosecka J (2006) Probabilistic location recognition using reduced feature set. In: ICRA
22. Li X, Wu C, Zach C, Lazebnik S, Frahm J-M (2008) Probabilistic location recognition using reduced feature set. In: ECCV
23. Li Y, Snavely N, Huttenlocher D, Fua P (2012) Worldwide pose estimation using 3D point clouds. In: ECCV
24. Li Y, Snavely N, Huttenlocher D (2010) Location recognition using priotized feature matching. In: ECCV
25. Lowe DG (2004) Distinctive image features from scale-invariant keypoints. In: IJCV
26. Ni K, Kannan A, Criminisi A, Winn J (2009) Epitomic location recognition. In: TPAMI
27. Nistér D, Stewénius H (2006) Scalable recognition with a vocabulary tree. In: CVPR
28. Olsson C, Eriksson AP, Kahl F (2007) Solving large scale binary quadratic problems: spectral methods vs. semidefinite programming. In: CVPR
29. Duncan R, Roberto C (2004) An image-based system for urban navigation. In: BMVC
30. Schindler G, Brown M, Szeliski R (2007) City-scale location recognition. In: ICCV
31. Se S, Lowe D, Little J (2002) Global localization using distinctive visual features. In: IROS
32. Simon I, Snavely N, Seitz SM (2007) Scene summarization for online image collections. In: ICCV
33. Snavely N, Seitz SM, Szeliski R (2006) Photo tourism: exploring photo collections in 3D. In: TOG(SIGGRAPH)
34. Snavely N, Seitz SM, Szeliski R (2008) Skeletal graphs for efficient structure from motion. In: CVPR
35. Torralba A, Murphy KP, Freeman WT, Rubin MA (2008) Context-based vision system for place and object recognition. In: ICCV
36. Zamir AR, Shah M (2010) Accurate image localization based on Google Maps street view. In: ECCV
37. Zhang W, Kosecka J (2006) Image based localization in urban environments. In: 3DPVT

Chapter 11
Image-Based Large-Scale Geo-localization in Mountainous Regions

Olivier Saurer, Georges Baatz, Kevin Köser, L'ubor Ladický and Marc Pollefeys

Abstract Given a picture taken somewhere in the world, automatic geo-localization of such an image is an extremely useful task especially for historical and forensic sciences, documentation purposes, organization of the world's photographs and intelligence applications. While tremendous progress has been made over the last years in visual location recognition within a single city, localization in natural environments is much more difficult, since vegetation, illumination, seasonal changes make appearance-only approaches impractical. In this chapter, we target mountainous terrain and use digital elevation models to extract representations for fast visual database lookup. We propose an automated approach for very large-scale visual localization that can efficiently exploit visual information (contours) and geometric constraints (consistent orientation) at the same time. We validate the system at the scale of Switzerland (40 000 km^2) using over 1000 landscape query images with ground truth GPS position.

O. Saurer (✉) · L. Ladický · M. Pollefeys
ETH Zürich, Zürich, Switzerland
e-mail: saurero@inf.ethz.ch

L. Ladický
e-mail: lubor.ladicky@inf.ethz.ch

M. Pollefeys
e-mail: marc.pollefeys@inf.ethz.ch

G. Baatz
Google Inc., Zürich, Switzerland
e-mail: gbaatz@google.com

K. Köser
GEOMAR, Kiel, Germany
e-mail: kkoeser@geomar.de

A.R. Zamir et al. (eds.), *Large-Scale Visual Geo-Localization*,
Advances in Computer Vision and Pattern Recognition,
DOI 10.1007/978-3-319-25781-5_11

11.1 Introduction and Previous Work

In intelligence and forensic scenarios as well as for searching archives and organizing photo collections, automatic image-based location recognition is a challenging task that would be extremely useful when solved. In such applications, GPS tags are typically not available in the images requiring a fully image-based approach for geo-localization. During the last years, progress has been made in urban scenarios, in particular with stable man-made structures that persist over time. However, recognizing the camera location in natural environments is substantially more challenging, since vegetation changes rapidly during seasons, and lighting and weather conditions (e.g. snow lines) make the use of appearance-based techniques (e.g. patch-based local image features [6, 25]) very difficult. Additionally, dense street-level imagery is limited to cities and major roads, and for mountains or countryside only aerial footage exists, which is much harder to relate with terrestrial imagery.

In this chapter, we give a more in-depth discussion on camera geo-localization in natural environments. In particular, we focus on recognizing the skyline in a query image, given a digital elevation model (DEM) of a country—or ultimately, the world. In contrast to previous work of matching e.g. a peak in the image to a set of mountains known to be nearby, we aggregate shape information across the whole skyline (not only the peaks) and search for a similar configuration of basic shapes in a large-scale database that is organized to allow for query images of largely different fields of view. The method is based on sky segmentation (either automatic or easily supported by an operator for challenging pictures such as those with reflection, occlusion, or taken from inside a cable car).

11.1.1 Contributions

The main contributions are a novel method for robust contour encoding as well as two different voting schemes to solve the large- scale camera pose recognition from contours. The first scheme operates only in descriptor space (it verifies where in the model a panoramic skyline is most likely to *contain* the current query picture) while the second one is a combined vote in descriptor and rotation space. We validate the whole approach using a public digital elevation model of Switzerland that covers more than 40 000 km^2 and a set of over 1000 images with ground truth GPS position. In particular, we show the improvements of all novel contributions compared to a baseline implementation motivated by classical bag-of-words [28]—based techniques like [6]. Furthermore, we demonstrate that the skyline is highly informative and can be used effectively for localization (Fig. 11.1).

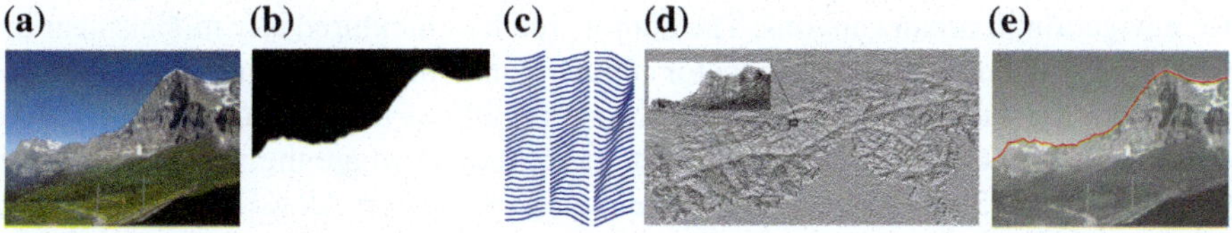

Fig. 11.1 Different stages in the proposed pipeline: **a** Query image somewhere in Switzerland, **b** sky segmentation, **c** sample set of extracted 10° contourlets, **d** recognized geo-location in digital elevation model, **e** overlaid skyline at retrieved position

11.1.2 Previous Work

To the best of our knowledge this is the first attempt to localize photographs at large-scale based on a digital elevation model.[1] The closest works to ours are smaller scale navigation and localization in robotics [29, 33], and building/location recognition in cities [1, 6, 25] or with respect to community photo collections of popular landmarks [17]. These, however, do not apply to landscape scenes of changing weather, vegetation, snowlines, or lighting conditions. The robotics community has considered the problem of robot navigation and robot localization using digital elevation models for quite some time. Talluri et al. [30] reason about intersection of known viewing ray directions (north, east, south, west) with the skyline and relies thus on the availability of 360° panoramic query contours and the knowledge of vehicle orientation (i.e. north direction). Thompson et al. [31] suggest general concepts of how to estimate pose and propose a hypothesize and verification scheme. They also rely on known view orientation and match viewpoint-independent features (peaks, saddle points, etc.) of a DEM to features found in the query image, ignoring most of the signal encoded in the skyline. In [9], computer vision techniques are used to extract mountain peaks which are matched to a database of nearby mountains to support a remote operator in navigation. However, we believe that their approach of considering relative positions of absolute peaks detected in a DEM is too restrictive and would not scale to our orders of magnitude larger problem, in particular with respect to less discriminative locations. Reference [22] proposes to first match three features of a contour to a DEM and estimate an initial pose from that before doing a nonlinear refinement. Also here the initial step of finding three correct correspondences is a challenging task in a larger scale database. Reference [29] assumes panoramic query data with known heading, and computes super-segments on a polygon fit, however descriptiveness/robustness is not evaluated on a bigger scale, while [8] introduces a probabilistic formulation for a similar setting. The key point is that going from tens of potential locations to millions of locations requires a conceptually different approach, since exhaustive image comparison or trying all possible "mountain peaks" simply does not scale up to large-scale geo-localization problems. Similarly, for urban localization, in [24] an upward looking 180° field-of-view fisheye is used

[1] A preliminary version of this system has been presented in [2].

for navigation in urban canyons. The authors render untextured city models near the predicted pose and extract contours for comparison with the query image. Also this is meant as a local method for navigation. Most recently, in [3] the authors optimize the camera orientation given the exact position i.e. they estimate the viewing direction given a good GPS tag. None of the above-mentioned systems considered recognition and localization at large scale.

On the earth scale, the authors of [11] source photo collections and aim at learning location probability based on colour, texture and other image-based statistics. Conceptually, this is not meant to find an exact pose based on geometric considerations but rather discriminates landscapes or cities with different (appearance) characteristics on a global scale. The authors of [16] exploit the position of the sun (given the time) for geo-localization. In the same work, it is also shown that identifying a large piece of clear sky without haze provides information about the camera pose (although impressive given the data, over 100 km mean localization error is reported). Both approaches are appealing for excluding large parts of the earth from further search but do not aim at exactly localizing the camera within a few hundred meters.

Besides attacking the DEM-based, large-scale geo-localization problem we propose new techniques that might also be transferred to bag-of-words approaches based on local image patches (e.g. [6, 25, 28]). Those approaches typically rely on pure occurrence-based statistics (visual word histogram) to generate a first list of hypotheses and only for the top candidates geometric consistency of matches is verified. Such a strategy fails in cases where pure feature coocurrence is not discriminative but where the relative locations of the features are important. Here, we propose to do a (weak) geometric verification already in the histogram distance phase. Furthermore, we show also a representation that tolerates largely different document sizes (allowing to compare a panorama in the database to an image with an order of magnitude smaller field of view).

11.2 Mountain Recognition Approach

The location recognition problem in its general form is a six-dimensional problem, since three position and three orientation parameters need to be estimated. We make the assumption that the photographs are taken not too far off the ground and use the fact that people rarely twist the camera relative to the horizon [5] (e.g. small roll). We propose a method to solve that problem using the outlines of mountains against the sky (i.e. the skyline). For the visual database we seek a representation that is robust with respect to tilt of the camera which means that we are effectively left with estimating the 2D position (latitude and longitude) on the digital elevation model and the viewing direction of the camera. The visible skyline of the DEM is extracted offline at regular grid positions (360° at each position) and represented by a collection of vector-quantized local contourlets (contour words, similar in spirit to visual words obtained from quantized image patch descriptors [28]). In contrast to visual word-based approaches, additionally an individual viewing angle α_d ($\alpha_d \in [0; 2\pi]$) relative to north direction is stored. At query time, a skyline segmentation technique is applied that copes with the often present haze and allows for user interaction in case of incorrect segmentation. Subsequently, the extracted contour is robustly described by a set of local contourlets plus their relative angular distance α_q with respect to the optical axis of the camera. The contour words are represented as an inverted file system, which is used to query the most promising location. At the same time the inverted file also votes for the viewing direction, which is a geometric verification integrated in the bag-of-words search.

11.2.1 Processing the Query Image

11.2.1.1 Sky Segmentation

The estimation of the visible skyline can be cast as a foreground-background segmentation problem. As we assume almost no camera roll and since overhanging structures are not modelled by the 2.5D DEM, finding the highest foreground pixel (foreground height) for each image column provides an good approximation and allows for a dynamic programming solution, as proposed in [18]. To obtain the data term for a candidate height in a column we sum all foreground costs below the candidate contour and all sky costs above the contour. The assumption is, when traversing the skyline, there should be a local evidence in terms of an orthogonal gradient (similar in spirit to flux maximization [32] or contrast sensitive smoothness assumptions [4, 13] in general 2D segmentation).

We express the segmentation problem in terms of an energy:

$$E = E_d(1) + \sum_{x=2}^{\text{width}} (E_d(x) + \lambda E_s(x-1, x)), \qquad (11.1)$$

where E_d represents the data term, E_s the smoothness term, λ a weighting factor and x runs over all image columns. The data term $E_d(x)$ in one column x evaluates the cost of all pixels below it to be assigned a foreground label, while all pixels above it are assigned a background (sky) label. The cost is incorporated into the optimization framework as a standard negative-log-likelihood:

$$E_d = \sum_{i=1}^{k-1} - \log h(\mathscr{F}|z_i) + \sum_{i=k}^{\text{height}} - \log h(\mathscr{B}|z_i), \qquad (11.2)$$

where $h(\mathscr{F}|z_i)$ denotes the probability of pixel z_i to be assigned to the foreground $\mathscr{F}$ model. While $h(\mathscr{B}|z_i)$ denotes the probability of a pixel z_i to be assigned to the background $\mathscr{B}$ model. The likelihoods $h(z|\mathscr{F})$ and $h(z|\mathscr{B})$ are computed by the pixel-wise classifier, jointly trained using contextual and super-pixel-based feature representations [15].

Fig. 11.2 Superpixel-based segmentation: **a** Input image. **b** MeanShift filtered image. **c** MeanShift region boundaries. **d** Final segmentation

The contextual part of the feature vector [14, 27] consists of a concatenation of bag-of-words representations over a fixed random set of 200 rectangles, placed relative to the corresponding pixel. These bag-of-words representations are built using four dense features-textons [20], local ternary patterns [12], self-similarity [26] and dense SIFT [19], each one quantized to 512 clusters using standard K-means clustering. The super-pixel part of the feature vector consists of concatenation of a bag-of-words representations of the super-pixels, the pixel belongs to [15]. Four superpixel segmentations are obtained by varying the parameters of the MeanShift algorithm [7], see Fig. 11.2. Pixels, belonging to the same segment, share a large part of the feature vector, and thus tend to have the same labels, leading to segmentations, that follow semantic boundaries.

The most discriminative weak features are found using AdaBoost [10]. The contextual feature representations are evaluated on the fly using integral images [27], the super-pixel part is evaluated once and kept in memory. The classifier is trained independently for five colour spaces—Lab, Luv, Grey, Opponent and Rgb. The final likelihood is calculated as an average of these 5 classfiers.

The pairwise smoothness term is formulated as:

$$E_s(x, x+1) = \sum_{i \in C} (|\frac{\mathbf{disp}}{||\mathbf{disp}||} \mathbf{R} g_i|)^{-1}, \qquad (11.3)$$

where C is the set of pixels connecting two pixels (p_n and p_m) in column x and column $x+1$ along the Manhattan path. Then **disp** denotes the Euclidean vector between p_n and p_m, g_i is the image gradient at pixel i and $\mathbf{R}$ represents a 90° rotation matrix. The intuition is, that all pixels on the contour should have a gradient orthogonal to the skyline.

Given the energy terms defined in Eqs. (11.2) and (11.3), the segmentation is obtained by minimizing Eq. (11.1) using dynamic programming. Our framework also allows for user interaction, where simple strokes can mark foreground or background (sky) in the query image. In case of a foreground labelling this forces all pixel below the stroke to be labels as foreground and in case of a backround stroke, the stroke pixel and all pixels above it are marked as background (sky). This provides a simple and effective means to correct for very challenging situations, where buildings, trees partially occlude the skyline.

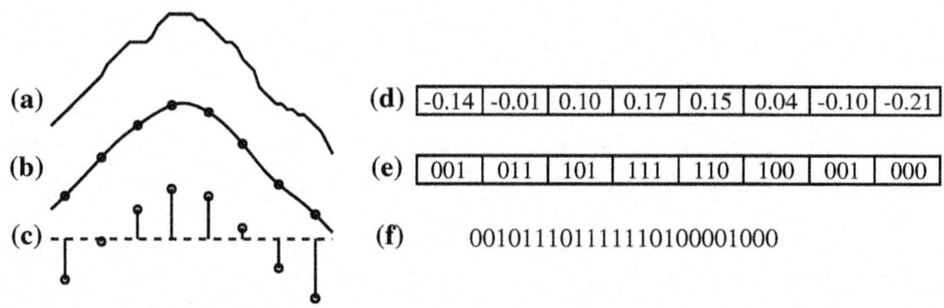

Fig. 11.3 Contour word computation: **a** raw contour, **b** smoothed contour with n sampled points, **c** sampled points after normalization, **d** contourlet as numeric vector, **e** each dimension quantized to 3 bits, **f** contour word as 24-bit integer

11.2.1.2 Contourlet Extraction

In the field of shape recognition, there are many shape description techniques that deal with closed contours, e.g. [21]. However, recognition based on partial contours is still a largely unsolved problem, because it is difficult to find representations invariant to viewpoint. For the sake of robustness to occlusion, to noise and systematic errors (inaccurate focal length estimate or tilt angle), we decided to use local representations of the skyline (see [34] for an overview on shape features).

To describe the contour, we consider overlapping curvelets of width w (imagine a sliding window, see Fig. 11.1). These curvelets are then sampled at n equally spaced points, yielding each an n-dimensional vector $\tilde{y}_1, \ldots, \tilde{y}_n$ (before sampling, we low-pass filter the skyline to avoid aliasing). The final descriptor is obtained by subtracting the mean and dividing by the feature width (see Fig. 11.3a–d):

$$y_i = \frac{\tilde{y}_i - \bar{y}}{w} \text{ for } i = 1, \ldots, n \quad \text{where} \quad \bar{y} = \frac{1}{n}\sum_{j=1}^{n} \tilde{y}_j \tag{11.4}$$

Mean subtraction makes the descriptor invariant w.r.t. vertical image location (and therefore robust against camera tilt). Scaling ensures that the y_i's have roughly the same magnitude, independently of the feature width w.

In a next step, each dimension of a contourlet is quantized (Fig. 11.3e–f). Since the features are very low-dimensional compared to traditional patch-based feature descriptors like SIFT [19], we choose not to use a vocabulary tree. Instead, we directly quantize each dimension of the descriptor *separately*, which is both faster and more memory-efficient compared to a traditional vocabulary tree. In addition the best bin is guaranteed to be found. Each y_i falls into one bin and the n associated bin numbers are concatenated into a single integer, which we refer to as *contour word*. For each descriptor, the viewing direction α_q, relative to the camera's optical axis is computed using the camera's intrinsics parameters and is stored together with the visual word. We have verified that an approximate focal length estimate is sufficient. In case of an unknown focal length, it is possible to sample several tentative focal length values, which we evaluate in in Sect. 11.3.

11.2.2 Visual Database Creation

The digital elevation model we use for validation is available from the Swiss Federal Office of Topography, and similar datasets exist also for the US and other countries. There is one sample point per $2\,\text{m}^2$ and the height quality varies from 0.5 m (flat regions) to 3–8 m (above 2000 m elevation) average error.[2] This data is converted to a triangulated surface model with level-of-detail support in a scene graph representation.[3] At each position on a regular grid on the surface (every 0.001° in N–S direction and 0.0015° in E–W direction, i.e. 111 m and 115 m respectively) and from 1.80 m above the ground,[4] we render a cube-map of the textureless DEM (face resolution 1024×1024) and extract the visible skyline by checking for the rendered sky color. Overall, we generate 3.5 million cubemaps. Similar to the query image, we extract contourlets, but this time with *absolute* viewing direction. We organize the contourlets in an index to allow for fast retrieval. In image search, inverted files have been used very successfully for this task [28]. We extend this idea by also taking into account the viewing direction, so that we can perform rough geometric verification on-the-fly. For each word we maintain a list that stores for every occurrence the panorama ID and the azimuth α_d of the contourlet.

11.2.3 Recognition and Verification

11.2.3.1 Baseline

The baseline for comparison is an approach borrowed from patch-based systems (e.g. [6, 23, 25]) based on the (potentially weighted) L1-norm between normalized visual word frequency vectors:

$$D^E(\tilde{\mathbf{q}}, \tilde{\mathbf{d}}) = \|\tilde{\mathbf{q}} - \tilde{\mathbf{d}}\|_1 = \sum_i |\tilde{q}_i - \tilde{d}_i| \quad \text{or} \quad D^{E_w}(\tilde{\mathbf{q}}, \tilde{\mathbf{d}}) = \sum_i w_i |\tilde{q}_i - \tilde{d}_i| \tag{11.5}$$

$$\text{with } \tilde{\mathbf{q}} = \frac{\mathbf{q}}{\|\mathbf{q}\|_1} \quad \text{and} \quad \tilde{\mathbf{d}} = \frac{\mathbf{d}}{\|\mathbf{d}\|_1} \tag{11.6}$$

where q_i and d_i is the number of times visual word i appears in the query or database image respectively, and $\tilde{q}_i$, $\tilde{d}_i$ are their normalized counterparts. w_i is the weight of visual word i (e.g. as obtained by the term frequency—inverse document frequency (tf-idf) scheme). This gives an ideal score of 0 when both images contain the same visual words at the same proportions, which means that the L1-norm favours images that are *equal* to the query.

[2] http://www.swisstopo.admin.ch/internet/swisstopo/en/home.

[3] http://openscenegraph.org.

[4] Synthetic experiments verified that taking the photo from 10 or 50 m above the ground does not degrade recognition besides very special cases like standing very close to a small wall.

Reference [23] suggested transforming the weighted L1-norm like this

$$D^{E_w}(\tilde{\mathbf{q}}, \tilde{\mathbf{d}}) = \sum_i w_i \tilde{q}_i + \sum_i w_i \tilde{d}_i - 2 \sum_{i \in Q} w_i \min(\tilde{q}_i, \tilde{d}_i) \tag{11.7}$$

in order to enable an efficient method for evaluating it by iterating only over the visual words present in the query image and updating only the scores of database images containing the given visual word.

11.2.3.2 "Contains"-Semantics

In our setting, we are comparing 10°–70° views to 360° panoramas, which means that we are facing a 5×–36× difference of magnitude. Therefore, it seems ill-advised to implement an "equals"-semantics, but rather one should use a "contains"-semantics. We modify the weighted L1-norm as follows:

$$D^C(\mathbf{q}, \mathbf{d}) = \sum_i w_i \max(q_i - d_i, 0) \tag{11.8}$$

The difference is that we are using the raw contour word frequencies, q_i and d_i without scaling and we replace the absolute value $|\cdot|$ by $\max(\cdot, 0)$. Therefore, one only penalizes contour words that occur in the query image, but not in the database image (or more often in the query image than in the database image). An ideal score of 0 is obtained by a database image that contains every contour word at least as often as the query image, plus any number of other contour words. If the proposed metric is transformed as follows, it can be evaluated just as efficiently as the baseline:

$$D^C(\mathbf{q}, \mathbf{d}) = \sum_{i \in Q} w_i q_i - \sum_{i \in Q} w_i \min(q_i, d_i) \tag{11.9}$$

This subtle change makes a huge difference, see Fig. 11.6a and Table 11.1: (B) versus (C). Note that this might also be applicable to other cases where a "contains"-semantics is desirable.

11.2.3.3 Location and Direction

We further refine retrieval by taking geometric information into account already during the voting stage. Earlier bag-of-words approaches accumulate evidence purely based on the frequency of visual words. Voting usually returns a shortlist of the top n candidates, which are reranked using geometric verification (typically using the number of geometric inliers). For performance reasons, n has to be chosen relatively small (e.g. $n = 50$). If the correct answer already fails to be in this shortlist, then no amount of reordering can bring it back. Instead, we check for geometric consistency

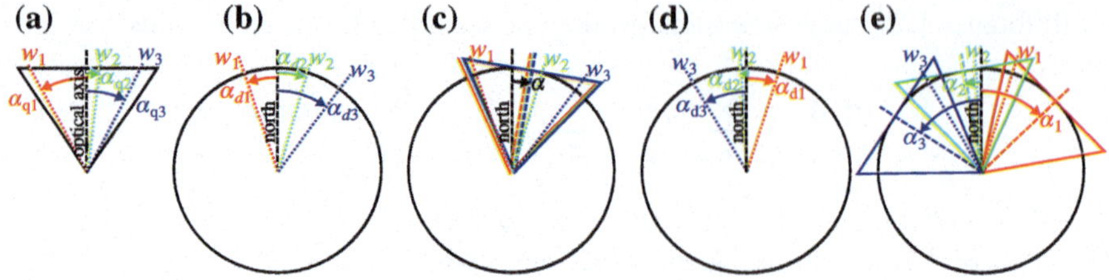

Fig. 11.4 Voting for a direction is illustrated using a simple example: We have a query image **a** with contour words w_i and associated angles α_{qi}. We consider a panorama **b** with contour words in the same *relative* positions α_{di} as the query image. Since the contour words appear in the same order, they all vote for the same viewing direction α **c**. In contrast, we consider a second panorama **d** with contour words in a different order. Even though the contour words occur in close proximity they each vote for a different direction α_i, so that none of the directions gets a high score **e**

already at the voting stage, so that fewer good candidates get lost prematurely. Not only does this increase the quality of the shortlist, it also provides an estimated viewing direction, which can be used as an initial guess for the full geometric verification. Since this enables a significant speedup, we can afford to use a longer shortlist, which further reduces the risk of missing the correct answer.

If the same contour word appears in the database image at angle α_d (relative to north) and in the query image at angle α_q (relative to the camera's optical axis), the camera's azimuth can be calculated as $\alpha = \alpha_d - \alpha_q$. Weighted votes are accumulated using soft binning and the most promising viewing direction(s) are passed on to full geometric verification. This way, panoramas containing the contour words in the right order get many votes for a single direction, ensuring a high score. For panoramas containing only the right mix of contour words, but in random order, the votes are divided among many different directions, so that none of them gets a good score (see Fig. 11.4). Note that this is different from merely dividing the panoramas into smaller sections and voting for these sections: Our approach effectively requires that the order of contour words in the panorama matches the order in the query image. As an additional benefit, we do not need to build the inverted file for any specific field-of-view of the query image.

11.2.3.4 Geometric Verification

We verify the top 1000 candidates with a geometric check. Verification consists in calculating an optimal alignment of the two visible skylines using iterative closest points (ICP). While we consider in the voting stage only one angle (azimuth), ICP determines a full 3D rotation. First, we sample all possible values for azimuth and keep the two other angles at zero. The most promising one is used as initialization for ICP. In the variants that already vote for a direction, we try only a few values around the highest ranked ones. The average alignment error is used as a score for re-ranking the candidates.

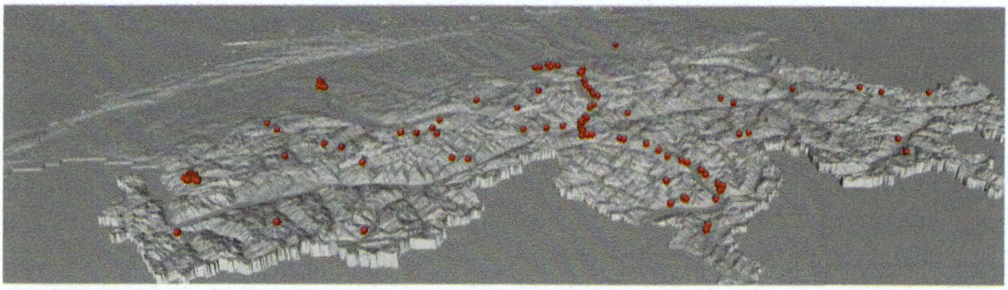

Fig. 11.5 Oblique view of Switzerland, spanning a total 40 000 km^2. *Spheres* indicate the query images' of the CH1 dataset at ground truth coordinates (size reflects 1 km tolerance radius). Source of DEM: Bundesamt für Landestopografie swisstopo (Art. 30 GeoIV): 5704 000 000

11.3 Evaluation

In this section, we evaluate the proposed algorithm on two real datasets consisting of a total of 1151 images. We further give a detailed evaluation of the algorithm under varying roll and tilt angles, and show that in cases where the focal length parameter is unknown it can effectively be sampled.

11.3.1 Query Set

In order to evaluate the approaches we assembled two datasets, which we refer to as CH1 and CH2. The CH1 dataset consists of 203 photographs obtained from different sources such as online photo collections and on site image capturing. The CH2 dataset consists of 948 images which were solely captured on site. For all of the photographs, we verified the GPS tag or location estimate by comparing the skyline to the surface model. For the majority of the images the information was consistent. For a few of them the position did not match the digital elevation model's view. This can be explained by a wrong cell phone GPS tag, due to bad/no GPS reception at the time the image was captured. For those cases, we use dense geometric verification (on each 111 × 115 m grid position up to a 10 km radius around the tagged position) to generate hypotheses for the correct GPS tag. We verify this by visual inspection and removed images in case of disagreement. The complete set of query images used is available at the project website.[5] The distribution of the CH1 dataset is drawn on to the DEM in Fig. 11.5. For all of the query images FoV information is available (e.g. from EXIF tag). However, we have verified experimentally that also in case of fully unknown focal length the system can be applied by sampling over this parameter, see Fig. 11.10 as example and subsection 11.3.6.

[5]http://cvg.ethz.ch/research/mountain-localization.

11.3.2 Query Image Segmentation

We used the CH1 query images which were already segmented in [2] as training set and apply our segmentation pipeline to the CH2 dataset. Out of the 948 image 60 % of the images were segmented fully automatically, while 30 % required little user interaction, mainly to correct for occluders such as trees or buildings. 10 % of the images required a more elaborate user interaction, to correct for snow fields, (often confused as sky), clouds hiding small parts of the mountain or for reflections appearing when taking pictures from inside a car, cable car or train.

11.3.3 Parameter Selection

The features need to be clearly smaller than the images' field-of-view, but wide enough to capture the geometry rather than just discretization noise. We consider descriptors of width $w = 10°$ and $w = 2.5°$. The number of sample points n should not be so small that it is uninformative (e.g. $n = 3$ would only distinguish concave/convex), but not much bigger than that otherwise it risks being overly specific, so we choose $n = 8$. The curve is smoothed by a Gaussian with $\sigma = \frac{w}{2n}$, i.e. half the distance between consecutive sample points. Descriptors are extracted every σ degrees.

Each dimension of the descriptor is quantized into k bins of width 0.375, the first and last bin extending to infinity. We chose k as a power of 2 that results in roughly 1 million contour words, i.e. $k = 8$. This maps each y_i to 3 bits, producing contour words that are 24 bit integers. Out of the 2^{24} potential contour words, only 300–500k (depending on w) remain after discarding words that occur too often (more than a million) or not at all.

11.3.4 Recognition Performance

The recognition pipeline using different voting schemes and varying descriptor sizes is evaluated on the *CH1* dataset published in [2], which consists of 203 images, see Table 11.1. All of the tested recognition pipelines return a ranked list of candidates. We evaluate them as follows: For every $n = 1, \ldots, 100$, we count the fraction of query images that have at least one correct answer among the top n candidates. We consider an answer correct if it is within 1 km of the ground truth position (see Fig. 11.5).

In Fig. 11.6a, we compare different voting schemes: (B) voting for location only, using the traditional approach with normalized visual word vectors and L1-norm ("equals"-semantics); (C) voting for location only, with our proposed metric ("contains"-semantics); (E) voting for location and direction simultaneously (i.e.

Table 11.1 Overview of tested recognition pipelines, evaluated on the CH1 dataset

	Voting scheme	Descriptor width	Dir. bin size	Geo. ver.	Top 1 correct (%)
(A)	random	N/A	N/A	No	0.008
(B)	"equals"	10°	N/A	No	9
(C)	"contains"	10°	N/A	No	31
(D)	loc.&dir.	10°	2°	No	45
(E)	loc.&dir.	10°	3°	No	43
(F)	loc.&dir.	10°	5°	No	46
(G)	loc.&dir.	10°	10°	No	42
(H)	loc.&dir.	10°	20°	No	38
(I)	loc.&dir.	2.5°	3°	No	28
(J)	loc.&dir.	10° and 2.5°	3°	No	62
(K)	loc.&dir.	10° and 2.5°	3°	Yes	88

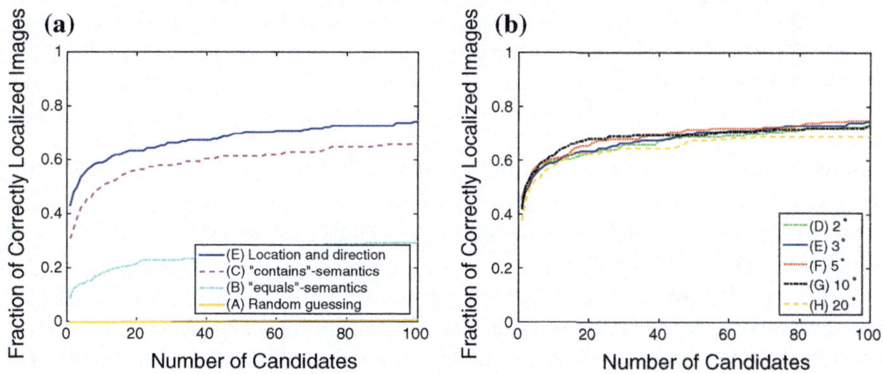

Fig. 11.6 Retrieval performance for different: **a** voting schemes, **b** bin sizes in direction voting. Evaluated on the CH1 dataset

taking order into account). All variants use 10° descriptors. For comparison, we also show (A) the probability of hitting a correct panorama by random guessing (the probability of a correct guess is extremely small, which shows that the tolerance of 1 km is not overly generous). Our proposed "contains"-semantics alone already outperforms the baseline ("equals"-semantics) by far, but voting for a direction is even better!

In Fig. 11.6b, we analyse how different bin sizes for direction voting affects results. (D)–(H) correspond to bin sizes of 2°, 3°, 5°, 10°, 20° respectively. While there are small differences, none of the settings outperforms all others consistently: Our method is quite insensitive over a large range of this parameter.

In Fig. 11.7a, we study the impact of different descriptor sizes: (E) only 10° descriptors; (I) only 2.5° descriptors; (J) both 10° and 2.5° descriptors combined.

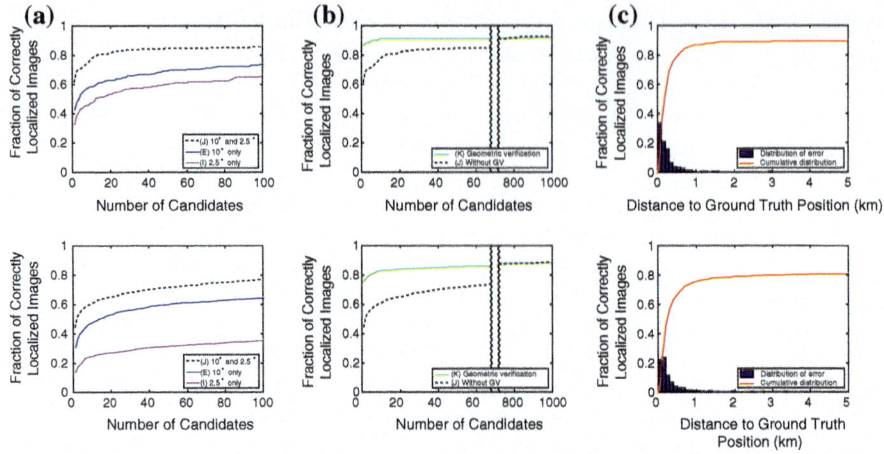

Fig. 11.7 Retrieval performance for CH1 (*top*) and CH2 (*bottom*) dataset: **a** Different descriptor sizes. **b** Retrieval performance before and after geometric verification. **c** Fraction of queries having at most a given distance to the ground truth position. Not shown: 21 images (9.9 %) with an error between 7 and 217 km from the CH1 dataset and 177 images (18.6 %) between 13 and 245 km from the CH2 dataset

All variants vote for location and direction simultaneously. While 10° descriptors outperforms 2.5° descriptors, the combination of both is better than either descriptor size alone. This demonstrates that different scales capture different information, which complement each other.

In Fig. 11.7b, we show the effect of geometric verification by aligning the full countours using ICP: (J) 10° and 2.5° descriptors voting for location and direction, without verification; (K) same as (J) but with geometric verification. We see that ICP-based reranking is quite effective at moving the best candidate(s) to the beginning of the short list: On the CH1 dataset the top ranked candidate is within a radius of 1 km with a probability of 88 %. On the CH2 dataset we achieve a recognition rate of 76 % for a maximum radius of 1 km. See Fig. 11.7c for other radii. In computer-assisted search scenarios, an operator would choose an image from a small list which would further increase the percentage of correctly recovered pictures. Besides that, from geometric verification we not only obtain an estimate for the viewing direction but the full camera orientation which can be used for augmented reality. Figures 11.8 and 11.9 show successful and unsuccessful examples.

11.3.5 Field-of-View

In Fig. 11.10 we illustrate the effect of inaccurate or unknown field-of-view (FoV). For one query image, we run the localization pipeline (K) assuming that the FoV is 11° and record the results. Then we run it again assuming that the FoV is 12°

Fig. 11.8 Sample results: *first and fourth column* are input images. *Second* and *fifth column* show the segmentations and *third* and *sixth column* show the query images augmented with the skyline, retrieved from the database. The images in the *last five rows* were segmented with help of user interaction

Fig. 11.9 Some incorrectly localized images. This usually happens to images with a relatively smooth skyline and only few distinctive features. The pipeline finds a contour that fits somewhat well, even if the location is completely off

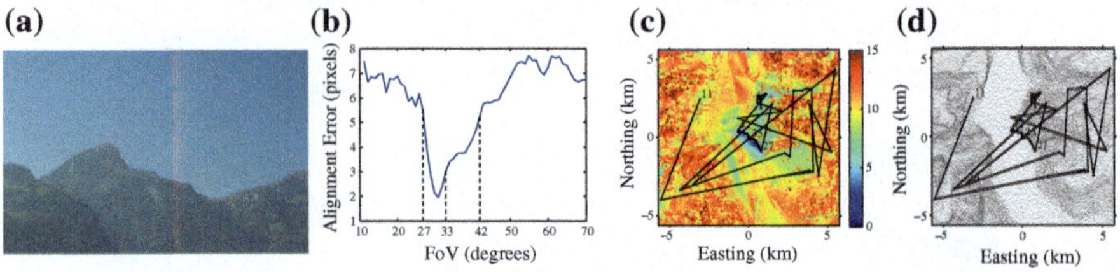

Fig. 11.10 **a** Query image. **b** Alignment error of the best position for a given FoV. *Dashed lines* indicate the limits of the stable region and the FoV from the image's EXIF tag. **c** Alignment error of the best FoV for a given position. For an animated version, see http://cvg.ethz.ch/research/mountain-localization. **d** Shaded terrain model. The overlaid *curve* in (**c**) and (**d**) starts from the best location assuming 11° FoV and continues to the best location assuming 12°, 13°, etc. Numbers next to the markers indicate corresponding FoV

etc., up to 70°. Figure 11.10 shows how the alignment error and estimated position depend on the assumed FoV.

In principle, it is possible to compensate a wrong FoV by moving forward or backward. This holds only approximately if the scene is not perfectly planar. In addition, the effect has hard limits because moving too far will cause objects to move in or out of view, changing the visible skyline. Between these limits, changing the FoV causes both the alignment error and the position to change smoothly. Outside of this stable range, the error is higher, fluctuates more and the position jumps around wildly.

This has two consequences: First, if the FoV obtained from the image's metadata is inaccurate it is usually not a disaster, the retrieved position will simply be slightly inaccurate as well, but not completely wrong. Second, if the FoV is completely unknown, one can get a rough estimate by choosing the minimum error and/or looking for a range where the retrieved position is most stable.

The field-of-view (FoV) extracted from the EXIF data may not always be 100 % accurate. This experiment studies the effects of a slight inaccuracy. We modify the FoV obtained from the EXIF by ±5 % and plot it against the recognition rate obtained over the entire query set CH1. We observe in Fig. 11.11 that even if the values are off by ±5 %, we still obtain a recognition rate of 70–80 %.

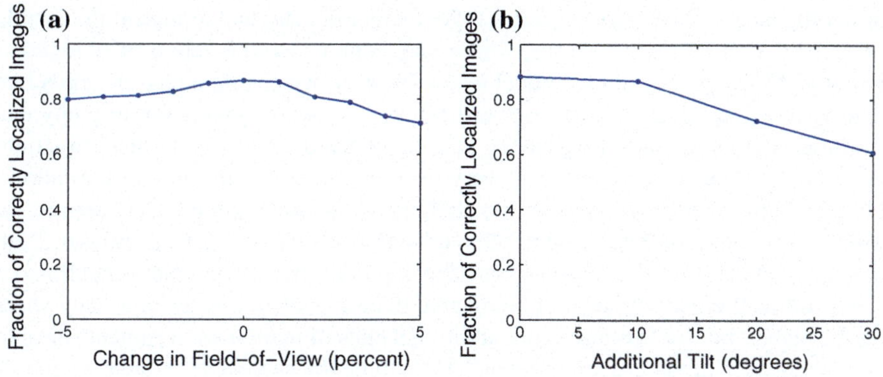

Fig. 11.11 **a** Recognition rate under varying FoV. **b** Recognition rate under varying tilt angle

11.3.6 Tilt Angle

Our algorithm assumes that landscape images usually are not subject to extreme tilt angles. In the final experiment evaluated in Fig. 11.11b, we virtually rotate the extracted skyline of the query images by various angles in order to simulate camera tilt and observe how recognition performance is affected. As shown in Fig. 11.11b with 30° tilt we still obtain a recognition rate of 60 % on the CH1 dataset. This is a large tilt angle, considering that the skyline is usually straight in front of the camera and not above or below it.

11.3.7 Runtime

We implemented the algorithm partly in C/C++ and partly in Matlab. The segmentation runs at interactive frame rate and gives direct visual feedback to the operator, given the unary potential of our segmentation framework. Given the skyline it takes 10 s to find the camera's position and rotation in an area of 40 000 km^2 per image. Exhaustively computing an optimal alignment between the query image and each of the 3.5 M panoramas would take on the order of several days. For comparison, the authors of [3] use a GPU implementation and report 2 min computation time to determine the rotation only, assuming the camera position is already known.

11.4 Conclusion and Future Work

We have presented a system for large-scale location recognition based on digital elevation models. This is very valuable for geo-localization of pictures when no GPS information is available (for virtually all video or DSLR cameras, archive pictures, in intelligence and military scenarios). We extract the sky and represent the visible skyline by a set of contour words, where each contour word is represented together with its offset angle from the optical axis. This way, we can do a bag-of-words like approach with integrated geometric verification, i.e. we are looking for the panorama (portion) that has a similar frequency of contour words with a consistent direction. We show that our representation is very discriminative and the full system allows for excellent recognition rates on two challenging datasets. On the CH1 dataset we achieve a recognition rate close to 90 % and almost 80 % on the CH2 dataset. Both datasets include different seasons, landscapes and altitudes. We believe that this is a step towards the ultimate goal of being able to geo-localize images taken anywhere on the planet, but for this also other additional cues of natural environments have to be combined with the given approach. This will be the subject of future research.

Acknowledgments This work has been supported through SNF grant 127224 by the Swiss National Science Foundation. We also thank Simon Wenner for his help to render the DEMs and Hiroto Nagayoshi for providing the CH2 dataset.

References

1. Baatz G, Köser K, Chen D, Grzeszczuk R, Pollefeys M (2012) Leveraging 3d city models for rotation invariant place-of-interest recognition. IJCV, special issue on mobile vision 96
2. Baatz G, Saurer O, Köser K, Pollefeys M (2012) Large scale visual geo-localization of images in mountainous terrain. In: Proceedings of the 12th european conference on computer vision - volume part II, ECCV'12, pp 517–530, Berlin, Heidelberg, 2012. Springer
3. Baboud L, Cadík M, Eisemann E, Seidel H.-P (2011) Automatic photo-to-terrain alignment for the annotation of mountain pictures. In: CVPR, pp 41–48
4. Blake A, Rother C, Brown M, Perez P, Torr P (2004) Interactive image segmentation using an adaptive gmmrf model. In: ECCV, pp 428–441
5. Brown M, Lowe DG (2007) Automatic panoramic image stitching using invariant features. Int J Comput Vis 74:59–73
6. Chen D, Baatz G, Köser K, Tsai S, Vedantham R, Pylvanainen T, Roimela K, Chen X, Bach J, Pollefeys M, Girod B, Grzeszczuk R (2011) City-scale landmark identification on mobile devices. In: CVPR
7. Comaniciu D, Meer P, Member S (2002) Mean shift: a robust approach toward feature space analysis. IEEE Trans Pattern Anal Mach Intell 24:603–619
8. Cozman F (1997) Decision making based on convex sets of probability distributions: quasi-bayesian networks and outdoor visual position estimation. Ph.D. thesis, Robotics Institute, Carnegie Mellon University, Pittsburgh, PA, December 1997
9. Cozman F, Krotkov E (1996) Position estimation from outdoor visual landmarks for teleoperation of lunar rovers. In: WACV'96, pp 156–161
10. Friedman J, Hastie T, Tibshirani R (2000) Additive logistic regression: a statistical view of boosting. Ann Stat
11. Hays J, Efros AA (2008) im2gps: estimating geographic information from a single image. In: Proceedings of CVPR 2008
12. Hussain SU, Triggs B (2012) Visual recognition using local quantized patterns. In: European conference on computer vision
13. Kolmogorov V, Boykov Y (2005) What metrics can be approximated by geo-cuts, or global optimization of length/area and flux. In: Proceedings of ICCV 2005, pp 564–571, Washington, DC, USA, 2005. IEEE Computer Society

14. Ladicky L, Russell C, Kohli P, Torr P (2014) Associative hierarchical random fields. IEEE Trans Pattern Anal Mach Intell, 36(6):1056–1077
15. Ladicky L, Zeisl B, Pollefeys M (2014) Discriminatively trained dense surface normal estimation. In: European conference on computer vision
16. Lalonde JF, Narasimhan SG, Efros AA (2010) What do the sun and the sky tell us about the camera? Int J Comput Vis 88(1):24–51
17. Li Y, Snavely N, Huttenlocher DP (2010) Location recognition using prioritized feature matching. In: Proceedings of the 11th european conference on computer vision: part II, ECCV'10, pp 791–804, Berlin, Heidelberg, 2010. Springer
18. Lie WN, Lin TCI, Lin TC, Hung KS (2005) A robust dynamic programming algorithm to extract skyline in images for navigation. Pattern Recog Lett 26(2):221–230
19. Lowe DG (2004) Distinctive image features from scale-invariant keypoints. Int J Comput Vis 60(2):91–110
20. Malik J, Belongie S, Leung T, Shi J (2001) Contour and texture analysis for image segmentation. Int J Comput Vis 43(1):7–27
21. Manay S, Cremers D, Hong BW, Yezzi A, Soatto S (2006) Integral invariants for shape matching. Pattern Anal Mach Intell
22. Naval PC, Mukunoki M, Minoh M, Ikeda K (1997) Estimating camera position and orientation from geographical map and mountain image. In: 38th pattern sensing group research meeting, society of instrument and control engineers, pp 9–16
23. Nistér D, Stewénius H (2006) Scalable recognition with a vocabulary tree. In: CVPR (2), pp 2161–2168
24. Ramalingam S, Bouaziz S, Sturm P, Brand M (2010) Skyline2gps: localization in urban canyons using omni-skylines. In: IROS 2010, pp 3816–3823
25. Schindler G, Brown M, Szeliski R (2007) City-scale location recognition. Comput Vis Pattern Recog CVPR '07, pp 1–7
26. Shechtman E, Irani M (2007) Matching local self-similarities across images and videos. In: Conference on computer vision and pattern recognition
27. Shotton J, Winn J, Rother C, Criminisi A (2006) Textonboost: joint appearance, shape and context modeling for multi-class object. In: IN ECCV, pp 1–15
28. Sivic J, Zisserman A (2003) Video Google: a text retrieval approach to object matching in videos. In: Proceedings ICCV 2003, vol 2, pp 1470–1477
29. Stein F, Medioni G (1995) Map-based localization using the panoramic horizon. IEEE Trans Robot Autom 11(6):892–896
30. Talluri R, Aggarwal J (1992) Position estimation for an autonomous mobile robot in an outdoor environment. Trans Robot Autom 8(5):573–584
31. Thompson WB, Henderson TC, Colvin TL, Dick LB, Valiquette CM (1993) Vision-based localization. In: Image understanding workshop, pp 491–498
32. Vasilevskiy A, Siddiqi K (2002) Flux maximizing geometric flows. IEEE Trans Pattern Anal Mach Intell 24:1565–1578
33. Woo J, Son K, Li T, Kim GS, Kweon IS (2007) Vision-based uav navigation in mountain area. In: MVA, pp 236–239
34. Yang M, Kpalma K, Ronsin J (2008) A survey of shape feature extraction techniques. In: Yin PY (ed) Pattern recognition, pp 43–90. IN-TECH, November 2008

Chapter 12
Adaptive Rendering for Large-Scale Skyline Characterization and Matching

Jiejie Zhu, Mayank Bansal, Nick Vander Valk and Hui Cheng

Abstract We propose an adaptive rendering approach for large-scale skyline characterization and matching with applications to be automated geo-tagging of photos and images. Given an image, our system automatically extracts the skyline and then matches it to a database of reference skylines generated from rendered images using digital elevation data (DEM). The sampling density of these rendering locations determines both the accuracy and the speed of skyline matching. The proposed approach successfully combines global planning and local greedy search strategies to select new rendering locations incrementally. We report quantitative and qualitative results from synthesized and real experiments, where we achieve a computational speedup of around 4X.

12.1 Introduction

Skylines, especially in mountainous areas, provide robust and often unique features to characterize an area. By looking at the skyline in a photo, such as the half dome in Yosemite, one can recognize the area where the photo was taken. However, to recognize where a photo is taken over a large area of tens or hundreds thousands of square kilometers is very difficult. Based on this observation and previous work [1, 2] on skyline analysis, we developed a large-scale skyline characterization and matching system for automated geo-tagging of photos and images.

J. Zhu (✉) · M. Bansal · N. Vander Valk · H. Cheng
Vision Technologies Lab, SRI International, Princeton, NJ 08540, USA
e-mail: jiejie.zhu@sri.com

M. Bansal
e-mail: mayank.bansal@sri.com

N. Vander Valk
e-mail: nicholas.vandervalk@sri.com

H. Cheng
e-mail: hui.cheng@sri.com

A.R. Zamir et al. (eds.), *Large-Scale Visual Geo-Localization*,
Advances in Computer Vision and Pattern Recognition,
DOI 10.1007/978-3-319-25781-5_12

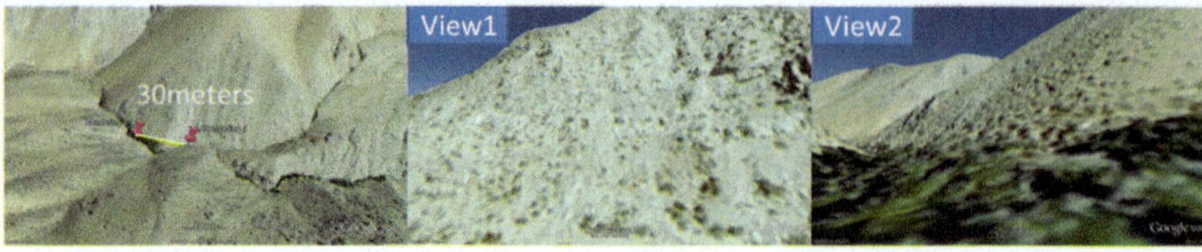

Fig. 12.1 Dramatic skyline change caused by occlusions in complex mountainous terrain

Given an image, our system automatically extracts the skyline and then matches it to a database of reference skylines generated from digital elevation data (DEM). The reference skylines are generated through rendering using DEM at locations over the entire area of interest. The selection of these rendering locations determines both the accuracy and the speed of the skyline matching. In addition, it determines the time needed to build the reference skyline database.

We can use a grid on the ground-plane as rendering locations—in this case the distance between two adjacent grid locations is the sampling distance. The larger the sampling distance is, the faster the rendering and the matching algorithms run, but less accurate the skyline matching is. The smaller the sampling distance is, the more precise the matching is, but the slower the rendering and the matching algorithms run. For example, skyline rendering and extraction over a 10,000 km^2 using 50–100 m sampling distance can take months to complete. On the other hand, if the sampling distance is large, photos taken close to a mountain or inside a mountain, such as those taken on a hiking trail or a road, passing through a mountain cannot be found. Figure 12.1 depicts such an example where two views that are separated by only 30 m have very different skylines.

In this chapter, we propose an adaptive rendering approach to select rendering locations. By modeling both the rendering process and the skyline matching process, our system can compute the set of optimal rendering locations based on the DEM and a predefined matching threshold. The viewpoint selection is optimal in the sense that for a given matching accuracy, it requires far fewer numbers of renderings than if the rendering locations were uniformly sampled on a dense grid. In the example of Fig. 12.1 above, a good viewpoint selection algorithm should automatically render densely when close to the mountain and coarsely when far away.

12.2 Related Work

In recent literature, there has been increasing interest in geo-localization of ground-level imagery using visible skylines by matching them to a database of known skyline shapes. For urban geo-localization, Ramalingam et al. [2, 3] focused on matching omni-skylines from an upward facing camera to skyline renderings generated on-the-fly from 3D building models of the scene. Our focus on this chapter is on geo-localization in natural terrain—in this case, the problem is much harder since we have to match to a specific viewpoint (i.e. the query camera is not upward facing), and we cannot pre-render on sparse road networks like in the urban case. It, therefore, becomes important to devise a scheme that will render at the fewest number of viewpoints without reducing the localization accuracy achievable.

Baboud et al. [1] recently addressed the problem of automatic photo-to-terrain alignment with a goal of annotating mountain pictures. In their work, high resolution elevation maps are rendered to create synthesized panoramic views, which are compared to the mountain picture using a robust edge matching algorithm. However, their approach assumes that the GPS location and FOV of the query picture are known and then they solve for the unknown camera pose relative to the terrain. In contrast, we would like to address the problem of localization of the query picture itself by matching to a set of terrain renderings obtained from sampled viewpoints. To achieve this, we have to either sample these viewpoints very densely everywhere, or we can adopt an adaptive rendering strategy to minimize the search required on the matching stage. In this chapter, we focus on such an approach.

Russell et al. [4] address the problem of automatically aligning historical architectural paintings with 3D models obtained from modern photographs. A key step in their approach is the "view-sensitive retrieval" that aims to find a 3D viewpoint that is sufficiently close to the painting viewpoint. To achieve this, they sample a large (dense) pool of virtual viewpoints around the 3D model and then use a matching procedure to retrieve a small set of nearby matching candidate viewpoints. In this chapter, we propose an approach to allow a non-dense rendering of the 3D scene for query localization in the context of natural terrain where dense sampling quickly becomes a computational bottleneck.

Stein [5] used fixed spacing to render a DEM map corresponds to an area of approximately 4×3.5 km^2. Each spacing corresponding to a size of 5×5 m^2 within the map. Outside the map, the spacing is 75×75 m^2. Similar to [5], Baatz also used fixed spacing to render a DEM map corresponds to a large area approximately 40,000 km^2. The spacing in this approach is 111 and 115 m respectively. This generates 3.5 million renders.

There has been some work on the literature on efficient rendering of large terrains. However, the majority of this work has focused on the graphics aspect including adaptive means to enable faster Google-Earth like renderings served to an end user. For example, Lerbour et al. [6, 7] describe a generic data structure to adaptively serve data to the client rendering system and to improve the database loading and rendering speeds independent of the database size. Similarly, a hardware accelerated terrain rendering approach is outlined in [8]. In this chapter, we assume that a terrain rendering algorithm is available to us as a black-box and we can use it to render the terrain at any specified location and viewpoint.

12.3 Skyline Rendering and Matching

In this section, we briefly introduce our skyline rendering and matching algorithms. In skyline rendering, the system takes an area of geo-localized DEM data as input. A ground-level camera location inside this area is specified along with its intrinsic and extrinsic parameters. The system renders the DEM into a depth image, which is used to extract a skyline corresponding to the specified viewpoint e.g. Fig. 12.2.

Given all the rendered skylines, the system extracts features from each of them and saves them in a database for future matching purposes. In feature extraction, a skyline is first approximated by polylines and then the endpoints of the line-segments composing these polylines are used as feature points.

Given a query image like in Fig. 12.3, we use a skyline extraction algorithm [2] to extract the skyline. The polyline approximation-based algorithm described above is then used to extract the query skyline features. For each extracted key feature from the query skyline, the matching process finds a key feature from the rendered skylines that best matches it. The matching score between a keypoint pair is computed as the Chamfer distance between the local skylines centered at these keypoints. This establishes a correspondence between the keypoints of the two skylines following which RANSAC is used to find inliers corresponding to an affine transformation

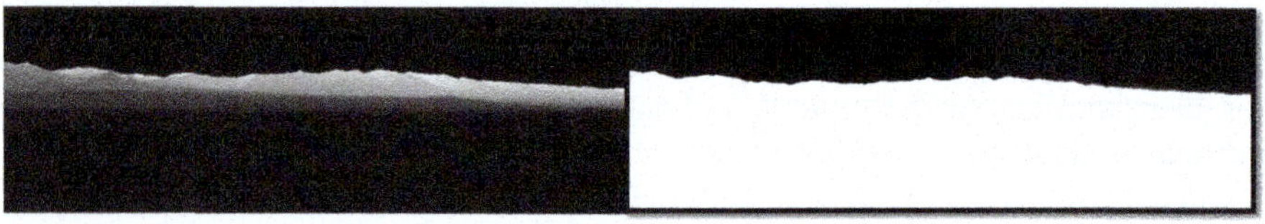

Fig. 12.2 Example of a rendered depth image and its extracted skyline. To extract the skyline, the system looks for the depth image from *top* to *bottom* in each column, and labels as skyline pixel the first pixel with a non-infinite depth

Fig. 12.3 Example of a query image and its skyline. The contours of the trees are false skyline pixels automatically computed from the skyline extraction model—the RANSAC-based matching process labels them correctly as outliers

between the two skylines. The overall matching score between the two skylines is computed as the Euclidean distance of the matched keypoints after warping the query skyline using the computed affine transformation.

12.4 Problem Formulation

We assume that we start the adaptive rendering algorithm from a set of prerendered viewpoints which are on a uniform but coarsely sampled grid, e.g. at 1 km spacing. The goal of adaptive rendering is to automatically predict optimal viewpoints c_k (with spacing finer than 1 km) at which the rendered skyline looks sufficiently different from existing renderings of the same 3D feature (mountain). To achieve this goal, we approach the viewpoint selection process as an incremental algorithm that adds new viewpoints to the set of existing renderings, *but without rendering the new viewpoints first.* Around each already rendered skyline from a viewpoint c_0, we can explicitly compute a "tolerance area" within which the skyline projection from another candidate viewpoint c_k looks similar to c_0. This similarity is defined by an image projection distance between the skylines at c_k and c_0. Thus, for a given threshold on the similarity metric, we can estimate the tolerance area around each existing rendering. Intuitively, the tolerance area specifies the extent within which no additional rendering is required. Thus, for an existing viewpoint inside a complex terrain, we can expect to obtain a much smaller tolerance area than for a simple flat terrain. The tolerance area computation algorithm is described in Sect. 12.4.2. Given the tolerance areas (and their shape) for each of the existing renderings in our set, in Sect. 12.4.3, we propose and compare several novel planning strategies to determine the next-best viewpoint that avoids overlap with the existing tolerance areas.

12.4.1 Camera Configuration

We assume a distortion-free ideal pin-hole camera model for the rendering camera with square pixels and a camera center coincident with the image center. This leads to an ideal camera intrinsic matrix. For our experiments, we cover the 360° view surrounding each candidate rendering location using four camera viewpoints $v_1, \ldots, v_4$, each with a horizontal field-of-view of $\theta = 90°$ and image resolution of $w = h = 640$. The optical axis for each viewpoint v_i can be described by a predefined rotation matrix R_i which fixes a single look-at direction for this viewpoint independent of the rendering location.

In our framework, we represent the 3D points corresponding to the skylines (mountain silhouettes) in world-coordinates by variable $X_i \in \Re^3$. The skyline projection at any viewpoint c_k is then given by:

$$\mathbf{P}_k X_i = \mathbf{K}[\mathbf{R}|\mathbf{t}_k] X_i$$

where $\mathbf{P}_k = \mathbf{K}[\mathbf{R}|\mathbf{t}_k]$ is the camera projection matrix for the camera located at displacement $\mathbf{t}_k$ w.r.t the world coordinate system origin. The rotation $\mathbf{R}$ is one of $\mathbf{r}_1, \ldots, \mathbf{r}_4$ depending on the viewpoint's look-at direction.

In the following, we will follow the convention that the world coordinate system is defined with XZ as the ground-plane and the Y-axis pointing upwards.

12.4.2 Tolerance Area Computation

The squared difference between projected skylines I_0 and I_k (at locations c_0 and c_k respectively) can serve as a simple error metric for the estimation of tolerance area. However, the skyline matching algorithm (Sect. 12.3) accounts for any small distortions of the skyline by an affine transformation model. It therefore makes sense to measure the projection error between skylines visible at two viewpoints after allowing for an affine transformation. Figure 12.4 illustrates an example of two projections of a synthetic skyline where the distance between the original skylines does not correctly reflect the difference between their shape. Warping one of the skylines using an affine transformation, however, leads to a much more accurate error metric.

Estimating affine transformation between two images has been studied extensively. Most techniques rely on first applying RANSAC [9] to determine the corresponding pixels and then to recover their geometrical relationships by rejecting outliers. Here, we can apply a similar approach but the computation cost is high considering the huge number of such pair comparisons required in the system.

We propose an analytical approximation approach to compute the affine transformation between the two cameras given their projection matrix. It has merits of

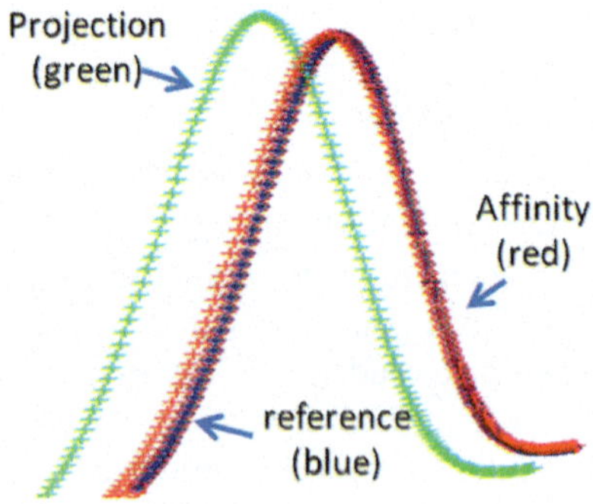

Fig. 12.4 Example of skyline projection improvement using an affine transformation model from a synthetic experiment. The *green* and *blue curves* are two candidate skylines; the *red curve* is the *green curve* after warping by an affine transformation that best aligns the candidate skylines. The direct Euclidean distance between the *green* and *blue* skylines is around 26 pixels. The corresponding distance between the *red* and *blue curves* is 3.6, which is a more accurate estimate of the difference between the skyline shapes than the direct distance

requiring much less computation (since it is analytical) while preserving the accuracy compared with the traditional correspondence-based estimation method.
Affine Transformation Given two projection matrices $\mathbf{P}_1 = \mathbf{K}_1[\mathbf{R}_1|\mathbf{t}_1]$ and $\mathbf{P}_2 = \mathbf{K}_2[\mathbf{R}_2|\mathbf{t}_2]$, without loss of generality, we can assume the first camera to be in a canonical form such that $\mathbf{R}_1 = I$, $\mathbf{t}_1 = 0$. At each location, there are four viewing directions. Instead of using four renders, we apply adaptive rendering for each viewing direction separately. This gives us $\mathbf{R}_2 = \mathbf{R}_1$. We can also assume $\mathbf{K}_2 = \mathbf{K}_1 = \mathbf{K}$. If a 3D point X is visible to both cameras, its projection can be expressed by $\mathbf{u}_1 = \mathbf{K}X$ and $\mathbf{u}_2 = \mathbf{K}X + \mathbf{K}\mathbf{t}$, where $\mathbf{t} = \mathbf{t}_2 - \mathbf{t}_1$. By substituting X using $\mathbf{K}^{-1}\mathbf{u}_1$, we have

$$\mathbf{u}_2 = \mathbf{u}_1 + \mathbf{K}\mathbf{t} \tag{12.1}$$

By representing $\mathbf{u}_1$ in homogeneous coordinates $\mathbf{u}'_1$, $\mathbf{K} = [\mathbf{k}_1, \mathbf{k}_2, \mathbf{k}_3]'$, we can write the above equation in a matrix formulation,

$$\mathbf{u}_2 = \begin{pmatrix} \mathbf{k}_3X & 0 & \mathbf{k}_1\mathbf{t} \\ 0 & \mathbf{k}_3X & \mathbf{k}_2\mathbf{t} \\ 0 & 0 & \mathbf{k}_3X + \mathbf{k}_3\mathbf{t} \end{pmatrix} \begin{pmatrix} \frac{\mathbf{k}_1X}{\mathbf{k}_3X} \\ \frac{\mathbf{k}_2X}{\mathbf{k}_3X} \\ 1 \end{pmatrix} \tag{12.2}$$

If we represent $\mathbf{u}_2$ in homogeneous coordinates $\mathbf{u}'_2$, we have

$$\mathbf{u}'_2 = \frac{\mathbf{u}_2}{\mathbf{k}_3X + \mathbf{k}_3T} = \mathbf{A}\mathbf{u}'_1 \tag{12.3}$$

$$= \begin{pmatrix} \frac{\mathbf{k}_3X}{\mathbf{k}_3X+\mathbf{k}_3T} & 0 & \frac{\mathbf{k}_1\mathbf{t}}{\mathbf{k}_3X+\mathbf{k}_3\mathbf{t}} \\ 0 & \frac{\mathbf{k}_3X}{\mathbf{k}_3X+\mathbf{k}_3\mathbf{t}} & \frac{\mathbf{k}_2\mathbf{t}}{\mathbf{k}_3X+k_3\mathbf{t}} \\ 0 & 0 & 1 \end{pmatrix} \mathbf{u}'_1 \tag{12.4}$$

In the above equation, we should note that $\mathbf{k}_3X$ is the displacement between the 3D point X and the reference ($\mathbf{P}_1$) camera center, $\mathbf{k}_3\mathbf{t}$ is the displacement between the 3D point X and the candidate ($\mathbf{P}_2$) camera center. $\mathbf{k}_3X$ will be different for each 3D point located on the skyline if their Z-coordinates are different. Building a per-point affine transformation generates the exact transformation between $\mathbf{u}_1$ and $\mathbf{u}_2$, but it is computationally expensive given the large number of 3D points on skylines. Instead, our system chooses a z-value that is representative of the skyline such as the mean or median distance of the skyline points.

Given the above formulation, we now outline our tolerance area computation algorithm: We sample a 10×10 neighborhood around the reference location c_0. At each sampled location, we project the skyline to $\mathbf{P}_1$ by computing $\mathbf{u}_1$. We then compute $\mathbf{u}_2$ using Eq. 12.4. If the distance between $\mathbf{u}_1 - \mathbf{u}_2$ is within a given threshold, the sampled location is included in the tolerance area (corresponding to the reference location c_0), otherwise it is excluded.

12.4.3 Optimal Viewpoint Planning and Selection

Without loss of generality, we assume that we have already computed the tolerance areas for the four corners of a 1 km $\times$ 1 km on the terrain using the algorithm in the previous section. The next step in our pipeline is to select a new location at which the renderer should render so as to cover maximum uncovered ground. In the following, we describe five different strategies to plan the next viewpoint. In the experiments section, we will discuss how each of these strategies performed on our simulation data followed by results of the best performing strategy on real data.

Random Optimistic Viewpoint (ROV). ROV randomly chooses a number of candidate viewpoint locations from the uncovered area. The number of candidate viewpoint locations is computed as the number of connected components in the uncovered area. Their tolerance area are computed and ranked based on their overlap with the uncovered area. ROV selects the location with the highest coverage and performs rendering at the selected location. This process will iterate until the coverage reaches an acceptable number, e.g., 90 or 98 % used in the experiments.

Approximated Random Optimistic Viewpoint (AROV). The tolerance area computation in ROV is expensive since it requires projecting each 3D skyline point at multiple candidate locations (please refer to the computation cost in the last row of Table 12.1). To make this process more efficient, AROV interpolates a candidate viewpoint's tolerance area bilinearly from its nearest rendered viewpoints whose tolerance areas are already computed.

Hierarchy Optimistic Viewpoint (HOV). Different from ROV and AROV, HOV searches optimal viewpoint using a coarse to fine process. At each level, AROV is used to select the next-best viewpoint; when coverage is satisfied at a level, HOV will move to the next finer level.

Shape-assisted Optimistic Viewpoint (SOV). SOV investigates how the shape of the tolerance area may be used in assisting optimal viewpoint selection. From ROV, AROV, and HOV, we noticed that most of tolerance areas can be best described using 2D ellipses. Thus, instead of randomly picking the candidate locations, in SOV we pack them along the minor axes of the ellipse-like tolerance areas so as to achieve a tighter packing over the uncovered area.

Line Planning Optimistic Viewpoint (LPOV). SOV uses independent tolerance areas to select the optimal viewpoint and this strategy may not be optimal for covering the whole area. In order to include more global information in selecting an optimal viewpoint, LPOV pre-locates optimal viewpoints on each line joining a pair of sampled viewpoints. The location of a candidate viewpoint is predicted as a convex combination of the locations of the endpoints; the weights in this convex combination are proportional to the size of each end viewpoint's tolerance area. LPOV encourages sparse optimal viewpoint selection in flat areas since the tolerance areas of the rendered locations are larger in comparison with the rendered viewpoints in clutter areas, such as inside mountains. This line-wise planning and local greedy combined approach contributes to less number of rendered viewpoints with a fast coverage.

Table 12.1 Simulation results for viewpoint planning methods using criteria (A)/(B)

Parameters	ROV	AROV	HOV	SOV	LPOV
Non-Affine ($p = 1, t = 15,$ $c = 90\,\%, r = 0$)	**240**/1545	286/1636	324/**1210**	275/1941	325/1515
Affine ($p = 1, t = 3, c = 98\,\%,$ $r = 0$)	81/6664	**80**/5824	98/6723	100/10043	86/**5489**
Affine ($p = 2, t = 3, c = 98\,\%,$ $r = 30$)	239/9028	217/6991	252/7226	245/9148	**191/5943**
Computation cost (rounded) (min)	45	10	5	30	10

p number of peaks on the skyline; t reprojection error threshold in pixels
r skyline rotation angle; $r = 0$ represents a skyline that is parallel to the camera plane
c coverage percentage at which the adaptive rendering process is terminated

12.5 Experimental Results

This section exhibits our quantitative and qualitative experimental results on (i) a synthetic dataset generated by sampling skylines from a parametric model, and (ii) a DEM dataset covering around 50 km^2 on a mountainous terrain.

12.5.1 *Simulation Results*

We evaluated the five viewpoint selection techniques described in Sect. 12.4.3 on a synthetic dataset consisting of an area with 100×100 potential (dense) viewpoint locations. All the experiments start with four rendered viewpoints at the corners initially. We evaluate the performance of the algorithms using two criteria:

(A) Total number of viewpoints selected for rendering and,
(B) Size of overlapped area—computed as the number of times the same viewpoint location is included in any of the tolerance areas.

The goal of a good adaptive rendering algorithm is to cover as much area as possible using minimal number of rendered viewpoints where the coverage is defined as the union of tolerance areas from all rendered viewpoints. Thus, our objective is to achieve smaller numbers for both criteria (A) and (B).

Table 12.1 reports the results of all the methods for the 100×100 grid using the convention x/y where x represents the value of criterion (A) and y the value of criterion (B). We include results from experiments with skylines of different complexities with the reprojection error evaluated with and without the affine transformation model proposed in this chapter.

We can conclude the following from the simulation results:

1. Overall, the affine transformation model requires fewer number of renderings, but produces more overlaps, as shown by a comparison between row 1 and rows 2, 3.
2. Overall, the number of selected viewpoints is proportional to the complexity of the skylines as observed by comparing row 2 and row 3.
3. SOV consistently produces the largest size of overlaps which shows that locally optimal algorithms do not achieve good results.
4. ROV requires the highest computation cost. Other algorithms require less computational since the tolerance area is approximated using nearest neighbors. Among them, HOV requires the least computation because a fixed pattern of optimal viewpoint selection is used.
5. LPOV obtains the best performance in the test using skylines with two peaks. This suggests that a planning strategy with a local greedy objective may give good results.

12.5.2 Experiments Using DEM Data

For experiments with real data, we used an area approximately 5 km × 10 km in size in a mountainous terrain. We select the best performing viewpoint selection method from the simulation (LPOV) for experiments with real data and compare its performance with a hierarchical uniform approach. To characterize the performance of the generated renderings using either approach, we use a set of query images (with known ground-truth geo-location information) shown in Fig. 12.5 and match each of them to the generated renderings. We selected three groups of query images: close to the mountain (Fig. 12.5, column-1), medium range from the mountain (Fig. 12.5, column-2), and far from the mountain (Fig. 12.5, column-3). We measure the distance between the location of the best matching rendering and the known ground-truth location of the query to showcase the improvement in geo-location using renderings from our adaptive algorithm versus renderings on a uniform grid at four different levels of resolution (1024, 512, 256, and 128 m).

Fig. 12.5 Six query images used in all the experiments, along with their geo-locations on the overhead view

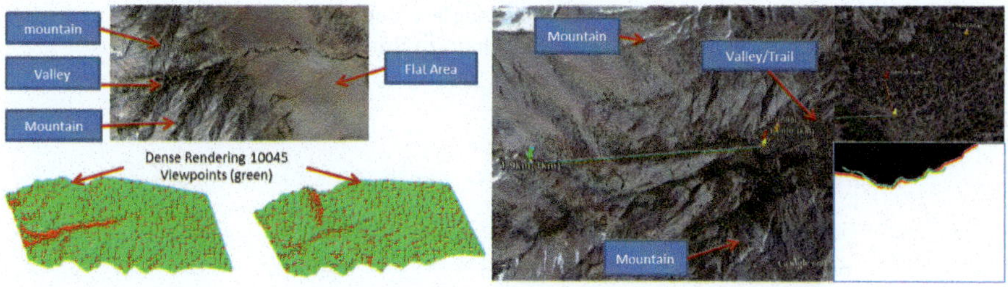

Fig. 12.6 *Left* Example of viewpoints selected from the proposed adaptive rendering approach for heading 0° and 90°. Looking at the terrain area shown on the *top*, we can see that a large number of viewpoints are selected in the valleys and areas with clutter while a fewer number of viewpoints are selected by our algorithm in flat areas of the terrain. It is also interesting to see that different heading directions will result in different adaptive rendering results given the terrain structure. *Right* Result of query 3976. In uniform viewpoint sampling, a 1024 m spacing generates best geo-location around 1900 m away, while a 128 m spacing gives the best geo-localization around 64 m away. Adaptive rendering has an error of 174 m, but with a 400 % system performance improvement. The inlay on the *bottom right* illustrates the skyline matching result for this query. The *green curve* indicates the skyline in the query image. The *red curve* shows the skyline in the rendered image

Figure 12.6 shows an example of the viewpoints selected from the proposed adaptive rendering approach with two headings at 0° and 90° (similar results are obtained from the other two headings). We can see that dense viewpoint sampling is required in cluttered terrain areas such as valley and trails. Incidentally, these areas are critical to skyline matching because occlusions introduce a large number of mismatches.

Skyline matching tests are performed for query images shown in Fig. 12.5. One of the geo-localization results is shown in Fig. 12.6. The numerical results are reported in Table 12.2.

Figure 12.7 highlights the important fact that using the proposed adaptive rendering algorithm, one can achieve a given matching accuracy using only a small fraction of the number of renderings required by a uniform fixed renderer.

Table 12.2 Query geo-localization accuracy results with and without adaptive rendering

	Rendering resolution/Number of renderings				
	1024 m/182	512 m/665	256 m/2662	128 m/10648	Adaptive/2800
Close to mountain (m)	1906	800	379	64	127
Medium range (m)	1609	703	402	102	98
Far from mountain (m)	1247	604	453	202	125

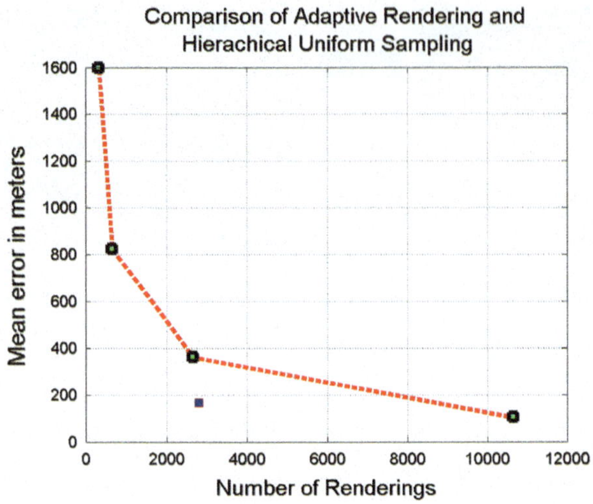

Fig. 12.7 Comparison of skyline matching accuracy using adaptive rendering and uniform viewpoint sampling at multiple resolution levels. In terms of accuracy, adaptive rendering leads to an improvement of almost 200 %. It has a mean error of 200 m with the number of renderings around 2500, while the error increases to 400 m with the same number of renderings in the case uniform sampling. In terms of the system performance, adaptive rendering achieves 300 % improvement as uniform sampling requires more than 7000 renderings to bring the mean error down to 200 m

12.6 Conclusion

We have proposed an adaptive rendering-based approach to enhance geo-localization from 2D skyline images. Using affinity error metric, viewpoint optimality, and line-wise planning presented in this chapter, our method can successfully reduce overall 4X computational cost while preserving the geo-localization accuracy compared with a uniform sampling approach. In the near future, we would like to explore testing on a larger area. We expect to see even further improvements since terrain surfaces in large areas will likely contain more scattered complex areas where a uniform rendering solution will be infeasible for a desired level of geo-localization accuracy.

References

1. Baboud L, Cadik M, Eisemann E, Seidel H (2011) Automatic photo-to-terrain alignment for the annotation of mountain pictures. In: 2011 IEEE conference on computer vision and pattern recognition (CVPR), pp 41–48. IEEE
2. Ramalingam S, Bouaziz S, Sturm P, Brand M (2010) Skyline2gps: localization in urban canyons using omni-skylines. In: 2010 IEEE/RSJ international conference on intelligent robots and systems (IROS), pp 3816–3823. IEEE
3. Ramalingam S, Bouaziz S, Sturm P, Brand M (2009) Geolocalization using skylines from omni-images. In: 2009 IEEE 12th international conference on computer vision workshops (ICCV workshops), pp 23–30. IEEE

4. Russell B, Sivic J, Ponce J, Dessales H (2011) Automatic alignment of paintings and photographs depicting a 3d scene. In: 2011 IEEE international conference on computer vision workshops (ICCV workshops), pp 545–552. IEEE
5. Stein F, Medioni GG (1995) Map-based localization using the panoramic horizon. IEEE Trans Robot Autom 11:892–896
6. Lerbour R, Marvie J, Gautron P (2009) Adaptive streaming and rendering of large terrains: a generic solution. In: Proceedings of WSCG
7. Lerbour R, Marvie J, Gautron P (2010) Adaptive real-time rendering of planetary terrains. In: Proceedings of WSCG
8. Röttger S, Ertl T (2001) Hardware accelerated terrain rendering by adaptive slicing. In: Workshop on vision, modelling, and visualization VMV, vol 1, pp 159–168
9. Choi S, Kim T, Yu W (2009) Performance evaluation of ransac family. In: BMVC

Chapter 13
User-Aided Geo-location of Untagged Desert Imagery

Eric Tzeng, Andrew Zhai, Matthew Clements, Raphael Townshend and Avideh Zakhor

Abstract We propose a system for user-aided visual localization of desert imagery without the use of any metadata such as GPS readings, camera focal length, or field-of-view. The system makes use only of publicly available datasets—in particular, digital elevation models (DEMs)—to rapidly and accurately locate photographs in nonurban environments such as deserts. Our system generates synthetic skyline views from a DEM and extracts stable concavity-based features from these skylines to form a database. To localize queries, a user manually traces the skyline on an input photograph. The skyline is automatically refined based on this estimate, and the same concavity-based features are extracted. We then apply a variety of geometrically constrained matching techniques to efficiently and accurately match the query skyline to a database skyline, thereby localizing the query image. We evaluate our system using a test set of 44 ground-truthed images over a 10,000 km^2 region of interest in a desert and show that in many cases, queries can be localized with precision as fine as 100 m^2.

E. Tzeng (✉) · A. Zhai · M. Clements · A. Zakhor
UC Berkeley, Berkeley, CA, USA
e-mail: etzeng@eecs.berkeley.edu

A. Zhai
e-mail: azhai@eecs.berkeley.edu

M. Clements
e-mail: clements@eecs.berkeley.edu

A. Zakhor
e-mail: avz@eecs.berkeley.edu

R. Townshend
Stanford University, Stanford, CA, USA
e-mail: raphtown@stanford.edu

A.R. Zamir et al. (eds.), *Large-Scale Visual Geo-Localization*,
Advances in Computer Vision and Pattern Recognition,
DOI 10.1007/978-3-319-25781-5_13

13.1 Introduction

Automatic geo-location of imagery has many exciting use cases. For example, such a tool could semantically organize large photo collections by automatically adding location information. Additionally, real-time solutions would serve as an alternative method of localization in instances where GPS systems are typically unreliable, such as urban canyons.

Researchers have attempted to solve this localization problem on the global scale. The authors of [4] use large image databases to identify probable query locations based on image properties such as color and texture. Additionally, [5] shows that using multiple queries with temporal information can improve results. However, because of their reliance on existing imagery, such systems are suited to situations in which ground-level imagery is abundant in the region of interest.

Most of the previous work achieving localization precision on the order of meters has focused on urban environments, relying on distinctive man-made landmarks. In particular, researchers have had great success using standard feature descriptors such as SIFT and street-view databases to localize imagery in cities [10, 13, 14]. These approaches typically detect salient keypoints using a feature descriptor of their choosing, then use a bag-of-visual-words matching scheme to retrieve a corresponding view from a database of street-level imagery.

In this chapter however, we are interested in localizing of imagery in natural environments such as deserts. Techniques that work well in urban environments fail to produce usable results in these settings, since they lack the abundance of discriminative keypoints that is characteristic of urban settings. Additionally, even if typical descriptors did work well, available ground-based imagery datasets are too sparse in these environments to result in reliable ground-to-ground image matching.

Thus, in our work we turn to the boundary between land and sky, or the skyline, as our main source of discriminative information. This approach has been explored many times previously. Ramalingam et al. [9] use this information in urban settings, where the angular skylines differ significantly from the smooth skylines present in natural settings. For geo-location of imagery in natural environments, Stein and Medioni [11] rely on panoramic skylines as their primary piece of discriminative information. In contrast, we focus on single images taken with a camera with no additional stitching or postprocessing.

The recent work by Baatz et al. [1] on geo-localization of images in mountainous terrain is perhaps the most similar work on large-scale localization in natural environments. In their work, they focus on the setting with a known camera field-of-view for a given query image, and provide minimal results in the unknown setting. However, in many practical scenarios, there is no knowledge of camera parameters or additional image metadata whatsoever. There has also been much work in the context of robot localization [11, 12]. However, in addition to making use of prior knowledge of camera parameters, these systems generally operate on a much smaller scale.

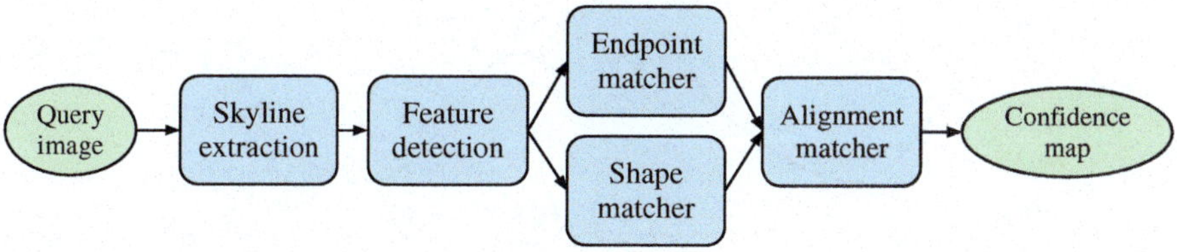

Fig. 13.1 Block diagram for the query localization system

In this chapter we propose a system to solve the large-scale localization problem in desert terrain by building upon existing works while overcoming many of their limitations. Similar to prior work, we focus on the boundary between land and sky as our main source of discriminative information. However, inspired by the previous work of Lamdan et al. [6], we use a different feature descriptor based on concavities in the skyline. These features have stable endpoints even when scales and in-plane rotations are applied, allowing us to operate in cases where camera parameters such as the field-of-view are unknown.

Armed with this feature descriptor, we use a digital elevation model (DEM) to synthesize skylines at a regular sampling grid within our region of interest, then build a database out of these skylines and their detected features. Our choice to use DEMs is due in part to their high availability across the world. When a query photograph needs to be localized, the image can then be sent through our processing pipeline, outlined in Fig. 13.1. First, a user marks a rough estimate of the skyline location, which is then automatically refined by our system. We then extract features from the query skyline to be matched to database skyline features in order to locate a corresponding view in the database. Using geometric hashing [6] and nearest-neighbor techniques [8], we rapidly and aggressively prune the space of candidate matches. We then compute alignments between the query skyline and a reduced subset of the database skylines in order to determine which database skyline matches the query image, thus completing the localization process.

We begin with a discussion of our concavity and convexity features in Sect. 13.2, before moving onto generation of a synthetic skyline database in Sect. 13.3. We then outline the query recognition process in Sect. 13.4, and discuss our experimental results in Sect. 13.5. Finally, Sect. 13.6 concludes with a summary of our contributions and a discussion of the future work.

13.2 Concavity Features and Feature Detection

In order to cope with unknown camera parameters of the query photograph, it is necessary to build scale-invariance into our system. Lamdan et al. use scale-invariant concavity and convexity features for object recognition from boundary curves [6]. We utilize features that are similar in concept for skyline matching. However, since the

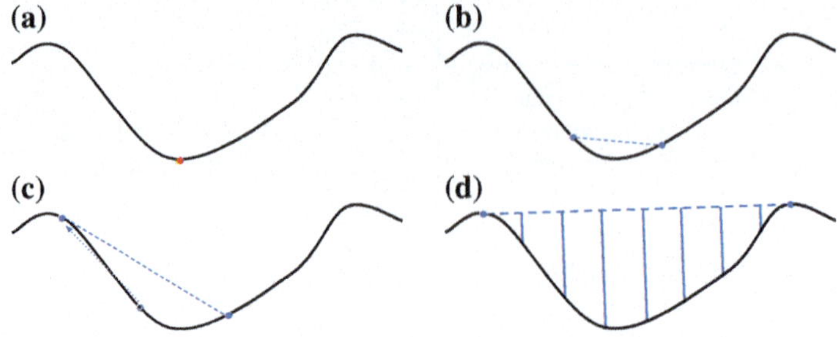

Fig. 13.2 **a** The input curve to feature extraction and the detected point of extreme curvature. **b** Initial estimates of endpoint locations. **c** One iteration of the refinement process. **d** The final refined endpoint locations shown by the *dashed line* and the sectors used to form the shape vector

notion of a concavity or convexity is poorly defined on open curves such as skylines, our method differs significantly in its details. For the remainder of this chapter, we use "concavity" to refer to both concavities and convexities, unless otherwise noted. A high-level overview of feature extraction is provided in Fig. 13.2. We now outline the process in greater detail.

First, we use the method developed by Fischler and Wolf [3] to detect points of extreme curvature on the input skyline, as shown in Fig. 13.2a. These points are found almost exclusively within concavities on the skyline. However, in practice they are too unstable to be used for localization by themselves. Rather, we use them to initialize the locations of our concavity features as follows: for each detected point of extreme curvature, we select a point to the left and to the right as a rough estimate of the concavity's endpoints, as seen in Fig. 13.2b. We then refine this estimate by iteratively and alternately moving the endpoints away from each other. Specifically, we push each endpoint out until the slope formed between the initial curvature point and the endpoint reaches a local maximum. Once this occurs, we continue examining an additional δ points. If a higher slope is found within these additional points, we continue pushing the endpoint outwards; otherwise, we leave the endpoint at the local maximum and begin pushing out the other endpoint. A sample iteration of this refinement process is shown in Fig. 13.2c. This process repeats alternately between the two endpoints until neither endpoint can move any further. For the sake of efficiency, we also impose an iteration limit, although in practice this limit is rarely reached. The final, refined concavity feature can be seen as the dashed line in Fig. 13.2d.

Although this process results in an initial set of basic features, we can obtain additional, more complex features by examining overlapping features. Specifically, for any two overlapping features, we select their distant endpoints as an initial estimate of the concavity containing the two and refine as before. Further linking, e.g., linking nonoverlapping concavities is possible, but has been empirically found to negatively impact localization performance.

We perform an additional filtering step to remove features that are of low reliability. In particular, features with endpoints near the edges of the image are of dubious

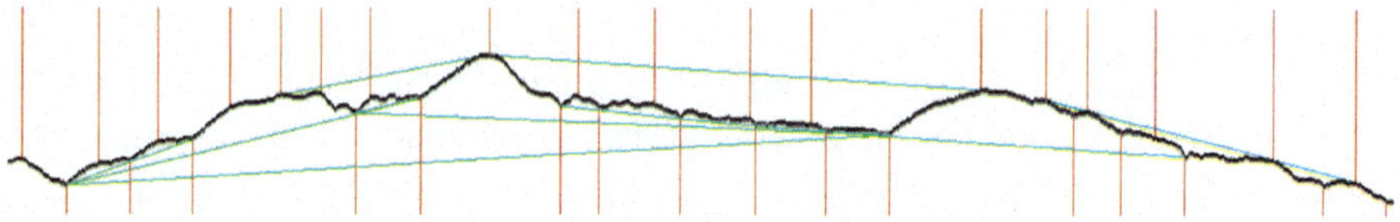

Fig. 13.3 A query skyline, the detected points of extreme curvature (indicated with *vertical lines*), and the refined concavity features (indicated with *straight lines* lying on the skyline) stretched vertically by a factor of 1.5 for illustrative purposes

quality, the concavity slope may continue to rise past the end of the image, or it may fall off steeply just beyond the edge of the image. To avoid false detections, features with at least one endpoint within a certain distance of the image edge are discarded.

In addition to the endpoints, we also characterize each feature as a d-dimensional vector as shown in Fig. 13.2d. In this step, we use feature curves, which are the portions of the skyline lying between a feature's endpoints. We apply a similarity transformation to the feature curve that maps its endpoints to $(-1, 0)$ and $(1, 0)$, thus normalizing each feature to a fixed length and orientation. This has two important effects. First, in normalizing the length of the feature, we account for any scale differences between feature curves, thereby achieving scale invariance. Second, because the orientation of each feature is normalized to lie along the x-axis, we achieve in-plane rotation invariance as well. Thus, barring any quantization effects, any two features curves with the same shape are normalized to the same final feature curve. After this normalization step, the d-dimensional feature vector is formed by computing the area between the normalized curve and the x-axis for d disjoint regions of equal width, as shown in Fig. 13.2d. Figure 13.3 shows an example output of feature detection, along with the sectors used to form the feature vector.

13.3 Database Generation

Localization of a query's skyline is performed via matching to a database of synthetic skylines. The skylines upon which we perform our tests are generated using DEMs with a resolution of $\frac{1}{3}$ arc-second, or 10 m, and which span a square region of 10,000 km^2. They were obtained from the National Elevation Dataset of the United States Geological Survey. The vertical heights of the dataset have a root mean square error of 2.44 m.

We evenly sample the DEM at a 1000 m resolution along both the north–south and east–west directions, forming a 2D grid of sample points as discussed by Baatz et al. [1]. This results in a 100×100 grid of samples. We then use a skyline generator to synthesize the skyline as seen from each sample point. We have implemented two versions of the skyline generator: a GPU-based approach where the full scene is rendered, and a CPU-based approach where only the primary skyline is extracted.

For the GPU version, at each sample point we render twenty-four $30° \times 20°$ images at 15° offsets, covering the full 360° panorama at the location. If the skyline is not completely captured by the viewport, we progressively pitch the camera upward until the full horizon is rendered. For our tests, we generate $100 \times 100 \times 24 = 240{,}000$ distinct images in total. Due to the large volume of images to render and the high resolution of the DEM, optimizations are needed in order to efficiently generate skyline databases. The DEM is initially tiled into chunks of 256×256 points, and an additional subsampled version of each tile is also created. The points within each tile are then formed into a mesh of triangles, which is saved along with the tile.

When rendering a viewpoint, we use a simple level of detail algorithm to only render the triangles for the tiles that are in view, and of those we only select the high-resolution version if the angular resolution of the skyline is finer than the angular diameter of neighboring DEM points within the lower-resolution tile. In practice, this translates to setting a distance threshold T where tiles further away than this threshold can be of lower resolution. Consider Fig. 13.5, in which A and B are adjacent points between output skylines. Using the fact that the angle θ swept between these points is very small, we approximate T as r/θ where r is the resolution in meters per point of the downsampled DEM tiles. Given that the horizontal field-of-view for our system is 30° and assuming that each view in the database is 1500 pixels wide, the angular resolution per pixel θ is 3.49×10^{-4} rad. Then, for a subsampled DEM resolution of 30 m between points, T can be estimated as $30/(3.49 \times 10^{-4})$ m, or about 86 km. This procedure allows for fine-grained selection of the DEM points to render. On an NVIDIA GeForce GT 650M, our skyline renderer generates over four 1500×1500-pixel images per second. Figure 13.4b shows an example rendered skyline. Once the 240,000 skylines have been synthesized, we extract features as outlined in Sect. 13.2. The final database then consists of all skylines and their features.

The CPU version directly generates the primary horizons from the DEM, allowing for faster performance. For each column in the output skyline, a ray passing through that column is shot outwards from the camera location. We trace each ray over the DEM, sampling points along it, in order to find the DEM point along each ray with the highest elevation angle to the camera point. This angle then determines how high

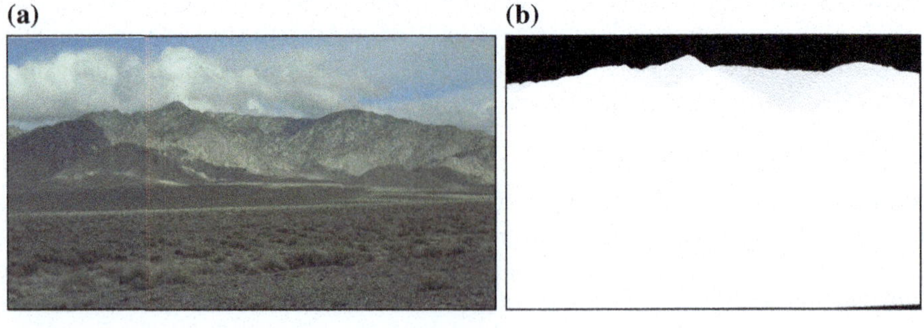

Fig. 13.4 **a** A query image. **b** Its corresponding database view

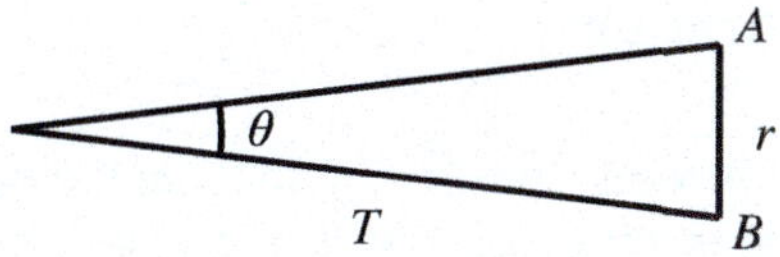

Fig. 13.5 From the spatial resolution of the downsampled DEM r and the angle θ swept between adjacent points A and B on the output skyline, we can determine at what distance T to begin using lower-resolution DEM tiles

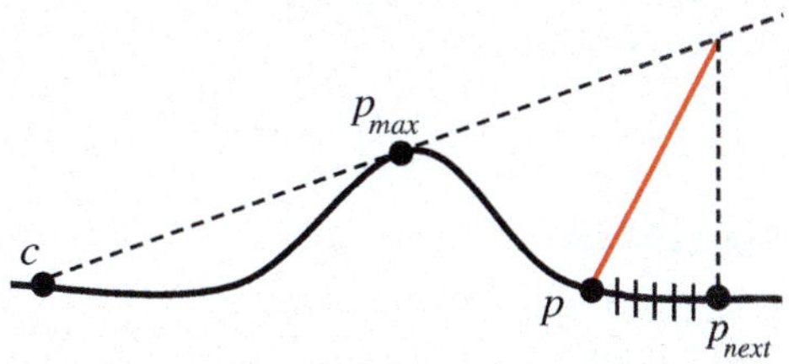

Fig. 13.6 A side view of the DEM during CPU-based skyline synthesis at camera location c. Using an estimate of the maximum slope of the DEM, indicated by the slope of the diagonal line extending from p, we allow ourselves to jump from p to p_{next} without sampling the points between

the skyline extends in that particular column. As we are only interested in the highest elevation angle along a ray, further optimizations are possible. A naïve approach would uniformly sample the elevation of every DEM point along the ray; however, if we assume an upper bound on the terrain's slope across the region, then we can sample points along the ray more sparsely, as shown in Fig. 13.6. In this figure, c denotes the camera position, p_{max} denotes the highest elevation point sampled so far, p denotes the last point sampled, and the slope of the diagonal line extending from p shows the estimate of maximum terrain slope. The naïve approach would sample at every point, represented by the tick marks, but under the maximum slope assumption, we can skip all points until p_{next} as we are guaranteed that no intermediate point has a higher elevation angle than p_{max}. Thus, the distance along each ray we can advance before sampling another point is a function of two factors: the maximum slope, and the difference between the maximum elevation angle seen thus far and the elevation angle of the point last sampled. In addition, we can use the determined maximum elevation angle of neighboring pixels in the skyline to estimate the elevation angle of the current pixel. Due to these optimizations, this method runs orders of magnitude faster than the previous version. Specifically, on a quad-core Intel Core i7, it generates over two hundred and fifty 2000-pixel-wide skylines per second, which is 60 times faster than the GPU based method.

13.4 Query Localization

The query localization process consists of four major steps, as diagrammed in Fig. 13.1. First, a skyline is extracted from a query photograph. Next, features are extracted from the query skyline. Then, a first round of matching feature shapes and endpoints occurs in order to rapidly narrow the set of candidate match locations. Finally, a more exhaustive alignment match occurs, producing a final ranked list of candidate database views. This list is used to generate a confidence map for the user to use in localizing the query photograph.

13.4.1 Skyline Extraction

Query processing begins with a user-directed skyline extraction process that follows a simple basic pattern: the user draws an approximate skyline on the query image, edge detection is run on the vicinity of that approximation, and finally a skyline is extracted from the resulting edge map.

For extraction of a skyline from a map of edge versus non-edge pixels, we use the strategy detailed in Lie et al. [7]. The dynamic programming algorithm first builds paths across the edge pixels in the image, then selects the one that is cheapest when evaluated on criteria such as smoothness, altitude, and number and size of gaps. For edge detection, we use a variant of Canny edge detection [2].

13.4.2 Endpoint Matcher

The first stage of skyline matching deals with matching endpoints of concavity features in the query and database skylines. It is done through a geometric hashing technique inspired by the line matching technique in [6]. In contrast to [6] which uses a line segment and an additional point to create an affine-invariant basis, we use a concavity feature as our basis, as explained in Sect. 13.2. We opt to forego affine-invariance because we have empirically found affine-invariant features to not possess as much discriminative power as similarity-invariant features. Additionally, the loss of affine-invariance is mitigated by the fact that our database is sampled uniformly, and thus usually has a view that is similar to the query view. We now outline the basic implementation of our geometric hashing method.

After generating our feature database as described in Sect. 13.3, we create a hash table to be used in the localization stage. Assume that a given skyline s has n features. For each feature $f \in s$, we find a 2D similarity transformation that maps its two endpoints to $(0, 0)$ and $(1, 0)$. We then apply this transformation to s and all of its features, generating a new configuration that we refer to as being normalized with respect to f. This transformation removes the effects of in-plane rotations and scales: any transformed version of this skyline s', when normalized with respect to its corresponding transformed feature f', results in the same configuration of feature endpoints as s when normalized to f. The new endpoints of the remaining $n-1$ features of s are used as indices to the hash table. At each index, we store the pair (s, f)—that is, the skyline and the feature used to normalize it.

This process is repeated for all skylines in the database to create a single hash table used for endpoint matching. Note that, since the building process does not require any knowledge of the query skyline or features, this hash table can be precomputed offline and saved for later use.

When a query skyline and its n features are provided as input to the endpoint matcher, we perform a similar normalization step. For each feature, we normalize the skyline and its other features with respect to it. We then use each of the other normalized $n-1$ features to index into the hash table built in the previous step. Each index operation produces a list of skyline/feature pairs, and each pair receives one vote. We then construct a list of votes for candidate skylines, where the number of votes a skyline receives is the maximum number of votes of any of its skyline/feature pairs.

After this process has been repeated for each of the n query features, we are left with n ranked lists of candidate database skylines. We combine these lists into a final list by simply summing all votes together. This final list is then resorted to generate the final, ranked list of database skylines for the endpoint matcher, where more votes translates into a better ranking. Figure 13.7 shows an example match detected by this matcher, where the features marked with arrows form the pair used for alignment, and the dashed lines denote features correspondences found by the matcher.

As previously discussed, by normalizing skylines with respect to their features, we achieve similarity invariance. Furthermore, our system is robust to partial occlusions; since the voting step uses every feature in every skyline, no single occluded feature significantly impacts the result. We now discuss a variety of strategies we employ to ensure the endpoint matcher is as robust as possible.

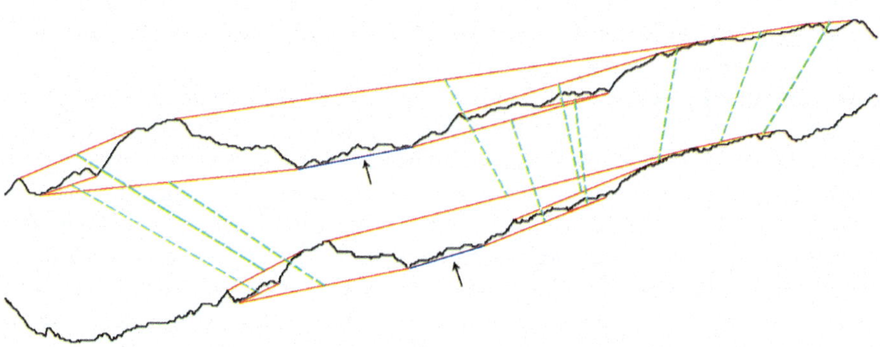

Fig. 13.7 A resulting match from the endpoint matcher between a database skyline (*top*) and a query skyline (*bottom*). The skylines have been stretched vertically by a factor of 3 for illustrative purposes

Soft Quantization in Matching Each time we query the database hash table with a normalized feature in the matching stage, we use a soft quantization of the normalized feature to account for noise in the query skyline. We achieve this soft quantization by querying a set of points rather than a single point for each hash table access. For a given query feature q and one of its endpoints x, the size of this set is determined by

$$s(q, x) = R + \min\left(\frac{||x||_2}{F + \sqrt[4]{q_{\text{len}}}}, M\right) \tag{13.1}$$

where $||x||_2$ indicates the distance of x from the origin, q_{len} indicates the distance between the endpoints of q and R, F and M are user-specified parameters to be described shortly. When indexing into the hash table, we index into all entries that fall within a box with side length $2s(q, x)$, centered at the endpoint x's location.

The above equation can be explained as follows. To account for noise, we allow the user to specify range R, ensuring that we always examine at least within a $2R \times 2R$ border from the original endpoint. Furthermore, since noise is amplified for points further away from the origin, we expand this window with an additional $||x||_2$ term. However, if left unchecked, this term can dominate. Thus, we attenuate its effect by F, a falloff term, and $\sqrt[4]{q_{\text{len}}}$. Larger features tend to be more stable after normalization, so we reduce the effect of the $||x||_2$ term by $\sqrt[4]{q_{\text{len}}}$. Finally, as a final precaution, we limit the contribution of this term to maximum value M. For our current system, we set $R = 20$, $F = 200$, $M = 10$.

Vote Threshold During the matching stage, n skyline vote lists are aggregated into a single final list. In order to reduce the influence of feature correspondences that occur by chance, we prevent an individual vote list from contributing to the final vote list for a given skyline unless the number of votes for that skyline in the individual list is above some threshold t. For our current system, $t = 4$.

Concavity/Convexity Alignment When querying the database hash table with a normalized feature, we ensure that the curve type of the normalized feature used matches the one in the database. In other words, if a convex normalized feature is used as a key to an entry to the hash table, we ensure that only a convex normalized feature can be used to get that entry in the hash table, since a convex-to-concave alignment is almost certainly erroneous.

Normalization by Features in Window If a database skyline has a large number of features, then the likelihood that the skyline receives feature correspondence votes simply by chance is higher. As a result, we need to normalize the number of votes a database skyline receives based on the number of features it contains. However, this normalization should not consider features that have no chance of matching query features, such as those that fall entirely outside the query skyline after alignment occurs. Thus, we normalize the skyline's votes by the number of features in the query's window, where the window is defined as the smallest axis-aligned bounding box that contains all query feature endpoints.

13.4.3 Shape Matcher

The matcher described in Sect. 13.4.2 considers only locations of feature endpoints without regards to the feature shapes. We now outline another basic matcher that considers only shape without regard to feature configurations.

Since each feature's shape is characterized as a d-dimensional vector shown in Fig. 13.2d, we can evaluate the similarity in shape between two features by computing the Euclidean distance of the shape vectors in $\mathbb{R}^d$. Thus, when processing a query skyline and its features, we use a k-d tree containing the database features for efficient retrieval of nearby shape vectors [8]. Since the k-d tree does not depend on the query features at all, it can be precomputed offline and stored. Doing so allows for matches to be performed in a matter of seconds.

To perform a match, we examine each query feature and index into the k-d tree. We retrieve all database skylines containing a feature within distance D from the query feature, and any database skyline with such a feature receives a vote. Since rare features are more discriminative than common features, we weigh a feature's vote using a function of the number of database skylines it votes for. More specifically, we weigh each vote by its inverse document frequency defined as

$$idf(f_q) = \ln \frac{|S|}{|\{s \in S \mid \exists f_d \in s : ||f_d - f_q||_2 < D\}|} \tag{13.2}$$

where f_q and f_d are query and database features, respectively, and S is the set of all database skylines.

Once voting is complete, the shape matcher outputs a ranked list of database skylines, ordered by descending vote count.

13.4.4 Alignment Matcher

As shown in Fig. 13.1, a small subset of the best-ranked results from the endpoint and shape matchers is selected and fed as input into the final alignment matcher. Specifically, to collect a set of N candidate database skylines, we collect the intersection of the top p endpoint matcher results and the top q shape matcher results. Any remaining slots are then split with a $1:2$ ratio between the top endpoint matcher and the shape matcher results respectively. This combination step is crucial to the performance of the system. The value of N can be set so as to balance accuracy and speed: higher values of N favor accuracy, whereas lower values favor speed. For our system, $N = 50{,}000$, $p = 30{,}000$, $q = 60{,}000$, resulting in an overall runtime of 2 hours on an 8-core 2.13 GHz Intel Xeon. The alignment step is by far the most computationally intense part of our entire system and as such dominates the run time in a significant way.

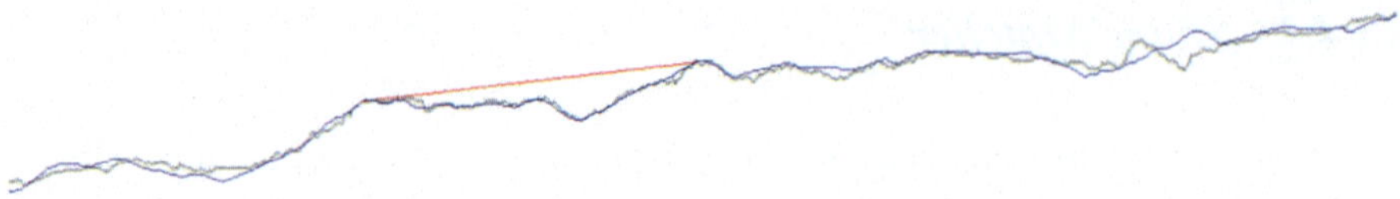

Fig. 13.8 An example alignment found by the alignment matcher. The query skyline is drawn in *black*, the database skyline is drawn in *blue*, and the feature used for alignment is drawn in *red*

When aligning a query skyline against a candidate database skyline, we consider the Cartesian set product of the skylines' features. Each query-database feature pair defines a potential alignment of the two skylines. In particular, for query and database skylines s_q, s_d and features f_q, f_d from each skyline respectively, we find a similarity transformation that maps the endpoints of f_d onto the endpoints of f_q. This transformation is applied to s_d to effectively "overlay" the two skylines onto each other.

We evaluate each potential alignment by sampling 1000 points from the overlapping region, then computing an error function

$$E(s_q, s_d) = \ln |O(s_q, s_d)|^{-1} \sum_{i=1}^{1000} (s_d[i] - s_q[i])^2 \tag{13.3}$$

where $|O(s_q, s_d)|$ denotes the size of the overlap between the skylines s_q and s_d. We divide the squared distance between the two curves by the logarithm of the overlap size, since a larger overlap should translate into a more confident match, i.e., a lower error score.

For a particular candidate database skyline, we find its best alignment and assign it that score. We can then produce a ranked list of candidate locations by sorting the skylines in ascending order of error score. Figure 13.8 shows an example top database match overlaid on its corresponding query.

To exclude degenerate alignments such as alignments in which only a small fraction of the skylines overlap, we require any alignment to result in at least a third of both skylines to overlap. We also exclude any alignments that match a concavity to a convexity or vice versa similar to the endpoint matcher, as well as any alignments with a scale difference above a threshold.

13.4.5 Confidence Map Generation

After the alignment matcher outputs the final rankings for each candidate database skyline, we generate a confidence map such as the one in Fig. 13.9 indicating how likely a match is for each grid location within the region of interest.

Although our localization system uses orientation information during the matching process, our confidence maps show only location. Thus, the very first step is to

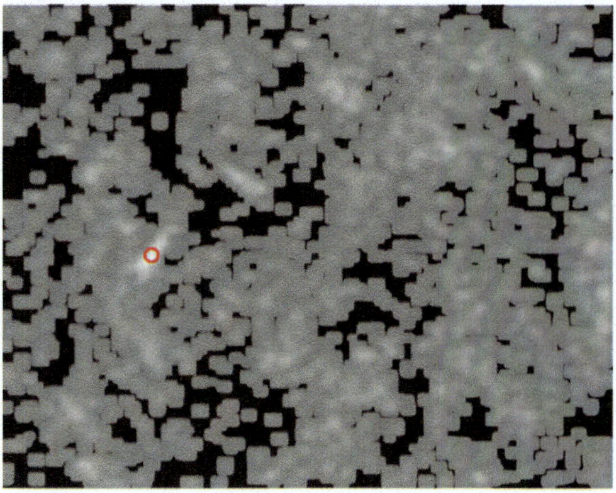

Fig. 13.9 An example confidence map, with the ground-truth location *circled*. Higher intensities denote locations with higher confidence

find the minimum error score for each grid location. This results in a ranked list of locations, rather than individual views. We now take the top k locations and plot them on a confidence map. The intensity value of each location is determined based on the error score; the top location has the lowest error score and thus the highest intensity, whereas the k-th location has the highest error score and thus the lowest intensity. The intensity of all other locations is determined through linear interpolation on their error scores.

As a final processing step on our confidence maps, we apply a uniform box filter with side lengths equal to that of two grid locations. This provides more lenience and allows grid locations with high intensity to count for their neighboring locations as well.

13.5 Evaluation and Results

Since our primary motivation is user-aided geo-location, typical metrics such as the percentage of top-1 matches are ill suited for evaluating our system. Rather, we use the geo-location area (GA), or the area of the region with confidence value greater than or equal to that of the ground-truth location, and compare it to the area of the entire region of interest (|ROI|). Intuitively, the GA represents the area over which a user would have to search in order to find the ground-truth location, assuming the search is done in order from the highest to the lowest confidence value. Figure 13.10 plots the fraction of queries correctly located against different GA/|ROI| values. As shown, our system correctly locates over 25 % of the queries with a GA/|ROI| of less than 0.01, and over 50 % of the queries with a GA/|ROI| of less than 0.10.

Figure 13.11 shows examples in which the system performs well or when it fails. As seen in Fig. 13.11a, the best performing query images are those of distant ridges

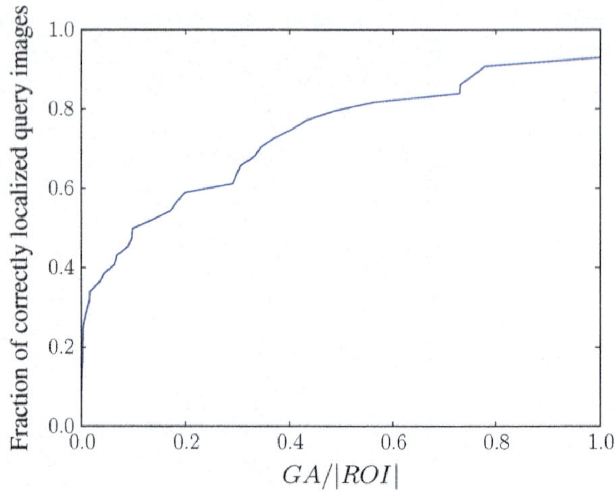

Fig. 13.10 Curve of system performance. We evaluate our system using the GA/|ROI| metric, which compares the area of the region with confidence value greater than or equal to the ground-truth location (GA) to the area of the entire region of interest |ROI|. We are able to locate one-third of all images with very high accuracy, and over half of the images to within just over 10 % of the region of interest

Fig. 13.11 **a** Two of the best performing query images, taken from flat ground with distant skylines. **b** Two of the worst performing query images, taken on a slope of mountainous terrain with nearby ridges and skylines

taken from flat ground, as these images are invariant to small shifts in location. When these conditions are met, we obtain GA/|ROI| scores as low as 1×10^{-5}. In contrast, photographs of nearby formations or photographs taken on mountainous terrain, as shown in Fig. 13.11b, are very sensitive to small changes in location. Often, this means that our database is not sampled densely enough to contain a good match.

13.6 Conclusions and Future Work

In this chapter we have presented a system for geo-location of untagged imagery using only digital elevation models. Our approach extracts the skyline from a query photograph and detects stable concavity features for use in matching against a database. We introduced a modular matching system that allows users to balance accuracy with speed as necessary and can be easily parallelized by simply distributing the database across multiple machines. Finally, we showed that this method is robust, attaining incredibly precise locations on many test queries.

By far the major weakness of our system is the relatively sparse sampling of our database. Our concavity features are sensitive to out-of-plane rotations, so denser sampling would ensure that a proper match exists in our database. However, since our current database generation uses a uniform grid across the ROI, increasing the sampling density causes a multiplicative increase in the number of database skylines and thus the runtime. If we can detect database sampling points at which the skyline is nearby or has sloped terrain—the two conditions under which the view is highly sensitive to precise location—then we can adaptively increase the sampling density at only those points, mitigating the multiplicative increase effect.

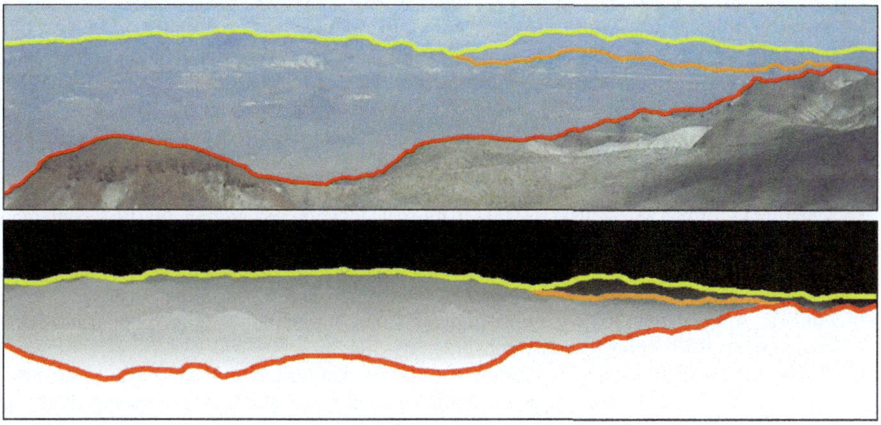

Fig. 13.12 A query (*top*) and corresponding database view (*bottom*) with multiple corresponding skylines identified

There is also an abundance of discriminative information in secondary "horizons" below the primary one, formed by the interplay between ridges of different heights, as shown in Fig. 13.12. The additional skylines provide more features with which to perform alignment, and might allow the system to recover in cases where the primary skyline is unclear or heavily occluded.

Acknowledgments Supported by the Intelligence Advanced Research Projects Activity (IARPA) via Air Force Research Laboratory, contract FA8650-12-C-7211. The U.S. Government is authorized to reproduce and distribute reprints for Governmental purposes notwithstanding any copyright annotation thereon. Disclaimer: The views and conclusions contained herein are those of the authors and should not be interpreted as necessarily representing the official policies or endorsements, either expressed or implied, of IARPA, AFRL, or the U.S. Government.

References

1. Baatz G, Saurer O, Koeser K, Pollefeys M (2012) Large scale visual geo-localization of images in mountainous terrain. In: Proceedings of the European conference on computer vision
2. Canny J (1986) A computational approach to edge detection. IEEE Trans Pattern Anal Mach Intell (6):679–698
3. Fischler MA, Wolf HC (1994) Locating perceptually salient points on planar curves. IEEE Trans Pattern Anal Mach Intell 16(2):113–112
4. Hays J, Efros A (2008) Im2gps: estimating geographic information from a single image. In: IEEE conference on computer vision and pattern recognition, CVPR 2008, pp 1–8. IEEE
5. Kalogerakis E, Vesselova O, Hays J, Efros A, Hertzmann A (2009) Image sequence geolocation with human travel priors. In: 2009 IEEE 12th international conference on computer vision, pp 253–260. IEEE
6. Lamdan Y, Schwartz JT, Wolfson HJ (1988) Object recognition by affine invariant matching. In: Computer society conference on computer vision and pattern recognition, Proceedings CVPR'88, pp 335–344. IEEE
7. Lie WN, Lin TCI, Lin TC, Hung KS (2005) A robust dynamic programming algorithm to extract skyline in images for navigation. Pattern Recog Lett 26(2):221–230
8. Muja M, Lowe DG (2009) Fast approximate nearest neighbors with automatic algorithm configuration. In: International conference on computer vision theory and application, VISSAPP'09, pp 331–340. INSTICC Press
9. Ramalingam S, Bouaziz S, Sturm P, Brand M (2010) Skyline2gps: localization in urban canyons using omni-skylines. In: 2010 IEEE/RSJ international conference on intelligent robots and systems (IROS), pp 3816–3823. IEEE
10. Schindler G, Brown M, Szeliski R (2007) City-scale location recognition. In: IEEE conference on computer vision and pattern recognition, CVPR'07, pp 1–7. IEEE
11. Stein F, Medioni G (1995) Map-based localization using the panoramic horizon. IEEE Trans Robot Autom 11(6):892–896
12. Talluri R, Aggarwal J (1992) Position estimation for an autonomous mobile robot in an outdoor environment. IEEE Trans Robot Autom 8(5):573–584
13. Zhang J, Hallquist A, Liang E, Zakhor A (2011) Location-based image retrieval for urban environments. In: 2011 18th IEEE international conference on image processing (ICIP), pp 3677–3680. IEEE
14. Zhang W, Kosecka J (2006) Image based localization in urban environments. In: Third international symposium on 3D data processing, visualization, and transmission, pp 33–40. IEEE

Chapter 14
Visual Geo-localization of Non-photographic Depictions via 2D–3D Alignment

Mathieu Aubry, Bryan Russell and Josef Sivic

Abstract This chapter describes a technique that can geo-localize arbitrary 2D depictions of architectural sites, including drawings, paintings, and historical photographs. This is achieved by aligning the input depiction with a 3D model of the corresponding site. The task is very difficult as the appearance and scene structure in the 2D depictions can be very different from the appearance and geometry of the 3D model, e.g., due to the specific rendering style, drawing error, age, lighting, or change of seasons. In addition, we face a hard search problem: the number of possible alignments of the depiction to a set of 3D models from different architectural sites is huge. To address these issues, we develop a compact representation of complex 3D scenes. 3D models of several scenes are represented by a set of discriminative visual elements that are automatically learnt from rendered views. Similar to object detection, the set of visual elements, as well as the weights of individual features for each element, are learnt in a discriminative fashion. We show that the learnt visual elements are reliably matched in 2D depictions of the scene despite large variations in rendering style (e.g., watercolor, sketch, and historical photograph) and structural changes (e.g., missing scene parts and large occluders) of the scene. We demonstrate that the proposed approach can automatically identify the correct architectural site as well as recover an approximate viewpoint of historical photographs and paintings with respect to the 3D model of the site.

M. Aubry (✉)
LIGM (UMR CNRS 8049), ENPC/Université Paris-Est,
77455 Marne-la-Vallée, France
e-mail: Mathieu.Aubry@imagine.enpc.fr

B. Russell
Adobe Research, Lexington, KY, USA
e-mail: brussell@adobe.com

J. Sivic
Inria, WILLOW Project-team, Département d'Informatique de l'Ecole
Normale Supérieure, ENS/INRIA/CNRS UMR, 8548 Paris, France
e-mail: Josef.Sivic@ens.fr

A.R. Zamir et al. (eds.), *Large-Scale Visual Geo-Localization*,
Advances in Computer Vision and Pattern Recognition,
DOI 10.1007/978-3-319-25781-5_14

14.1 Introduction

In this work, we seek to automatically geo-localize historical photographs and non-photographic renderings, such as paintings and line drawings, by matching them with a set of geo-referenced 3D models. Specifically, we wish to establish a set of point correspondences between local structures on the 3D models and their respective 2D depictions. The established correspondences will in turn allow us to identify the correct architectural site and find an approximate viewpoint of the 2D depiction with respect to the identified 3D model, thus geo-localizing the input depiction. We focus on depictions that are, at least approximately, perspective renderings of the 3D scene. Example results are shown in Fig. 14.1. We show that our alignment method works with complex textured 3D models obtained by recent multi-view stereo reconstruction systems [20] as well as with simplified models obtained from 3D modeling tools such as Trimble 3D Warehouse that often appear in geo-browsing tools such as Google Earth.

Why is this task important? First, non-photographic depictions are plentiful and comprise a large portion of our visual record. We wish to reason about them, and aligning such depictions to our 3D physical world is an important step towards this goal. Second, such depictions are often stored in archives and museums with limited access and search capabilities. Automatic large-scale geo-localization would change the way archivists access and organize such imagery. Finally, reliable

Fig. 14.1 Our system automatically geo-localizes paintings, drawings, and historical photographs by recovering their viewpoint with respect to a geo-referenced 3D model of the depicted architectural site. Here, geo-localized paintings of Notre Dame in Paris are visualized in the Google Earth geo-browser

automatic image-to-3D model matching is important in domains where geo-referenced 3D models are often available, but may contain errors or unexpected changes (e.g., something built/destroyed) [6], such as urban planning, civil engineering, or archeology.

The task of aligning 3D models to 2D non-photographic depictions is extremely challenging. As discussed in prior work [41, 47], local feature matching based on interest points (e.g., SIFT [35]) often fails to find correspondences across paintings and photographs. First, the rendering styles across the two domains can vary considerably. The scene appearance (colors, lighting, and texture) and geometry depicted by the artist can be very different from the rendering of the 3D model, e.g., due to the depiction style, drawing error, or changes in the geometry of the scene. Second, we face a hard search problem. The number of possible alignments of the depiction to a large 3D model, such as a partial reconstruction of a city, is huge. Which parts of the depiction should be aligned to which parts of the 3D model? How does one search over the possible alignments?

To address these issues, we introduce the idea of automatically discovering *discriminative visual elements* for a 3D scene. We define a discriminative visual element to be a mid-level patch that is rendered with respect to a given viewpoint from a 3D model with the following properties: (i) it is visually discriminative with respect to the rest of the "visual world" represented here by a generic set of randomly sampled patches, (ii) it is distinctive with respect to other patches in nearby views, and (iii) it can be reliably matched across nearby viewpoints. We employ modern representations and recent methods for discriminative learning of visual appearance, which have been successfully used in recent object recognition systems. Our method can be viewed as "multi-view geometry [26] meets part-based object recognition [17]"—here, we wish to automatically discover the distinctive object parts for a large 3D site.

We discover discriminative visual elements by first sampling candidate mid-level patches across different rendered views of the 3D model. We cast the image-matching problem as a classification task over appearance features with the candidate mid-level patch as a single positive example and a negative set consisting of a large set of "background" patches. Note that a similar idea has been used in learning per-exemplar distances [19] or per-exemplar support vector machine (SVM) classifiers [36] for object recognition and cross-domain image retrieval [47]. Here, we apply per-exemplar learning for matching mid-level structures between images.

For a candidate mid-level patch to be considered a discriminative visual element, we require that (i) it has a low training error when learning the matching classifier and (ii) it is reliably detectable in nearby views via cross-validation. Critical to the success of operationalizing the above procedure is the ability to efficiently train linear classifiers over Histogram of Oriented Gradients (HOG) features [12] for each candidate mid-level patch, which has potentially millions of negative training examples. In contrast to training a separate SVM classifier for each mid-level patch, we change the loss to a square loss, similar to [4, 22]. We show that the solution can be computed in closed form, which is computationally more efficient as it does not require expensive iterative training. In turn, we show that efficient training opens-up the possibility to evaluate the discriminability of millions of candidate visual elements densely sampled over all the rendered views. We further show how our formulation is related to recent work that performs linear discriminant analysis (LDA) by analyzing a large set of negative training examples and recovering the sample mean and covariance matrix that decorrelates the HOG features [22, 25].

The output for each discriminative visual element is a trained classifier. At run-time, for an input depiction (e.g., a painting), we run the set of trained classifiers in a sliding window fashion across different scales. Detections with high responses are considered as putative correspondences with the 3D model, from which camera resectioning is performed. The output is a geo-localization of the input depiction in the form of its approximate viewpoint with respect to the geo-referenced 3D model. We show that our approach is able to scale to a number of different 3D sites and handles different input rendering styles. To evaluate our alignment procedure, we use the publicly available dataset of [1]. First, we evaluate whether the proposed technique can coarsely localize the input depiction by correctly identifying the 3D model corresponding to the depicted architectural site. Second, for the correctly coarsely localized depictions we perform a user study where human subjects are asked to judge the goodness of the output alignment. Parts of this chapter were previously published in [1]. Here, we apply the 2D-to-3D alignment technique described in [1] to the task of automatic geo-localization of historical and non-photographic imagery.

14.2 Related Work

This section reviews prior work on visual geo-localization with a focus on non-photographic and historical imagery.

Visual geo-localization using local features Local invariant features and descriptors such as SIFT [35] represent a powerful tool for matching photographs of the same at least lightly textured scene despite changes in viewpoint, scale, illumination, and partial occlusion. Without explicitly representing the 3D structure of the scene, visual geo-localization can be cast as large-scale instance-level retrieval [37, 39, 48]. Local invariant features are extracted from each image in a geo-referenced image database. The query photograph is then localized despite changes in viewpoint or illumination by finding the best matching image in the database and transferring its geo-tag [7, 11, 24, 31, 43, 50, 51]. Large 3D scenes, such as a portion of a city [33], can be also represented as a geo-referenced 3D point cloud with associated local feature descriptors extracted from the corresponding photographs [30, 33, 42]. Geo-referenced camera pose of a given query photograph can be recovered from 2D to 3D correspondences obtained by matching appearance of local features verified using geometric constraints [26]. However, appearance changes beyond the modeled invariance, such as significant perspective distortions, non-rigid deformations, non-linear illumination changes (e.g., shadows), weathering, change of seasons, structural variations or a different depiction style (photograph, painting, sketch, drawing) cause local feature-based methods to fail [27, 41, 47]. Greater insensitivity to appearance variation can be achieved by matching the geometric or symmetry pattern of local image features [8, 27, 46], rather than the local features themselves. However, such patterns have to be detectable and consistent between the matched views.

Visual geo-localization via alignment of contours Contour-based 2D to 3D alignment methods [29, 34] rely on detecting edges in the image and aligning them with projected 3D model contours. Such approaches are successful if scene contours can be reliably extracted both from the 2D image and the 3D model. A recent example is the work on photograph localization using semi-automatically extracted skylines matched to clean contours obtained from rendered views of digital elevation models [2, 3]. Contour matching was also used for aligning paintings to 3D meshes reconstructed from photographs [41]. However, contours extracted from paintings and real-world 3D meshes obtained from photographs are noisy. As a result, the method requires a good initialization with a close-by viewpoint. In general, reliable contour extraction is a hard and yet unsolved problem.

Visual geo-localization with discriminative image representations Modern image representations developed for visual recognition, such as HOG descriptors [12], represent 2D views of objects or object parts [17] by a weighted spatial distribution of image gradient orientations. The weights are learnt in a discriminative fashion to emphasize object contours and de-emphasize non-object, background contours, and clutter. Such a representation can capture complex object boundaries in a soft manner, avoiding hard decisions about the presence and connectivity of imaged object edges. Learnt weights have also been shown to emphasize visually salient image structures matchable across different image domains, and have been used to coarsely geo-localize non-photographic depictions such as paintings or sketches using a global image descriptor [47]. Similar representation has been used to learn architectural elements that summarize a certain geo-spatial area by analyzing (approximately rectified) 2D street-view photographs from multiple cities [14] and to detect objects depicted in paintings which have been trained from images [10].

Building on discriminatively trained models for object detection, we develop a compact representation of 3D scenes suitable for alignment and visual geo-localization of arbitrary 2D depictions, such as paintings, drawings, or historical photographs. In contrast to [14, 47] who analyze 2D images, our method takes advantage of the knowledge and control over the 3D model to learn a set of mid-level 3D scene elements robust to a certain amount of viewpoint variation and capable of recovery of the (approximate) geo-referenced camera viewpoint. We show that the learnt mid-level scene elements are reliably detectable in 2D depictions of the scene despite large changes in appearance and rendering style.

14.3 Geo-localization by Matching Discriminative Visual Elements

The proposed method has two stages: first, in an offline stage we learn a set of discriminative visual elements representing one or more architectural sites; second, in an online stage a given unseen query depiction is aligned with the appropriate

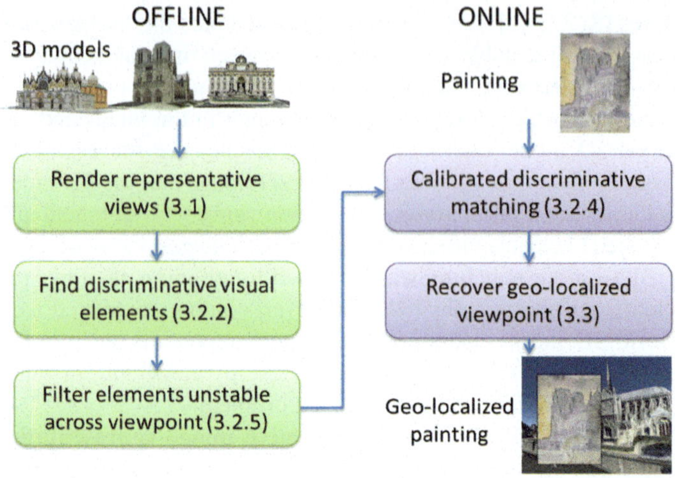

Fig. 14.2 Approach overview. In the offline stage, (*left*) we summarize a set of given geo-referenced 3D models using a collection of discriminative visual elements. In the online stage (*right*) we match the learnt visual elements to the input depiction and use the obtained correspondences to recover the camera viewpoint with respect to the best matching 3D model

3D model by matching with the learnt visual elements. The proposed algorithm is summarized in Fig. 14.2. In detail, the input to the offline stage are 3D models of multiple architectural sites. The output is a set of view-dependent visual element detectors able to identify specific structures of the different 3D models in various types of 2D imagery. The approach begins by rendering a set of representative views of each 3D model. Next, a set of visual element detectors is computed from the rendered views by identifying scene parts that are discriminative and can be reliably detected over a range of viewpoints. During the online stage, given an input 2D depiction, we match with the learnt visual element detectors and use the top scoring detections to recover a camera viewpoint with respect to the best matching 3D model.

14.3.1 Rendering Representative Views

We sample possible views of each 3D model in a similar manner to [2, 30, 41]. First, we identify the ground plane and corresponding vertical direction. The camera positions are then sampled on the ground plane on a regular grid. For each camera position, we sample 12 possible horizontal camera rotations assuming no in-plane rotation of the camera. For each horizontal rotation, we sample 2 vertical rotations (pitch angles). Views where less than 5 % of the pixels are occupied by the 3D model are discarded. This procedure results in 7,000–45,000 views for each model depending on the size of the 3D site. Note that the rendered views form only an intermediate representation and can be discarded after visual element detectors are extracted.

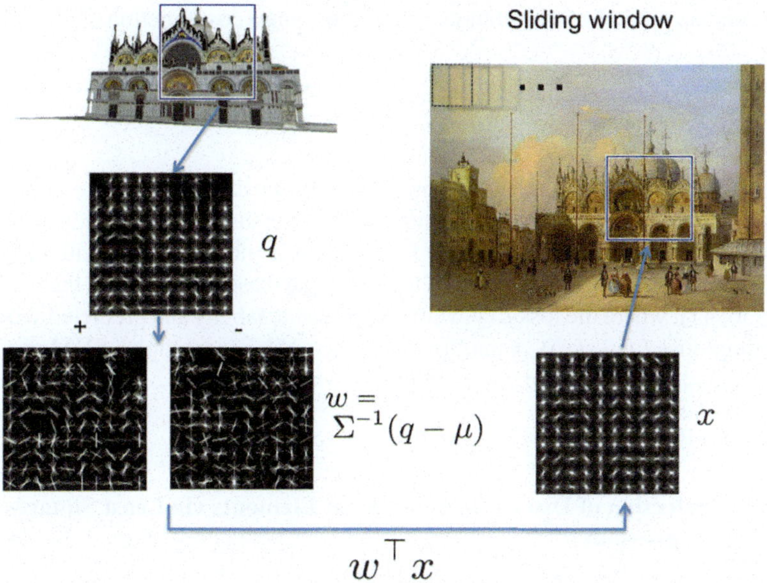

Fig. 14.3 Matching as classification. Given a region and its HOG descriptor q in a rendered view (*top left*), the aim is to find the corresponding region in a depiction (e.g., a painting, *top right*). This is achieved by training a linear HOG-based sliding window classifier using q as a single positive example and a large number of negative data. The classifier weight vector w is visualized by separately showing the positive (+) and negative (−) weights at different orientations and spatial locations. The best match x in the depiction is found as the maximum of the classification score

14.3.2 Finding and Matching Discriminative Visual Elements

14.3.2.1 Matching as Classification

The aim is to match a given rectangular image patch q (represented by a HOG descriptor [12]) in a rendered view to its corresponding image patch in the depiction, as illustrated in Fig. 14.3. Instead of finding the best match measured by the Euclidean distance between the descriptors, we train a linear classifier with q as a single positive example (with label $y_q = +1$) and a large number of negative examples x_i for $i = 1$ to N (with labels $y_i = -1$). The matching is then performed by finding the patch x^* in the depiction with the highest classification score

$$s(x) = w^\top x + b, \tag{14.1}$$

where w and b are the parameters of the linear classifier.

Parameters w and b can be obtained by minimizing a cost function of the following form

$$E(w,b) = L\left(y_q, w^T q + b\right) + \frac{1}{N}\sum_{i=1}^{N} L\left(y_i, w^T x_i + b\right), \tag{14.2}$$

where the first term measures the loss L on the positive example q (also called "exemplar") and the second term measures the loss on the negative data. A regularizer could be added to this cost E, but we found that was not necessary with our choice of loss functions. A particular case of the exemplar-based classifier is the exemplar-SVM [36, 47], where the loss $L(y, s(x))$ between the label y and predicted score $s(x)$ is the hinge-loss $L(y, s(x)) = \max\{0, 1 - ys(x)\}$ [5]. For exemplar-SVM cost (14.2) is convex and can be minimized using iterative algorithms [16, 45], but this remains computationally expensive.

14.3.2.2 Selection of Discriminative Visual Elements via Least Squares Regression

Using instead a square loss $L(y, s(x)) = (y - s(x))^2$, similarly to [4, 22], w_{LS} and b_{LS} minimizing (14.2) and the optimal cost E^*_{LS} can be obtained in closed form as

$$w_{\text{LS}} = \frac{2}{2 + \|\Phi(q)\|^2}\Sigma^{-1}(q - \mu), \tag{14.3}$$

$$b_{\text{LS}} = -\frac{1}{2}(q + \mu)^T w_{\text{LS}}, \tag{14.4}$$

$$E^*_{\text{LS}} = \frac{4}{2 + \|\Phi(q)\|^2}, \tag{14.5}$$

where $\mu = \frac{1}{N}\sum_{i=1}^{N} x_i$ denotes the mean of the negative examples, $\Sigma = \frac{1}{N}\sum_{i=1}^{N}(x_i - \mu)(x_i - \mu)^T$ their covariance and Φ is the "whitening" transformation such that

$$\|\Phi(x)\|^2 = (x - \mu)^\top \Sigma^{-1}(x - \mu). \tag{14.6}$$

We can use the value of the optimal cost (14.5) as a measure of the discriminability of a specific q. If the training cost (error) for a specific candidate visual element q is small, this visual element can be easily separated from the negative data and thus it is discriminative. This observation can be translated into a simple and efficient algorithm for ranking candidate element detectors based on their discriminability. Given a rendered view, we consider, as candidate visual elements, all patches that are local minima (in scale and space) of the training cost (14.5).

14.3.2.3 Relation to Linear Discriminant Analysis (LDA)

An alternative way to compute w and b is to use LDA, similarly to [22, 25]. This results in slightly different values of the parameters:

$$w_{\mathrm{LDA}} = \Sigma^{-1}(q - \mu_n), \tag{14.7}$$

and

$$b_{\mathrm{LDA}} = \frac{1}{2}\left(\mu^T \Sigma^{-1} \mu - q^T \Sigma^{-1} q\right). \tag{14.8}$$

Classifiers obtained by minimizing the least squares cost function (14.2) or satisfying the LDA ratio test can be used for matching a candidate visual element q to a 2D depiction as described in equation (14.1). Note that the decision hyperplanes obtained from the least squares regression, w_{LS}, and linear discriminant analysis, w_{LDA}, are parallel. As a consequence, for a particular visual element q the ranking of matches according to the matching score (14.1) would be identical for the two methods. In other words, in an object detection setup [12, 22, 25] the two methods would produce identical precision-recall curves. In our matching setup, for a given q the best match in a particular depiction would be identical for both methods. The actual value of the score, however, becomes important when comparing matching scores across different visual element detectors q. In object detection, the score of the learnt classifiers is typically calibrated on a held-out set of labeled validation examples [36].

14.3.2.4 Calibrated Discriminative Matching

We have found that calibration of matching scores across different visual elements is important for the quality of the final matching results. Below, we describe a procedure to calibrate matching scores without the need of any labeled data. First, we found [1] that the matching score obtained from LDA produces significantly better matching results than matching via least squares regression. Nevertheless, we found that the raw uncalibrated LDA score favors low-contrast image regions, which have an almost zero HOG descriptor. To avoid this problem, we further calibrate the LDA score by subtracting a term that measures the score of the visual element q matched to a low-contrast region, represented by zero (empty) HOG vector

$$s_{\mathrm{calib}}(x) = s_{\mathrm{LDA}}(x) - s_{\mathrm{LDA}}(0) \tag{14.9}$$

$$= (q - \mu)^T \Sigma^{-1} x. \tag{14.10}$$

This calibrated score gives much better results on the dataset of [27], as shown in [1], and significantly improves matching results.

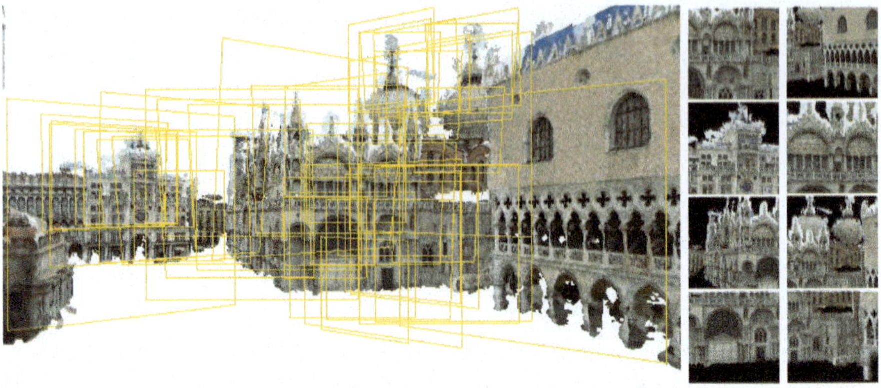

Fig. 14.4 Examples of selected visual elements for a 3D site. *Left* Selection of top ranked 50 visual elements visible from this specific view of the site. Each element is depicted as a planar patch with an orientation of the plane parallel to the camera plane of its corresponding source view. *Right* Subset of eight elements shown from their original viewpoints. Note that the proposed algorithm prefers visually salient scene structures such as the two towers in the *top-right* or the building in the *left part* of the view

14.3.2.5 Filtering Elements Unstable Across Viewpoint

We discard elements that cannot be reliably detected in close-by rendered views. This filtering criterion removes many unstable elements that are, for example, ambiguous because of repeated structures in the rendered view or cover large depth discontinuities and hence significantly change with viewpoint. To achieve that, we perform two additional tests on each visual element. First, to suppress potential repeated structures, we require that the ratio between the score of the first and second highest scoring detection in the image is larger than a threshold of 1.04, similar to [35]. Second, we run the discriminative elements in the views near the one where they were defined and keep only visual elements that are successfully detected in more than 80 % of the nearby views. The definition of what exactly is a nearby view is a difficult question, and the number of nearby views we consider varies greatly with the viewpoint. We refer the reader to [1] for more details. This procedure typically results in several thousand selected elements for each architectural site. Examples of the final visual elements obtained by the proposed approach are shown in Fig. 14.4.

14.3.3 Geo-localization by Recovering Viewpoint

In this section, we describe how, given the set of discriminative visual elements gleaned from the set of 3D models, we identify which 3D site is depicted in the input depiction, and recover the viewpoint of the input depiction with respect to the 3D model. We assume that the depictions are perspective scene renderings and seek to recover the camera center and the camera rotation matrix via camera resectioning [26]. As all 3D models are geo-referenced, the recovered camera position and viewpoint provide a geo-localization of the input depiction.

For detection, each discriminative visual element takes as input a 2D patch from the depiction and returns as output a 3D model ID, a 3D location **X** on the 3D model, a plane representing the patch extent on the 3D model centered at **X**, and a detector response score indicating the quality of the appearance match. Following the matching procedure described in Sect. 14.3.2.4, we form a set of putative discriminative visual element matches using the following procedure. First, we apply to the input depiction all visual element detectors from all 3D models and take the top 200 detections sorted according to the first to second nearest neighbor ratio [35], using the calibrated similarity score (14.9). This selects the least ambiguous matches. Second, we sort the 200 matches directly by score (14.9) and select the top 25 matches. This two step selection process choses putative matches that are both non-ambiguous (step 1) and have a high matching score (step 2). From each putative visual element match we obtain 5 putative point correspondences by taking the 2D/3D locations of the patch center and its four corners. The patch corners provide information about the patch scale and the plane location on the 3D model, which has been shown to work well for structure-from-motion with planar constraints [49]. At this point, the putative correspondences could still match to several different 3D models. To resolve this, we use RANSAC [18] to find the set of inlier correspondences to a camera model with a constraint that inliers must come from the same architectural site. Among the 125 putative correspondences derived from the 25 putative matches, at each RANSAC iteration we first select a correspondence at random, followed by a random selection of two correspondences from the same 3D model. We use the three points to estimate a camera matrix and then compute the number of inlier correspondences to the camera matrix. We use a restricted camera model where the intrinsics are fixed with the focal length set to the image diagonal length and the principal point set to the center of the image. The result of this RANSAC procedure is both a best matching 3D

Fig. 14.5 Illustration of alignment. We use the recovered discriminative visual elements to find correspondences between the input scene depiction (*left*) and a geo-referenced 3D model (*right*). Shown is the recovered viewpoint and inlier visual elements found via RANSAC

model and the corresponding camera matrix. The recovered viewpoint geo-localizes the input depiction as well as provides an alignment of the input depiction to the 3D model, as shown in Fig. 14.5.

14.4 Results

In this section, we evaluate the potential of our method for geo-localizing a given 2D depiction across different architectural sites and the quality of the recovered viewpoint. A detailed analysis of the 2D–3D instance alignment pipeline, as well as comparisons with other methods are given in [1]. Note that this prior work aligns a historical photograph or a non-photographic depiction to a 3D model of the depicted site assuming the identity of the depicted site is known. Here we are interested in the harder task of identifying the correct architectural site among a set of given 3D models of different architectural sites.

In the following, dataset and performance measures are described in Sect. 14.4.1, quantitative evaluation is given in Sect. 14.4.2 and qualitative results are shown in Sect. 14.4.3. Finally, the main failure modes are discussed in Sect. 14.4.4.

14.4.1 Dataset and Performance Measures

We consider a subset of the dataset introduced in [1] consisting of 3D models and historical photographs/non-photographic depictions of three architectural landmarks. The dataset contains 3D models downloaded from Trimble 3D Warehouse for the following architectural landmarks: Notre Dame of Paris, Trevi Fountain, and San Marco's Basilica. The 3D models consist of basic primitive shapes and have a composite texture from a set of images. The 2D depictions for the three sites were collected by [1] from the Internet and include 71 historical photographs and 224 non-photographic depictions, with 39 engravings, 58 drawings, and 127 paintings. The drawings category includes color renderings and the paintings category includes different rendering styles, such as watercolors, oil paintings, and pastels. Table 14.1 shows the number of images belonging to each category across the different sites.

We measure performance for the following two tasks. First, we evaluate the geo-localization accuracy, which measures the percentage of input depictions that are matched to the correct architectural site. Second, for the depictions assigned to the correct architectural site we evaluate the quality of the resulting alignment, which is measured by a user study via Amazon Mechanical Turk.

14.4.2 Quantitative Evaluation

We summarized the three Trimble 3D Warehouse models with 15,000 discriminative visual elements each. For each input depiction, we applied all of the 45,000 detectors

Table 14.1 Statistics of the dataset of historical photographs and non-photographic depictions used for our quantitative evaluation

	S. Marco Basilica	Trevi Fountain	Notre Dame	Total
Hist. photos	30	0	41	71
Paintings	41	34	52	127
Drawings	19	5	34	58
Engravings	9	10	20	39
Total	99	49	147	295

Table 14.2 The percentage of input depictions that were assigned to the correct architectural site split across different sites (rows) and depiction styles (columns)

	Paintings (%)	Historical photograph (%)	Engravings (%)	Drawings (%)	Average (%)
S. Marco Basilica	83	87	89	94	87
Trevi Fountain	82	–	90	80	84
Notre Dame	90	88	85	79	86
Average	86	87	87	84	86

Note that there are no historical photographs for Trevi Fountain in the database of [1]

corresponding to those elements, selected the 25 most confident ones, and performed camera resectioning using RANSAC as described in Sect. 14.3.3, with the constraint that only elements from the same site could be counted as inliers. Thus, our output is both a specific 3D model and a viewpoint.

We first report results on the task of identifying the 3D model of the architectural site. Table 14.2 shows the results separately for the three different sites and across different depiction styles. Despite the difficulty of the task due to the large variety of viewpoints and styles, our method identified correctly the architectural site for 86 % of the depictions, which is much larger than the 33 % chance performance.

We then evaluated the quality of the alignments for depictions that were assigned to the correct site. To quantitatively evaluate the goodness of our alignments, we have conducted a user study via Amazon Mechanical Turk. As in [1], the workers were asked to judge the viewpoint similarity of the resulting alignments to their corresponding input depictions by categorizing the viewpoint similarity as either a (a) Good match, (b) Coarse match, or (c) No match, illustrated in Fig. 14.6. We asked five different workers to rate the viewpoint similarity for each depiction and we report the majority opinion. Table 14.3 shows the performance of our algorithm for the different depiction styles. The performance varies across depiction style from 77 % of coarse/good matches for paintings to more than 90 % for historical photographs or engravings. Overall, 83 % of the input depictions are at least coarsely matched.

Fig. 14.6 Alignment evaluation criteria. We asked workers on Amazon Mechanical Turk to judge the viewpoint similarity of the resulting alignment to the input depiction. The workers were asked to categorize the viewpoint similarity into one of three categories: **a** Good match—the two images show a roughly similar view of the building; **b** coarse match—the view may not be similar, but the building is roughly at the same location in both images, not upside down, and corresponding building parts can be clearly identified; **c** no match—the views are completely different, e.g., upside down, little or no visual overlap

Table 14.3 Viewpoint similarity user study of our algorithm across different depiction styles

	Good match (%)	Coarse match (%)	No match (%)
Historical photographs	74	16	10
Paintings	57	20	23
Drawings	59	20	20
Engravings	65	29	6
Average	63	20	17

Fig. 14.7 Alignment of historical photographs of San Marco's Square (*top*) and Notre Dame of Paris (*bottom*) to their respective 3D models

14.4.3 Qualitative Evaluation

Figures 14.7 and 14.8 show example alignments of historical photographs and non-photographic depictions, respectively. Notice that the depictions are reasonably well aligned with the 3D models, with regions on the 3D model rendered onto the corresponding location for a given depiction. We are able to cope with a variety of viewpoints with respect to the 3D model as well as different depiction styles. Our approach succeeds in recovering the approximate viewpoint in spite of these challenging appearance changes and the varying quality of the 3D models.

Fig. 14.8 Example alignments of non-photographic depictions to 3D models. Notice that we are able to align depictions rendered in different styles and having a variety of viewpoints with respect to the 3D models

Note that Figs. 14.7 and 14.8, in addition to the Trimble 3D Warehouse models, also include alignment results using a 3D model of San Marco's Square that was reconstructed from a set of photographs using dense multi-view stereo [21]. While the latter 3D model has more accurate geometry than the Trimble 3D Warehouse models, it is also much noisier along the model boundaries. This model was excluded from the quantitative evaluation in Sect. 14.4.2 as it overlaps with the San Marco Basilica Trimble 3D Warehouse model, but we include it here to demonstrate alignment for different types of 3D models.

Fig. 14.9 Challenging examples successfully aligned by our method where the assumption of a perspective scene rendering is violated. Note that the drawing in (**c**) is a completely different cathedral. **a** Scene distortion, **b** drawing and 3D errors, **c** major structural differences

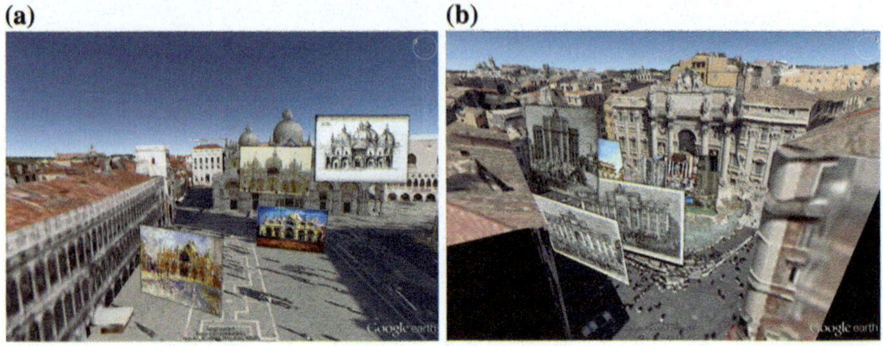

Fig. 14.10 Recovered viewpoints of some of the geo-localized depictions visualized in Google Earth. **a** San Marco's Basilica, **b** Trevi Fountain

In Fig. 14.9, we show alignments for a set of challenging examples where the assumption of a perspective rendering is significantly violated, but the proposed approach was still able to recover a reasonable alignment. Notice the severe non-perspective scene distortions, drawing errors, and major architectural differences (e.g., a part of the landmark may take a completely different shape).

Figures 14.1 and 14.10 show the recovered viewpoints of several different depictions for the three sites rendered in Google Earth. Figure 14.11 shows individual depictions rendered in Google Earth, which showcases a re-photography application by allowing a user to browse the depictions in the context of their modern environments. Please see additional qualitative results on the project webpage [28].

14.4.4 Failure Modes

We have identified three main failure modes of our algorithm, examples of which are shown in Fig. 14.12. The first is due to large-scale symmetries, for example when the front and side facade of a building are very similar. This problem is difficult to resolve with only local reasoning. For example, the proposed cross-validation step removes repetitive structures visible in the same view but not at different locations

Fig. 14.11 Examples of geo-localized depictions visualized in Google Earth. Note that the proposed method allows us to visualize the specific place across time and through the eyes of different artists. **a** Notre Dame, **b** San Marco's Basilica, **c** Trevi Fountain

of the site. The second failure mode is due to locally confusing image structures, for example, the vertical support structures on the cathedral in Fig. 14.12 (middle) are locally similar (by their HOG descriptor) to the vertical pencil strokes on the drawing. The learnt mid-level visual elements have a larger support than typical local invariant features (such as SIFT) and hence are typically more distinctive. Nevertheless, such mismatches can occur and in some cases are geometrically consistent with a certain view of the 3D model. The third failure mode is when the depicted viewpoint is not covered in the set of sampled views. This can happen for unusual viewpoints including extreme angles, large close-ups, or cropped views. Such unusual views are in some cases assigned to a wrong 3D site.

Fig. 14.12 Example failure cases. *Top* large-scale symmetry. Here, arches are incorrectly matched on a building with similar front and side facades. *Middle* locally confusing image structures. Here the vertical support structures on the cathedral (*right*) are locally similar by their HOG descriptor to the vertical pencil strokes on the drawing (*left*). *Bottom* two examples of paintings with unusual viewpoints

14.5 Conclusion

We have demonstrated that automatic geo-localization is possible for a range of non-photographic depictions and historical photographs, which represent extremely challenging cases for current local feature matching methods. To achieve this we have developed an approach to compactly represent 3D models of architectural sites by a set of visually distinct mid-level scene elements extracted from rendered views, and have shown that they can be reliably matched in a variety of photographic and non-photographic depictions. We have also shown an application of the proposed approach to computational re-photography to automatically geo-tag and find an approximate viewpoint of historical photographs and paintings, which allows for geo-browsing within Google Earth. This work is just a step towards computational reasoning about the content of non-photographic depictions. The developed approach for extracting visual elements opens-up the possibility of efficient indexing for visual search of paintings and historical photographs (e.g., via hashing of the HOG features as in [13]), or automatic fitting of complex non-perspective models used in historical imagery [40]. It would be also interesting to investigate learning our 3D mid-level visual elements with convolutional neural network descriptors [15, 23, 32, 38, 44, 52], which have recently shown promising results in object detection in non-photographic depictions [9].

Acknowledgments We are grateful to Guillaume Seguin, Alyosha Efros, Guillaume Obozinski and Jean Ponce for their useful feedback, and to Yasutaka Furukawa for providing access to the San Marco 3D model. This work was partly supported by the EIT ICT Labs, ANR project SEMAPOLIS (ANR-13-CORD-0003), and the ERC starting grant LEAP. The work was partly carried out at IMAGINE, a joint research project between Ecole des Ponts ParisTech (ENPC) and the Scientific and Technical Centre for Building (CSTB). Supported by the Intelligence Advanced Research Projects Activity (IARPA) via Air Force Research Laboratory. The U.S. Government is authorized to reproduce and distribute reprints for governmental purposes notwithstanding any copyright annotation thereon. Disclaimer: The views and conclusions contained herein are those of the authors and should not be interpreted as necessarily representing the official policies or endorsements, either expressed or implied, of IARPA, AFRL or the U.S. Government.

References

1. Aubry M, Russell B, Sivic J (2014) Painting-to-3D model alignment via discriminative visual elements. ACM Trans Graphics 33(2)
2. Baatz G, Saurer O, Köser K, Pollefeys M (2012) Large scale visual geo-localization of images in mountainous terrain. In: Proceedings of European conference on computer vision
3. Baboud L, Cadik M, Eisemann E, Seidel HP (2011) Automatic photo-to-terrain alignment for the annotation of mountain pictures. In: Proceedings of the conference on computer vision and pattern recognition
4. Bach F, Harchaoui Z (2008) Diffrac: a discriminative and flexible framework for clustering. In: Advances in neural information processing systems
5. Bishop CM (2006) Pattern recognition and machine learning. Springer
6. Bosché F (2010) Automated recognition of 3D CAD model objects in laser scans and calculation of as-built dimensions for dimensional compliance control in construction. Adv Eng Inf 24(1):107–118
7. Chen D, Baatz G et al (2011) City-scale landmark identification on mobile devices. In: Proceedings of the conference on computer vision and pattern recognition
8. Chum O, Matas J (2006) Geometric hashing with local affine frames. In: Proceedings of the conference on computer vision and pattern recognition
9. Crowley EJ, Zisserman A (2014) In search of art. In: Workshop on computer vision for art analysis, ECCV
10. Crowley EJ, Zisserman A (2014) The state of the art: object retrieval in paintings using discriminative regions. In: British machine vision conference

11. Cummins M, Newman P (2009) Highly scalable appearance-only SLAM—FAB-MAP 2.0. In: Proceedings of robotics: science and systems, Seattle, USA
12. Dalal N, Triggs B (2005) Histograms of oriented gradients for human detection. In: Proceedings of the conference on computer vision and pattern recognition
13. Dean T, Ruzon M, Segal M, Shlens J, Vijayanarasimhan S, Yagnik J (2013) Fast, accurate detection of 100,000 object classes on a single machine. In: Proceedings of the conference on computer vision and pattern recognition
14. Doersch C, Singh S, Gupta A, Sivic J, Efros AA (2012) What makes Paris look like Paris? ACM Trans Graphics (Proc SIGGRAPH) 31(4)
15. Donahue J, Jia Y, Vinyals O, Hoffman J, Zhang N, Tzeng E, Darrell T (2013) Decaf: a deep convolutional activation feature for generic visual recognition. arXiv:1310.1531
16. Fan R, Chang K, Hsieh C, Wang X, Lin C (2008) Liblinear: a library for large linear classification. J Mach Learn Res 9(1):1871–1874
17. Felzenszwalb P, Girshick R, McAllester D, Ramanan D (2010) Object detection with discriminatively trained part based models. IEEE Trans Pattern Anal Mach Intell 32(9)
18. Fischler MA, Bolles RC (1981) Random sample consensus: a paradigm for model fitting with applications to image analysis and automated cartography. Commun ACM 24(6):381–395
19. Frome A, Singer Y, Sha F, Malik J (2007) Learning globally-consistent local distance functions for shape-based image retrieval and classification. In: Proceedings of international conference on computer vision
20. Furukawa Y, Ponce J (2010) Accurate, dense, and robust multi-view stereopsis. IEEE Trans Pattern Anal Mach Intell 32(8)
21. Furukawa Y, Curless B, Seitz SM, Szeliski R (2010) Towards internet-scale multi-view stereo. In: Proceedings of the conference on computer vision and pattern recognition
22. Gharbi M, Malisiewicz T, Paris S, Durand F (2012) A Gaussian approximation of feature space for fast image similarity. Technical report, MIT
23. Girshick R, Donahue J, Darrell T, Malik J (2014) Rich feature hierarchies for accurate object detection and semantic segmentation. In: Proceedings of the conference on computer vision and pattern recognition
24. Gronat P, Obozinski G, Sivic J, Pajdla T (2013) Learning and calibrating per-location classifiers for visual place recognition. In: Proceedings of the conference on computer vision and pattern recognition
25. Hariharan B, Malik J, Ramanan D (2012) Discriminative decorrelation for clustering and classification. In: Proceedings of European conference on computer vision
26. Hartley RI, Zisserman A (2004) Multiple view geometry in computer vision, 2n edn. Cambridge University Press. ISBN: 0521540518
27. Hauagge D, Snavely N (2012) Image matching using local symmetry features. In: Proceedings of the conference on computer vision and pattern recognition
28. http://www.di.ens.fr/willow/research/painting_to_3d/
29. Huttenlocher DP, Ullman S (1987) Object recognition using alignment. In: International conference on computer vision
30. Irschara A, Zach C, Frahm JM, Bischof H (2009) From structure-from-motion point clouds to fast location recognition. In: Proceedings of the conference on computer vision and pattern recognition
31. Knopp J, Sivic J, Pajdla T (2010) Avoiding confusing features in place recognition. In: Proceedings of European conference on computer vision
32. Krizhevsky A, Sutskever I, Hinton GE (2012) Imagenet classification with deep convolutional neural networks. In: Advances in neural information processing systems
33. Li Y, Snavely N, Huttenlocher D, Fua P (2012) Worldwide pose estimation using 3D point clouds. In: Proceedings of European conference on computer vision
34. Lowe D (1987) The viewpoint consistency constraint. Int J Comput Vis 1(1):57–72
35. Lowe DG (2004) Distinctive image features from scale-invariant keypoints. Int J Comput Vis 60(2):91–110

36. Malisiewicz T, Gupta A, Efros AA (2011) Ensemble of exemplar-svms for object detection and beyond. In: Proceedings of international conference on computer vision
37. Nister D, Stewenius H (2006) Scalable recognition with a vocabulary tree. In: Proceedings of the conference on computer vision and pattern recognition
38. Oquab M, Bottou L, Laptev I, Sivic J (2014) Learning and transferring mid-level image representations using convolutional neural networks. In: Proceedings of the IEEE conference on computer vision and pattern recognition
39. Philbin J, Chum O, Isard M, Sivic J, Zisserman A (2007) Object retrieval with large vocabularies and fast spatial matching. In: Proceedings of the conference on computer vision and pattern recognition
40. Rapp J (2008) A geometrical analysis of multiple viewpoint perspective in the work of Giovanni Battista Piranesi: an application of geometric restitution of perspective. J Arch 13(6)
41. Russell BC, Sivic J, Ponce J, Dessales H (2011) Automatic alignment of paintings and photographs depicting a 3D scene. In: IEEE workshop on 3D representation for recognition (3dRR-11), associated with ICCV
42. Sattler T, Leibe B, Kobbelt L (2011) Fast image-based localization using direct 2d-to-3d matching. In: Proceedings of international conference on computer vision
43. Schindler G, Brown M, Szeliski R (2007) City-scale location recognition. In: Proceedings of the conference on computer vision and pattern recognition
44. Sermanet P, Eigen D, Zhang X, Mathieu M, Fergus R, LeCun Y (2013) Overfeat: integrated recognition, localization and detection using convolutional networks. arXiv:1312.6229
45. Shalev-Shwartz S, Singer Y, Srebro N, Cotter A (2011) Pegasos: primal estimated sub-gradient solver for SVM. Math Program Seri B 127(1):3–30
46. Shechtman E, Irani M (2007) Matching local self-similarities across images and videos. In: Proceedings of the conference on computer vision and pattern recognition
47. Shrivastava A, Malisiewicz T, Gupta A, Efros AA (2011) Data-driven visual similarity for cross-domain image matching. In: ACM Trans Graphics (Proc SIGGRAPH Asia)
48. Sivic J, Zisserman A (2003) Video Google: a text retrieval approach to object matching in videos. In: Proceedings of international conference on computer vision
49. Szeliski R, Torr P (1998) Geometrically constrained structure from motion: points on planes. In: European workshop on 3D structure from multiple images of large-scale environments (SMILE)
50. Torii A, Sivic J, Pajdla T, Okutomi M (2013) Visual place recognition with repetitive structures. In: Proceedings of the conference on computer vision and pattern recognition
51. Zamir A, Shah M (2010) Accurate image localization based on google maps street view. In: Proceedings of European conference on computer vision
52. Zeiler M, Fergus R (2013) Visualizing and understanding convolutional networks. arXiv:1311.2901

Part IV
Real-World Applications

Practical tools for geo-localization of web-scale databases and utilizing the extracted geo-location

Thus far in the book, we focused on the problem of geo-localization, defined as automatically estimating the location of an image. An important follow-up question would be: *what the location of an image can be used for?* The fact that modern cameras and cell phones are GPS-enabled adds to the importance of this question, as a considerable percentage of images captured nowadays are location-tagged at the time of collection. In the last part of the book, we discuss several real-world applications that actively utilize the location of an image, no matter extracted automatically or acquired from the GPS-chip, for better understanding its content. Also, as another real-world scenario, we will discuss a practical approach to geo-localization where a fully automated localization from scratch is inefficient, and the insights of a user in the loop can lead to a faster and more accurate localization.

Chapter 15
A Memory Efficient Discriminative Approach for Location-Aided Recognition

Sudipta N. Sinha, Varsha Hedau, C. Lawrence Zitnick and Richard Szeliski

Abstract In this chapter, we describe a visual recognition technique for fast recognition of urban landmarks on a GPS-enabled mobile device. Most existing methods offload their computation to a server by uploading the query image. Over a slow network, this can cause a latency of several seconds. In contrast, our approach requires uploading only the approximate GPS location to a server after which a compact, location-specific classifier is downloaded to the device and all subsequent computation is performed on it. Our approach is supervised and involves training compact random forest classifiers (RDF) on a database of geo-tagged images. The feature vector for the RDF is computed by densely searching the image for the presence of selective discriminative local image patches extracted from the training images. The images are rectified using detected vanishing points and binary descriptors allow for an efficient search for the discriminative patches, a step that is further accelerated using min-hash. We have evaluated the performance of our approach on representative urban datasets where it outperforms traditional methods based on bag-of-visual-words features or direct matching of local feature descriptors, neither of which are feasible approaches when processing must occur on a low-power mobile device.

S.N. Sinha (✉)
Microsoft Research, Redmond, WA, USA
e-mail: sudipsin@microsoft.com

V. Hedau
Apple, Cupertino, CA, USA
e-mail: varsha.hedau@gmail.com

C.L. Zitnick
Facebook AI Research, Palo Alto, CA, USA
e-mail: zitnick@fb.com

R. Szeliski
Facebook, Seattle, WA, USA
e-mail: szeliski@fb.com

A.R. Zamir et al. (eds.), *Large-Scale Visual Geo-Localization*,
Advances in Computer Vision and Pattern Recognition,
DOI 10.1007/978-3-319-25781-5_15

15.1 Introduction

The ubiquity of cameras on GPS-enabled mobile devices such as smartphones provides great utility for determining the location and contents of an image. It may be used to learn more about a specific landmark turning the phone into a device for visual queries or for automatic image-tagging allowing the images to be searched and organized later on. We address the problem of recognizing the landmark or location from a single image where a predefined set of locations and landmarks is available. In this chapter, we will describe a technique where the query processing can be implemented to run completely on a low-power mobile device.

The feasibility of location recognition has recently increased with the availability of large databases of dense geo-tagged imagery, e.g., Flickr photo collections and streetside imagery [24, 25] for cities. Recently, several approaches to location recognition have been proposed based on such databases [6, 15, 23, 31–33, 46, 53].

The main challenge in landmark or location recognition arises from variations in scale and scene appearance due to a diversity of camera viewpoints or due to illumination changes from time-of-day, weather, seasons etc. Although the intrinsic scene appearance does not change a lot, the visibility of landmarks covering a wide area can change dramatically with viewpoints. In addition, foreground objects such as people or cars can clutter the scene at busy urban locations. Finally, the appearance of a location may change temporally due to seasonal variation, or permanently due to construction, new store ownership, new billboards, etc.

Broadly speaking, most existing methods for landmark classification pose it as an image retrieval task [23, 43]. which requires ranking the database of geo-tagged images based on similarity to the query image. The similarity score is typically computed by comparing sets of local feature descriptors extracted at scale-invariant keypoints detected in the images. Although fast retrieval techniques exist for databases with millions of images, constant access to this huge database is necessary with consequent high storage and memory costs. For high precision, an expensive reranking step is often used for geometric postverification of feature matches over image pairs. Recently proposed alternatives to the retrieval approaches include direct image matching to sparse precomputed structure from motion (SfM) point clouds [33, 34, 36, 45]. Despite the higher compression in these methods, their storage costs and computational requirements are still quite significant.

Image categorization methods, on the other hand, use labeled data to train classifiers that are typically efficient to evaluate at query time and have lower storage costs [8, 32]. However, most of the existing techniques have traditionally focused on objects or scene categories. Large-scale and fine-grained classification has recently gained interest among researchers, however, with the exception of [8, 21, 32], landmark or location classification has received less attention from a purely classification or categorization standpoint.

15.1.1 Overview

With the advent of GPS-enabled modern mobile devices, one important and new aspect of location or landmark recognition is problem scoping, i.e., limiting the set of locations or landmarks that need to be searched. This is because GPS devices can easily narrow down a user's location to within approximately a hundred meters. In this chapter, we describe a memory-efficient discriminative approach [32] to location recognition that exploits approximate GPS coordinates. Instead of storing feature vectors for all the geo-tagged images in a database, we store compact, location-specific classifiers. Each classifier for a certain landmark is trained to distinguish it from only the nearby landmarks that lie in close proximity within the coarse location predicted using GPS. Our location classifiers are based on Random Decision Forests (RDF) [9] trained to predict the correct location or landmark within scope based on a global feature descriptor computed from the query image. Each dimension of this feature vector encodes the presence or absence of a specific discriminative local image patch within the query image.

During training, we identify potential candidates for such discriminative patches from the images of all the locations under consideration. Next, each candidate in this pool is densely matched to patches in all the training images and the matching score of the most similar patch found in the image is recorded in the global feature descriptor. All images are rectified prior to dense matching [6, 38, 44] to reduce the search space to just scale and 2D position and for improving robustness to perspective distortions. At query time, we propose to use a min-hash approach to speedup the feature vector computation. Specifically, this involves using approximate nearest neighbor search to identify the most similar patch within the query image, for each of the discriminative patches used by the RDF classifier.

We have evaluated our method on two public benchmarks for landmark recognition and two sets of urban images that we captured with a smartphone in a typical urban streetside scene. In the latter case, the query images were collected almost a year after the time the training imagery was acquired. Notable appearance changes were observed in the query images due to seasonal changes, viewpoints as well as structural changes due to construction. We show that even when approximately a hundred locations or landmarks must be discriminated from each other, our proposed method uses only about 120 KB for storage without sacrificing much accuracy and its performance is comparable to some of the most accurate methods that use significantly more storage. Thus, the approach appears to have several advantages. First, the RDF classifier achieves high accuracy despite the compactness constraints. Second, the min-hash-based dense matching and efficient feature vector computation step makes the query time computation feasible on a low-power mobile device. Image rectification helps to reduce the search space for dense matching and provides tolerance to large viewpoint changes that induce large perspective distortions in the image. Finally, RDF classifiers also implicity perform feature selection and this increases the overall likelihood of choosing more repeatable and temporally stable local dictionary features without additional postprocessing during the training stage.

15.1.2 Related Work

A number of location or landmark recognition approaches utilize local keypoint feature descriptors but improve their matching accuracy by employing offline learning. Knopp et al. [29] remove confusing features in a bag-of-words model, Torii et al. [49] exploit repeating features whereas Li et al. [33] prioritize the matching of repeatable and frequently occurring features. Turcot and Lowe [50] remove features with low repeatability and utilize features co-occuring in neighboring images. A tree-based descriptor quantization approach that maximizes the information gain of quantized visual words was proposed by Schindler et al. [46]. Zamir et al. [53] remove noisy matches between images using a variant of the descriptor distance ratio test [37] and Zhang et al. [54] uses a motion estimation technique that is robust to outliers. Our approach uses keypoint detection and local descriptor matching to identify a pool of candidates used for constructing the dictionary of discriminative patches. However, the global image descriptor or feature vector used by the RDF classifier is then computed using efficient min-hash-based dense matching techniques within the whole image rather than only at detected keypoints. Min hash, which is a common locality sensitive hashing technique, has been shown to be effective for efficient near-duplicate image search in large databases [41, 42].

The significance of image rectification using vanishing points as a preprocessing step for wide-baseline image matching and location recognition has been studied [6, 44]. This step provides higher invariance to perspective distortions when matching rectilinear structures in images of building facades in street scenes [38]. A heading dependent bag-of-features representation for panoramic images has been proposed by Guan et al. [22]. Efficient feature extraction and compactness can be obtained using compressed sensing techniques [22] or with low bit-rate compressed visual feature descriptors [14]. Another alternative approach to obtaining compact discriminative descriptors involves learning discriminative embeddings [26] or mapping the high-dimensional descriptors into the Hamming space to produce short binary codes that can be compared very efficiently [10].

For recognizing images of urban scenes, a number of recent methods employ offline structure from motion computation as a preprocessing step [17, 27, 35, 36, 39, 45]. Benchmarks for mobile image-based localization have been constructed to evaluate the performance of recognition systems [15, 16]. Clemens et al. [4] proposed such a method for localization on a mobile phone. Their approach is computationally efficient but the memory requirements grows linearly with scene size. Arth et al. [3] propose a similar technique that uses a panoramic image as a query and exploits inertial sensors available on most mobile devices. A number of recent augmented reality techniques related to image-based localization are discussed in [5].

For real-time image-based localization on images or video, efficient search index schemes have been used for matching image patch descriptors [36, 45]. Scale-invariant feature extraction at query-time can be slow but it can sometimes be avoided by increasing the storage redundancy [36]. For real-time performance on mobile devices, a part of the computation is often offloaded to a server [39]. Discriminative indexing strategies have been proposed recently for improving the accuracy of descriptor-based matching and classification of landmarks [12, 13]. Specifically, Cao and Snavely [12] exploits the implicit graph connectivity between known locations of the training images to learn more accurate image descriptors [13] leading to compact model representations without sacrificing area coverage or discriminative information.

Generic scene recognition can be achieved with approximate nearest neighbor search based on high-dimensional features consisting of color and texture histograms, line features, and global descriptors that encode image appearance, such as GIST [23]. Image-based features have also been shown to be effective for recognizing natural scenes and mountainous terrain [7]. In certain domains, hierarchical classification approaches [54] have been explored. However, as mentioned earlier, these methods all have high memory requirements as the feature vectors of the training set must be stored and accessed at query time.

Our work follows a different line of work that is more closely related to [8, 19, 21, 32]. Unlike localization methods, these techniques are not designed to estimate a metric camera pose or to register the image to a 3D scene model. Nevertheless, the predicted location label is sufficient for answering visual queries or automatically generating tags or useful metadata for the query images. Doersch et al. [19] propose a method to automatically discover discriminative mid-level visual elements from large-scale streetside imagery in urban scenes. These visual elements are shown to be extremely useful for geo-localization. The three methods [8, 21, 32] formulate location recognition as a multiclass classification problem and train compact linear classifiers for predicting the image label. Specifically, these methods train different variants of support vector machines (SVM) on high-dimensional global image feature vectors that are computed using a visual codebook and popular feature encoding techniques. Bergamo et al. [8] propose a method to learn a highly discriminative codebook by leveraging image correspondences recovered from Internet image collections using a modern structure from motion (SfM) technique. For large photo collections of landmarks and tourist scenes, their approach is an efficient alternative to [19] for discovering discriminative and repeatable visual features in the scene.

In contrast to [8, 21, 32], we use random forests for location classification. Random decision forests (RDF) [9] are applicable to a wide variety of tasks in computer vision, such as multiclass classification, regression, and density estimation as discussed in [18]. Some of its early applications include shape recognition [2], semantic segmentation [28], and efficient keypoint matching [30]. This chapter describes how they are also well suited for training memory-efficient location classifiers that can be compactly represented. Typically, RDFs have two advantages over SVMs. First, they extend naturally to multiclass classification tasks. Second, feature selection which is crucial for a compact representation, is performed implicity when training RDFs but may require expensive postprocessing when training SVMs [51].

15.2 Learning Location Classifiers

In this section, we describe our method for learning location classifiers using a RDF classifier. We assume that a training set of geo-referenced images has been collected within an area that can be predicted directly using GPS. We assume that the images of all the important locations or landmarks within this area in the training set have been labeled. Each location or landmark is treated as a different class in our method.

Classification is performed using a dictionary of highly discriminative image patches that are extracted from the geo-referenced images using a keypoint detector during offline training. The input to the RDF classifier is a high-dimensional image feature vector whose dimensionality is equal to the number of selected patches in the dictionary. Each dimension of the feature vector stores the matching cost between a specific dictionary patch and the patch in the input image that is most similar to it. The use of dense matching typically decreases the variance of the input features to the RDF compared to when sparse keypoint features are used [32]. However, to efficiently identify a set of discriminative dictionary patches, we first perform pairwise image matching [37] on the set of training images of each landmark or location and extract a subset of the most repeatable patches within these images. These typically correspond to the keypoints whose feature descriptors were matched reliably. The collection of such patches extracted from the different location classes serves as a pool of candidate features for the subsequent classifier training stage.

An overview of our location recognition system is illustrated in Fig. 15.1. The various stages of this system are described in the following sections. First, in Sect. 15.2.1, we describe our approach for rectifying the image using the extracted vanishing

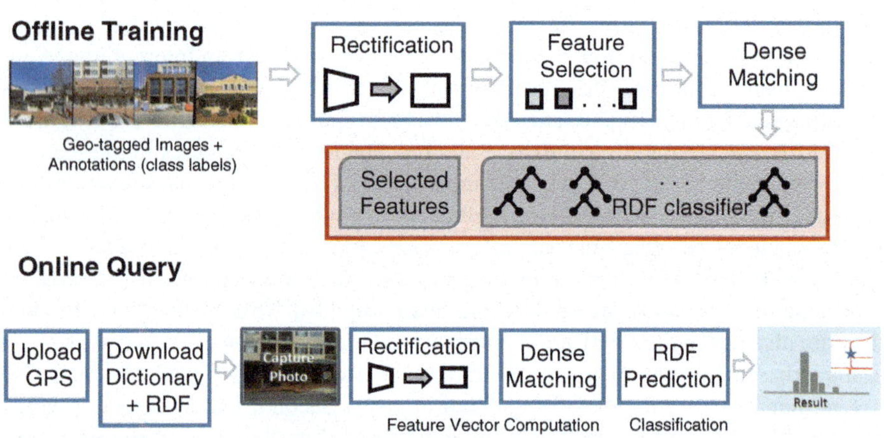

Fig. 15.1 An overview of our classification approach to location recognition. The training stage involves image rectification followed by the selection of discriminative local patches via image matching. Dense matching and search for each discriminative patch is used to construct feature vectors for a random decision forest (RDF) classifier. On a mobile device, at query time, only the GPS information needs to be uploaded to the server after which a compact, location-specific RDF classifier can be downloaded to the device for local computation and processing

points. Next, in Sect. 15.2.2, our approach for selecting a pool of candidate discriminative patches to be used as a dictionary for the classifier is described. Finally, Sect. 15.2.3 describes the details of training the random forests. Our method for efficiently constructing the feature vector using approximate nearest neighbor patch search is described in Sect. 15.2.4.

15.2.1 Image Rectification

Planar objects such as building facades may undergo severe perspective distortion in the image depending on the orientation and position of the camera. Searching over the full eight degrees of freedom provided by perspective distortions is computationally prohibitive for dense matching. If the focal length of the camera is known and vanishing points can be identified, the degrees of freedom can be reduced to three in a process called image rectification. Dense matching can then be performed by searching only in position and scale. Previous works have also used rectification to increase the repeatability and discriminability of interest points [6, 44].

We perform metric rectification by automatically detecting orthogonal vanishing points and using an approximate estimate of the focal length. We also assume that query images will have low camera roll (camera is mostly upright), thereby allowing us to identify the vertical vanishing point. As in *upright* SIFT [6], the absence of rotational invariance makes our features more discriminative.

We detect vanishing points in two stages. First, the vertical vanishing point in the image is estimated via RANSAC [20] on 2D line segments subtending a small angle to the vertical. For speed and accuracy, longer lines are sampled more frequently during the randomized hypothesis generation step. When the focal length is known, the vertical vanishing point determines the position of the horizon in the image. Horizontal vanishing points are found using a 1-line RANSAC on nonvertical lines that intersect the horizon. In the case of an unknown focal length, the RANSAC hypothesis also includes a random guess of the focal length, sampled from the normal distribution $N(f, \sigma)$ where $f = 1.5$ is the normalized focal length and $\sigma = 1.0$.

If reliable estimates of two orthogonal vanishing points are recovered, we rectify the image using the 2D homography, $H = \mathrm{KR}^{-1}K^{-1}$ where, the matrix $K = \mathrm{diag}([ff1])$ represents camera intrinsics with normalized focal length f and R represents the 3D camera rotation in the frame of the orthonormal vanishing directions. The rectified image is cropped whenever the local distortion caused by the perspective warp due to H exceeds a maximum threshold. Several examples of rectified image are shown in Fig. 15.2. An accurate metric rectification is desirable but not essential for our approach to work. When only the vertical vanishing point is detected, we perform roll correction, thereby eliminating one degree of freedom in camera rotation. Mobile devices are nowadays equipped with accelerometers which provide an approximate estimate of the vertical direction. When gyroscopes and magnetometers are available, they may provide alternative approaches to rectifying the image based on training data.

Fig. 15.2 Example query images from a mobile phone camera. The *top row* shows the original images, while the rectified versions of these images computed using our method are shown the *second row*

15.2.2 Selecting Discriminative Patches

In this section, we describe our approach to select a pool of discriminative image patches across all the location classes in the dataset under consideration. Let D denote the total number of candidate patches after resampling the original patches to 32×32 pixel each. These patches will be used to construct the D-dimensional feature vectors for training the Random Decision Forest (RDF).

To compute the k-th dimension of the feature vector, a dense search is performed within an image to find the patch most similar to the k-th patch in the candidate pool. This search is performed across all positions and a few discrete scales. For computational efficiency, we use BRIEF descriptors [11] to represent patches and compare their visual similarity. The BRIEF descriptor is a binary descriptor that is computed by randomly sampling m pixel pairs $\{(p_j, q_j)\}_{j=1}^{m}$ from the underlying 32×32 patch based on a 2D Gaussian distribution centered on the patch center. The j-th bit of the BRIEF descriptor is set to 1 when the intensity at pixel p_j is greater than the intensity at pixel q_j. Based on our experiments, we found $m = 192$ to provide a good compromise between accuracy and speed. The similarity between two patches is measured by the Hamming distance between their corresponding BRIEF descriptors. This can be computed very efficiently; on certain modern processors a dedicated *popcount* instruction exists for performing the computation in a single operation [11]. Finally, the k-th dimension of the feature vector is assigned to the minimum Hamming distance between the descriptor of the dictionary candidate and all descriptors extracted from the image.

An ideal set of candidate patches are ones that are both unique as well as repeatable within a location class. To select such patches, we first extract scale-invariant DoG [37] keypoints and DAISY descriptors [48, 52] from each image in the training set. Next, pairwise image matching is performed between all pairs of images from each location. The initial putative matching is done using approximate nearest neighbor search on the DAISY descriptors based on a standard kd-tree index [37]. The putative matches are then geometrically verified using RANSAC with an epipolar geometry model to prune outliers from the putative matching stage. The subset of two-view matches that satisfy the respective epipolar geometries are subsequently linked to form a track. First, a small number of tracks are randomly selected. The random sampling at this stage is biased toward favoring longer tracks over shorter tracks. Once a track has been selected, the specific patch within the set of patches in the track whose descriptor has the minimum distance to all the other descriptors in the track is chosen as the representative patch.

This process is repeated for all the location classes to select a uniform number of candidate patches from all the different locations. In our experiments we specify a budget of 4000 features uniformly divided across all the classes. However, for smaller datasets involving fewer classes, we expect that a lower number of candidates will be sufficient. For locations for which epipolar matching failed to produce long tracks on the training images, we select patches corresponding to DoG keypoints at random positions and scales from randomly chosen images and add them to candidate pool. All image patches are axis-aligned square patches and are resampled to 32×32 pixels before BRIEF descriptors are computed.

15.2.3 Random Decision Forests

After densely matching the D dictionary patches to all the images in the training set, the feature vectors required to train the RDF classifier are now available. The corresponding location labels for the training images are also available. In practice, these labels could be obtained from Flickr image-tags, by manually adding metadata to streetside imagery or by clustering the GPS coordinates of geo-tagged images in the training data.

Our random forest classifier consists of multiple binary decision trees that are trained independently. We use bagging [9] to create random subsets of the training data for training each decision tree. In our system, we randomly select 100 % of the training data with replacement, i.e., selecting n samples from the n original training samples with replacement which generates subsets with 63.2 % expected unique samples from the training set, the rest being duplicates.

Each node in a decision tree is optimized using the randomized greedy approach described in [18]. A fixed number of weak learners are randomly sampled. We refer to this hyperparameter as the *selection size*. Each of these weak learners correspond to a feature space partition along a randomly chosen dimension with the split occurring at a randomly chosen value. The final hyperplane at each node is found by selecting the split that resulted in the highest information gain which requires calculating the entropy of the empirical distributions before and after each split. Each axis-aligned weak learner can be represented compactly since only an index for the feature dimension and a threshold needs to stored. The leaf nodes of each tree store the resulting class distributions. During classification, the final result is obtained by averaging the probabilities of the classes predicted by the different trees in the forest. Unlike other implementations of random forests, no tree pruning is performed.

Note that the decision trees are already quite compact due to the choice of the underlying weak learners. However, another level of compactness is possible due to the following fact. In general, certain dimensions in the D-dimensional feature vector are never included in any of the weak learners used by the trained RDF. Let D^* denote the number of dictionary elements that are actually utilized in the RDF. This is then the effective dimensionality of the dictionary required at query time and dimensionality of the feature vector needed for evaluating the classifier at query time. In our experiments, we train forests with 50 trees and D^* typically ranges from 200 to 3000 depending on the number of classes in the dataset. The greedy approach for training random forests are controlled by the selection size hyperparameter which injects randomness and ensures that a wider variety of features are selected by the forest and this is known to encourage generalization. For example, the training images for a particular location may have a parked car seen in all the images. Even though the patches corresponding to the parked car may be the most discriminative given the training data, randomizing the node optimization of the random forest provides better generalization for the future when the scene appearance may have changed significantly.

15.2.4 Efficient Dense Matching

Computing the D^*-dimensional feature vector is the main computational bottleneck at query time. We now describe an efficient solution which is based on using min-hash for approximate nearest neighbor matching [41, 42]. Min-hash is used to compute a set of hashes for binary vectors that have a probability of collision or matching equal to the Jaccard similarity of the two vectors. When viewed as sets, the Jaccard similarity of the two vectors is just the cardinality of the intersection of the two sets divided by the set union. The min-hash of a binary string is the smallest index at which a bit is set (to 1) after a random permutation is applied to the original vector. Min-hash is appropriate for detecting exact duplicates very efficiently but can have low recall. To boost the recall, the selectivity of the hashes can be increased by concatenating multiple min-hashes into a *sketch*.

Let Q denote the set of BRIEF descriptors for the dictionary elements in our system. Let P denote all the BRIEF descriptors for patches densely sampled in the image. We compute t different sketches for each BRIEF descriptor, each of which consists of a concatenation of r min-hashes. A pair of descriptors (p_i, q_i) such that $p_i \in P$ and $q_i \in Q$ are deemed a potential match when s out of t sketches are found to be identical. Full Hamming distances are computed using the original descriptors only for the subset of descriptor pairs that are deemed potential matches. In our implementation, we set $t = 5$, $r = 5$, and $s = 2$, respectively.

Before the min-hash-based method can be used efficiently, a mechanism for efficient retrieval of descriptor sketches is required. This requires a small amount of precomputation on the mobile device after the dictionary is downloaded for the first time. The sketches for the set Q are first extracted and then an inverted index is constructed over those sketches to allow efficient retrieval of the descriptors in Q given any query sketch later on.

The actual dense matching is then performed using a brute-force linear scan. As described earlier, the min-hash technique is used to skip comparisons that are unlikely to yield the nearest neighbor in Hamming space. In practice, the set of descriptors for which the Hamming distance is computed is quite small and this is key to the speedup in the dense matching stage. On average, we found that Hamming distance calculations were skipped for 70–80 % of all patches extracted from query images.

Since all the bits of the BRIEF descriptor are not required for computing its min-hash sketch, another optimization is possible during dense matching. This involves computing the bits of the BRIEF descriptor on demand, i.e., only when needed by the min-hash function. Thus, for a majority of the image patches which get rejected using the min-hash technique, the full 192-dimensional BRIEF descriptor is never computed, rather partial descriptors are computed where only the bits essential to compute the min-hash sketches are evaluated.

Despite the various optimizations, in our experiments, we found that Hamming distance computation for potential matching pairs was often the bottleneck especially when the dictionary size grows larger (100 or above). However, Hamming distance can be computed very fast on certain processors [11] and such hardware acceleration is expected to be available on mobile platforms in the future making our approach feasible.

15.3 Experimental Results

To assess the feasibility of our approach for a mobile device, we analyze the download size of our classifiers and compare its accuracy to a method based on bag-of-visual words (BoW) as well as to a direct feature matching method. These results are shown in Fig. 15.4.

15.3.1 Datasets

We perform our evaluation on two public landmark datasets—ZuBuD [47] and CALTECH building datasets [1], each of which has five images of 200 and 50 buildings, respectively. For these benchmarks we created random query image sets using a leave one out strategy on each class. We also report results on two challenging streetside datasets—SUBURB-48 and TOWN-56, collected by us, where the training

Table 15.1 Summary of datasets used in our experiments and the performance of our proposed approach

Dataset	#Imgs	#Classes	Training			Query stats	
			#Features	#Trees	Download size (KB)	#Queries	Accuracy (%)
ZuBuD	804	200	4107	50	233	200	92
CalTech-50	200	50	2820	50	153	50	90
Suburb-48	504	48	2200	60	220	131	50
Town-56	464	56	3138	60	198	62	35.5

The number of features refers to the dictionary size and the download size refers to the storage used for the dictionary as well as the classifier

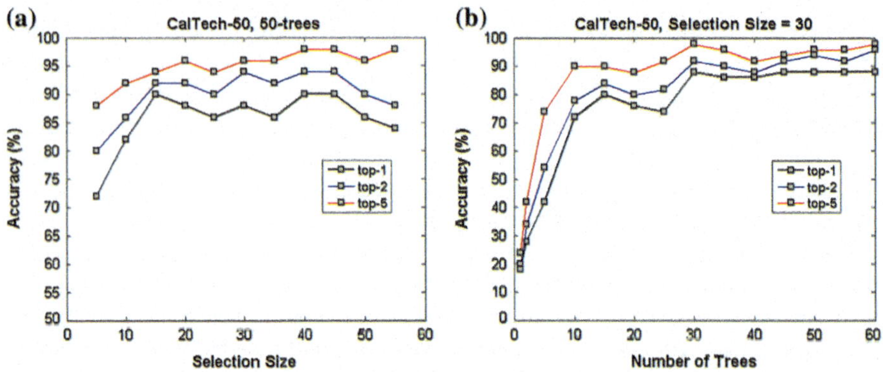

Fig. 15.3 The RDF's classification accuracy depends on the two hyperparameters—selection size and the number of decision trees in the forest as shown for the experiments on the CALTECH-50 building dataset. The top-k classification accuracy is shown here for $k = 1$, 2, and 5, respectively. **a** The selection size parameter controls the degree of randomness—increasing the value of this parameter first improves accuracy but using larger values hurts generalization. **b** Increasing the number of trees also consistently improves accuracy but also increases the storage cost. The accuracy remains almost the same beyond 40–50 trees. Similar trends are observed on the other datasets

images comprised of 504 and 464 images each of an area the size of four city blocks. For both datasets, prominent landmarks (classes) such as buildings, restaurants, and store fronts were manually identified and labeled. 200 images captured by pedestrians with mobile phones were used for queries. The database images were captured in a different season from when the query images were captured. Significant appearance variations caused by changing seasons and weather, and illumination and viewpoint variations make these datasets challenging for recognition (Table 15.1).

Our experiments demonstrate that compact classifiers and dictionaries can be constructed for tasks involving up to 200 locations or landmarks without sacrificing accuracy. Figure 15.3 shows the effect of RDF hyperparameters on the classification accuracy. The selection size parameter and number of decision trees affect the compactness of our classifiers. The accuracy improves as these parameters are increased but starts to converge after a selection size of about 20 and with 50 decision trees in the forest.

15.3.2 Comparison with traditional methods

We compare our method with SIFT feature matching [37] and BoW [40] methods, with different configurations that result in different storage size required by the different methods. The classification accuracy is the percentage of query images correctly recognized. For SIFT, the retrieved images for each query are ranked by the number of matches. The ranked image list is mapped to a list of classes by finding the first occurrence of each class in the sorted list. To impose memory/download constraints on SIFT matching, only a fraction of SIFT keypoints were sampled from the database image and used for matching. Further, the descriptor vectors were quantized to k-bits entries ($k = 1$–8). In a similar fashion, storage constraints were placed on the BoW method by choosing vocabularies with 1000–100,000 words and quantizing the histogram entries to use 1 to 16 bits. BoW histogram were represented as sparse vectors. Figure 15.4 shows that both SIFT and BoW accuracy degrades significantly when the memory/storage size is lowered. In our method, the download size is varied by choosing different RDF parameters. Note all comparisons were performed on rectified images computed based on automatic vanishing point detection.

For all three datasets on which the comparison was performed, our method outperforms both BoW and SIFT in accuracy even though our classifiers are one or more orders of magnitude more compact. For example, on TOWN-56 and SUBURB-48 datasets our method had an accuracy of 36 and 49.5 %, respectively, with about 200 KB storage size. While SIFT matching did not work at all, BoW methods had an accuracy of approximately 8 and 12 % for the two datasets. The best performance with BoW and SIFT was obtained with 2 MB+ and 40 MB+ storage size in both datasets. The comparison on CALTECH-50 is similar with the accuracy of our method being about 90 % with 100 KB of storage whereas BoW methods had an accuracy of about 50 % when compressed down to about 200 KB (Figs. 15.5 and 15.6).

This produces probability distributions at the leaf nodes which can be represented with a single integer, since all the training examples at a leaf node belong to the same class. Thus, with C classes, we need $\log_2(|C|)$ bits of storage at each leaf node. Each internal node requires roughly 5.5 bytes to represent a feature index, an integer threshold and two pointers. The total download size can be approximated as $24N + 6.5TH$ bytes (assuming we have fewer than 1024 classes), where the random forest selects N patches (each of which is represented using 32 bytes). T is the number of trees in the RDF and H is the average tree height. With more than a hundred classes, the storage size is typically dominated by the selected feature sets in our method.

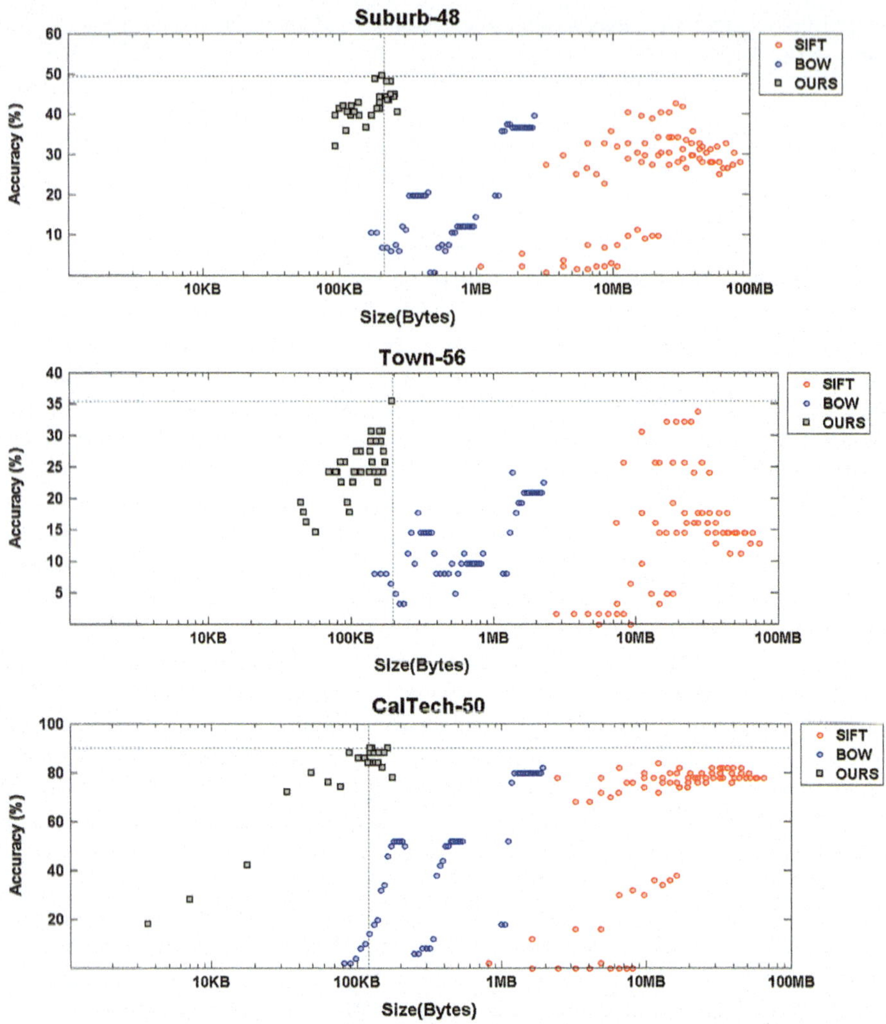

Fig. 15.4 COMPARISONS: This figure compares the classification accuracy of our method with that a BoW and a direct SIFT matching method which are considered state of the art for location recognition when memory footprint and storage is not an issue. For the two streetside datasets as well as the public CALTECH-50 dataset, our method outperforms both BoW and SIFT in accuracy even though our classifiers are one or more orders of magnitude more compact. Each method was configured to run with different storage/download sizes which generated the scatter plots shown here. The most accurate configuration for our method is indicated using dotted lines. Our method works reasonably well under 100 KB whereas BoW perform very poorly under such extreme compression and SIFT does not work at all. The X-axis in all plots is in log-scale

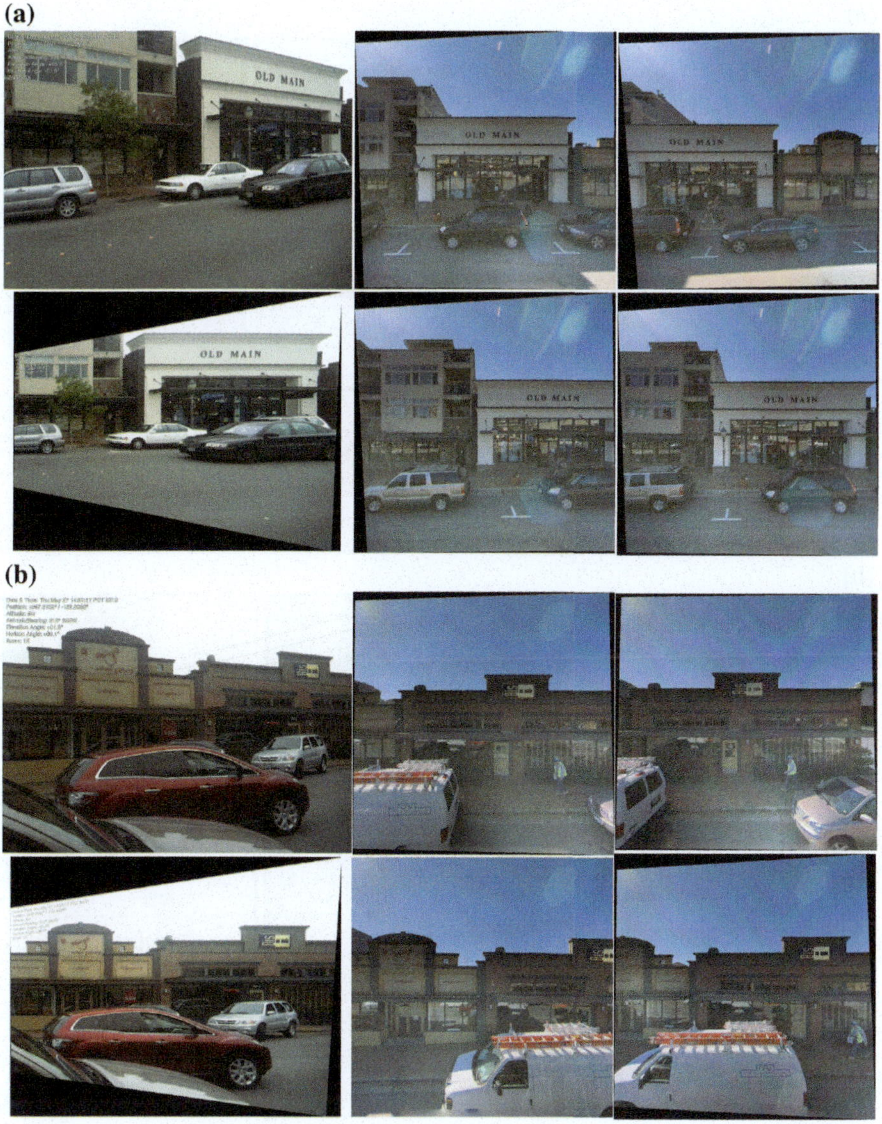

Fig. 15.5 Examples of successful location recognition queries using our method on the TOWN-56 and SUBURB-48 datasets. For each figure, the image in the *left column* and *top row* is the query image captured using a mobile device and the corresponding rectified image is shown in the *left column* and *bottom row*. The four images in the *middle* and *right columns* for all examples show the database images of the building which ranked highest among all the landmark or building classes for the particular dataset in question

15.3.3 Running Time

For the dense matching step, we resize the image such that its larger dimension is 512 pixels. When searching ten discrete levels of scale between a magnification factor of

Fig. 15.6 Examples of successful location recognition queries using our method on the TOWN-56 and SUBURB- 48 datasets. For each figure, the image in the *left column* and *top row* is the query image captured using a mobile device and the corresponding rectified image is shown in the *left column* and *bottom row*. The four images in the *middle* and *right columns* for all examples show the database images of the building which ranked highest among all the landmark or building classes for the particular dataset in question

$0.25\times$–$1.25\times$ of the size of the resized image, the running time for query processing varies between 0.5 and 2.0 s on a single CPU core for all our datasets. The time complexity of the dense matching step which dominates query processing is linear

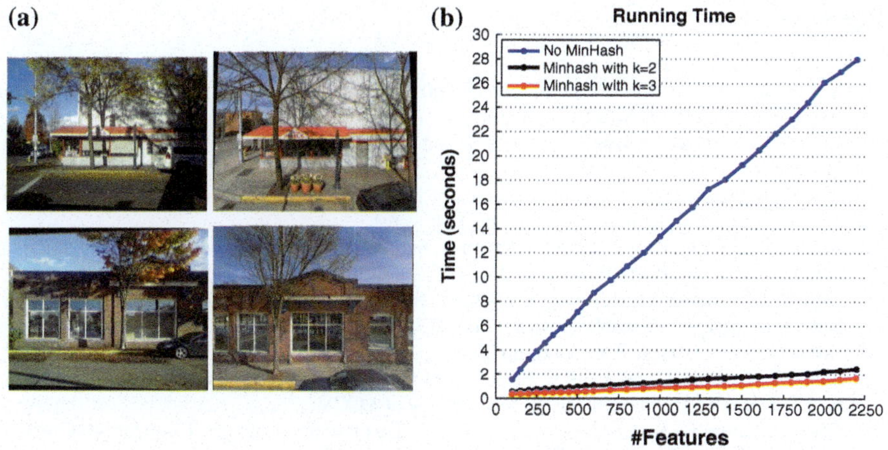

Fig. 15.7 **a** A couple of failure examples are shown (*query images on the left*). The database images shown on the *right contains shadows* and *parked cars* and was captured in a different season. **b** Running times for the dense matching step are reported. The *blue curve* shows timings when brute-force Hamming distance computation is performed during dense matching. The *red* and *black curves* shows timings for our min-hash-based approach, where Hamming distance is computed only when at least k min-hash sketches of the binary descriptors are identical

in the dimensionality of the feature vector, i.e., the number of patches in the dictionary. Figure 15.7b shows the running time for performing dense matching on a 512×420 pixel image on a laptop with a single core 2.66 GHz processor. For the min-hash-based speedup technique described here, we computed Hamming distance between a pair of descriptors only when k out of five sketches were identical. Figure 15.7b shows running time for $k = 2$ and 3 and shows how the min-hash strategy provides an order of magnitude speedup over the case when exact linear scan is performed using Hamming distance on BRIEF descriptors. In all our location recognition experiments, k was set to 2.

15.4 Conclusion

We have described a new discriminative approach for recognizing urban landmarks and locations that exploits approximate GPS location to train compact classifiers. The classifiers can also be efficiently evaluated at query time. The accuracy of our method is competitive with that of state of the art methods despite the compact representation. The small download footprint of the discriminative dictionary and the classifier as well as the efficient dense matching strategy makes the approach feasible for a location recognition system designed to run solely on a mobile device.

References

1. Aly M, Welinder P, Munich M, Perona P (2009) Towards automated large scale discovery of image families. CVPR Workshop Intern Vis 9–16
2. Amit YDG (1997) Shape quantization and recognition with randomized trees. Neural Comput 9
3. Arth C, Schmalstieg D (2011) Challenges of large-scale augmented reality on smartphones. Graz University of Technology, Graz, pp 1–4
4. Arth C, Wagner D, Klopschitz M, Irschara A, Schmalstieg D (2009) Wide area localization on mobile phones. In: ISMAR, pp 73–82
5. Arth C, Klopschitz M, Reitmayr G, Schmalstieg D (2011) Real-time self-localization from panoramic images on mobile devices. In: 2013 IEEE international symposium on mixed and augmented reality (ISMAR) vol 0, pp 37–46
6. Baatz G, Koser K, Grzeszczuk R, Pollefeys M (2010) Handling urban location recognition as a 2d homothetic problem. In: IEEE proceedings of ECCV
7. Baatz G, Saurer O, Köser K, Pollefeys M (2012) Large scale visual geo-localization of images in mountainous terrain. In: ECCV (2), pp 517–530
8. Bergamo A, Sinha SN, Torresani L (2013) Leveraging structure from motion to learn discriminative codebooks for scalable landmark classification. In: CVPR, pp 763–770
9. Breiman L (2001) Random forests. Machine Learn 45
10. Cstrecha AM, Bronstein MMB, Fua P (2012) LDAHash: improved matching with smaller descriptors. IEEE Trans Pattern Anal Mach Intell 34(1)
11. Calonder M, Lepetit V, Strecha C, Fua P (2010) BRIEF: Binary robust independent elementary features. In: ECCV 4:778–792
12. Cao S, Snavely N (2013) Graph-based discriminative learning for location recognition. In: CVPR, pp 700–707
13. Cao S, Snavely N (2014) Minimal scene descriptions from structure from motion models. In: CVPR
14. Chandrasekhar V, Takacs G, Chen D, Tsai S, Grzeszczuk R, Girod B (2009) CHoG: compressed histogram of gradients a low bit-rate feature descriptor. In: IEEE conference on computer vision and pattern recognition (2009), pp 2504–2511
15. Chen DM, Baatz G, Koser K, Tsai SS, Vedantham R, Pylvanainen T, Roimela K, Chen X, Bach J, Pollefeys M, Girod B, Grzeszczuk R (2011) City-scale landmark identification on mobile devices. In: 2013 IEEE conference on computer vision and pattern recognition, vol 0, pp 737–744
16. Cheng Z, Ren J, Shen J, Miao H (2013) Building a large scale test collection for effective benchmarking of mobile landmark search. In: Advances in multimedia modeling, pp 36–46. Springer
17. Crandall D, Owens A, Snavely N, Huttenlocher D (2011) Discrete-continuous optimization for large-scale structure from motion. In: CVPR, pp 3001–3008
18. Criminisi A, Shotton J, Konukoglu E (2012) Decision forests: a unified framework for classification, regression, density estimation, manifold learning and semi-supervised learning. Found Trends Comput Graph Vis 7(2–3):81–227
19. Doersch C, Singh S, Gupta A, Sivic J, Efros AA (2012) What makes paris look like paris? ACM Trans Graph 31(4)
20. Fischler MA, Bolles RC (1981) Random sample consensus: a paradigm for model fitting with applications to image analysis and automated cartography. Commun ACM 24:381–395
21. Gronat P, Obozinski G, Sivic J, Pajdla T (2013) Learning and calibrating per-location classifiers for visual place recognition. In: Proceedings of the IEEE conference on computer vision and pattern recognition
22. Guan T, Fan Y, Duan L, Yu J (2014) On-device mobile visual location recognition by using panoramic images and compressed sensing based visual descriptors. PloS one 9(6):e98,806
23. Hays J, Efros A (20078) IM2GPS: estimating geographic information from a single image. In: IEEE proceedings of CVPR

24. http://maps.google.com/help/maps/streetview/
25. http://www.bing.com/maps/
26. Hua G, Brown M, Winder S (2007) Discriminant embedding for local image descriptors. In: IEEE proceedings of ICCV
27. Irschara A, Zach C, Frahm JM, Bischof H (2009) From structure-from-motion point clouds to fast location recognition. In: CVPR, pp 2599–2606. IEEE
28. Jshotton M, Johnson RC (2008) Semantic texton forests for image categorization and segmentation. In: IEEE proceedings of CVPR
29. Knopp J, Sivic J, Pajdla T (2010) Avoiding confusing features in place recognition. In: IEEE proceedings of ECCV
30. Lepetit V, Fua P (2006) Keypoint recognition using randomized trees. PAMI 28:1465–1479
31. Li X, Wu C, Zach C, Lazebnik S, Frahm JM (2008) Modeling and recognition of landmark image collections using iconic scene graphs. In: IEEE proceedings of ECCV
32. Li Y, Crandall D, Huttenlocher D (2009) Landmark classification in large-scale image collections. In: IEEE Proceedings of ICCV
33. Li Y, Snavely N, Huttenlocher D (2010) Location recognition using prioritized feature matching. In: IEEE Proceedings of ECCV
34. Li Y, Snavely N, Huttenlocher D, Fua P (2012) Worldwide pose estimation using 3d point clouds. In: Computer Vision–ECCV 2012, pp 15–29. Springer
35. Li Z, Yap KH (2012) Content and context boosting for mobile landmark recognition. IEEE Sig Process Lett 19(8):459–462
36. Lim H, Sinha SN, Cohen MF, Uyttendaele M (2012) Real-time image-based 6-dof localization in large-scale environments. In: 2012 IEEE conference on computer vision and pattern recognition (CVPR), pp 1043–1050. IEEE
37. Lowe DG (2004) Distinctive image features from scale-invariant keypoints. Int J Comput Vis 60
38. Micusík B, Wildenauer H, Kosecka J (2008) Detection and matching of rectilinear structures. In: IEEE Proceedings of CVPR
39. Middelberg S, Sattler T, Untzelmann O, Kobbelt L (2014) Scalable 6-dof localization on mobile devices. In: Computer vision ECCV 2014, lecture notes in computer science, vol 8690, pp 268–283
40. Nister D, Stewenius H (2006) Scalable recognition with a vocabulary tree. In: CVPR, pp 2161–2168
41. Ondrej Chum JP, Zisserman A (2008) Near duplicate image detection: min-hash and tf-idf weighting. In: BMVC
42. Perdoch OCM, Matas J (2009) Geometric min-hashing: Finding a (thick) needle in a haystack. In: IEEE Proceedings of CVPR
43. Philbin J, Chum O, Isard M, Sivic J, Zisserman A (2007) Object retrieval with large vocabularies and fast spatial matching. In: IEEE proceedings of CVPR
44. Robertson D, Cipolla R (2004) An image based system for urban navigation. In: BMVC, pp 819–828
45. Sattler T, Leibe B, Kobbelt L (2012) Improving image-based localization by active correspondence search. In: ECCV 2012, pp 752–765. Springer
46. Schindler G, Brown M, Szeliski R (2007) City-scale location recognition. In: IEEE proceedings of CVPR
47. Shao H, Svoboda T, Gool LV (2003) ZUBUD-Zurich buildings database for image based recognition. Tech. rep., No. 260, Swiss Federal Inst. of Technology
48. Tola E, Lepetit V, Fua P (2010) DAISY: an efficient dense descriptor applied to wide baseline stereo. IEEE transactions on pattern analysis and machine intelligence 32(5):815–830
49. Torii A, Sivic J, Pajdla T, Okutomi M (2013) Visual place recognition with repetitive structures. In: Proceedings of the IEEE conference on computer vision and pattern recognition
50. Turcot P, Lowe DG (2009) Better matching with fewer features: the selection of useful features in large database recognition problems. In: ICCV workshop on emergent issues in large amounts of visual data (WS-LAVD)

51. Weston J, Mukherjee S, Chapelle O, Pontil M, Poggio T, Vapnik V (2000) Feature selection for SVMs. In: Advances in neural information processing systems, vol 13, pp 668–674. MIT Press
52. Winder SAJ, Hua G, Brown M (2009) Picking the best daisy. In: CVPR, pp 178–185
53. Zamir A, Shah M (2010) Accurate image localization based on google maps street view. In: IEEE proceedings of ECCV
54. Zhang W, Kosecka J (2007) Hierarchical building recognition. Image Vis Comput 25(5):704–716

Chapter 16
A Real-World System for Image/Video Geo-localization

Himaanshu Gupta, Yi Chen, Minwoo Park, Kiran Gunda, Gang Qian, Dave Conger and Khurram Shafique

Abstract Determining where an image was taken and geo-locating depicted structures are important tasks from a surveillance and intelligence standpoint. For example, the image might show terrorist training facilities or the vicinity of a safe house. To geo-localize, the user must combine prior knowledge of the area with subtle clues from the image in order to mitigate the tedious manual search of GIS reference data. This process is extremely challenging, time-consuming, and often yields poor accuracy. In this chapter, we describe WALDO (Wide Area Localization of Depicted Objects), a system that solves this challenging problem by combining the insight of analysts with the power of automated analysis for Internet-scale, geo-location-driven data mining. WALDO's goal-driven constrained resource management leverages a full spectrum of data-driven, semantic, and geometric geo-localization experts and user tools.

16.1 Introduction

16.1.1 Issues and Challenges

Geo-localization of an image or video is a challenging task that has received significant attention in recent years, largely due to the increased availability of geo-tagged images on social web sites as well as the improved computational power of modern computers. However, the state of the art on automated geo-localization has been largely limited to finding matches with geo-tagged ground imagery. Unfortunately, such approaches are only applicable to areas that are frequently visited and photographed and therefore are well represented in available ground-imagery databases. The problem gets even more complicated when one takes a closer look at the content of these images. Many of the images in these databases depict objects or scenes that

H. Gupta (✉) · Y. Chen · M. Park · K. Gunda · G. Qian · D. Conger · K. Shafique
Object Video, Inc., 11600 Sunrise Valley Drive, Suite 210, Reston, VA 20191, USA
e-mail: hgupta@objectvideo.com

A.R. Zamir et al. (eds.), *Large-Scale Visual Geo-Localization*,
Advances in Computer Vision and Pattern Recognition,
DOI 10.1007/978-3-319-25781-5_16

are not relevant for the purposes of image geo-localization, for example, close-up shots of people, indoor scenes, scenes with no salient objects, etc. Furthermore, even when the representative images are available in the database, they may have significantly different image properties, for example, viewing directions, illumination, weather conditions, etc., that make them very hard to match automatically.

Some of the typical challenges encountered in the real world are:

- Reference data reliability
- Lack of GIS data
- Diverse reference data sources
- Lack of identifiable and distinct features in queries
- Wide variety of images/videos and geo-location cues
- Sparse geo-tagged image data
- System scalability
- Low-resolution reference maps
- Lack of user expertise and subject matter knowledge

Previous chapters in this book have studied specific algorithms, mostly automatic algorithms, but these have largely been academic pursuits. To solve the grand real-world problem we need combinations of automated techniques and human feedback. We believe that a successful solution to this problem hinges upon creating an effective blend of the power of automated systems to mine the vast amounts of available data sources, and perception of the human user, to identify the subtle patterns and shepherd the search. The design decisions for WALDO are guided by the interaction between the user and novel automated tools for knowledge discovery, finding geo-informative features, semantic scene understanding, data mining and fusion, and searching large databases of ground-level images and GIS datasets. These automated and interactive tools are combined together by a Bayesian theoretic decision and control engine that updates and maintains belief models, schedules, and triggers various tools based on data and belief state, and optimizes the candidate geo-locations for the given query conditioned on user requirements.

The WALDO system includes several technical innovations in each of the core research areas that are unique compared with the state of the art. These include:

- Data and resource-dependent selection of appropriate geo-localization tools from a pool of matchers: A wide variety of images, geo-localization cues, and data sources preclude the use of a single approach for geo-localization. Therefore, the WALDO system exploits a pool of matchers with different operating conditions and data expectations. It uses a Boosting-like decision theoretic framework to mesh together different automated and interactive tools by optimizing the performance under a cost-benefit resource valuation.
- Matcher Combination and Anytime Prediction: Multiple predictions from different matchers need to be fused into a geo-distribution. Commonly used Naïve Bayesian approach treats all predictions as equally good—and independent—often resulting in inaccurate predictions. WALDO refines this approach by implementing a

logarithmic opinion pooling of the mixture of experts algorithm to produce a distribution that is consistent with the constraints and correlations between the estimates output by different matchers. The approach provides a mechanism for "anytime prediction", that is, the system can produce a distribution at any time given the available set of constraints.

- Maintaining and Updating Belief Models regarding the geo-location of query image/video: WALDO combines results from the above-mentioned methods (matchers) to maintain a probability distribution of the camera location of the query image/video. The system uses this probability distribution to generate a candidate list of geo-spatial regions that are likely to contain the camera position of the query.
- Bottom-up discovery of the relationship between visual features and geographic attributes from reference data: State-of the-art methods require dense ground level imagery in world regions, restricting applicability in the regions where such imagery is not easily available. The WALDO system exploits matchers that exploit widely available GIS data and available geo-tagged imagery (not necessarily from the world region) to determine the relationship between visual features and geographic attributes. In addition, it estimates location-specific information content of features for improved matching efficiency and avoidance of confusing features.
- Automated Discovery and Exploitation of Semantic Attributes: For efficient geo-localization, the system must be able to match queries with diverse data sources and quickly reject large regions in the world. WALDO enables this by automated discovery and exploitation of scene categorization (e.g., coast, forest, rural) and semantic attribute identification (e.g., object labels, weather, scene elements, text, etc.) for geo-localization. To enable large-scale semantic matching, it uses a hierarchical scene similarity assessment and matching framework for coarse-to-fine estimation of geo-location.
- Virtual Subject Matter Experts: Automated methods do not always provide categorization resolutions necessary for geo-localization. Moreover, analysts cannot be expected to have subject matter expertise. WALDO uses a collection of visual tools—Virtual Subject Matter Experts—exploiting taxonomies of geo-informative scene attributes and their visual representations for fine-grained categorization.
- Interactive Tools for Knowledge Discovery and Geo-localization Improvement: WALDO exploits various interactive tools aimed toward improving knowledge discovery and geo-localization of automated methods. These include tools for image search, image tagging and markup, estimation of camera/scene geometry (camera calibration, surface orientation) from a single view, and image-based mensuration.
- Exploitation and Fusion of Information from Large Geo-spatial Datasets: WALDO uses state-of-the-art techniques for rapid fusion of geo-spatial imagery and other data sources such as landcover databases, DEM, LIDAR imagery, etc.

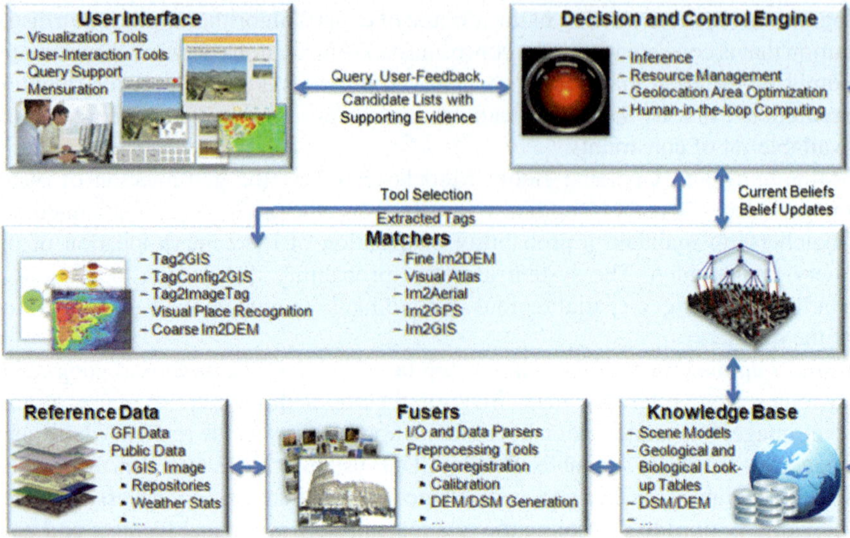

Fig. 16.1 High-level block diagram of the WALDO system

16.1.2 System Overview

A high-level block diagram of the proposed WALDO system is shown in Fig. 16.1. WALDO provides a powerful mix of state-of-the-art automated and user-in-the-loop tools for geo-locating images, videos, and depicted objects of interest. This mix is realized by innovative user interfaces, automated knowledge discovery and geo-localization modules (Matchers), and a Bayesian Decision and Control Engine (DCE) that optimizes resource allocation according to the gestalt of expected-benefit, resource, time, and data constraints. The constrained task-driven resource management framework of the DCE calls upon a set of matchers. Each matcher performs a mapping from the query images/videos to a probability distribution (PDF) over geo-locations, and is treated as an individual resource with its own data prerequisites, costs and benefits. The DCE continuously evaluates the belief model, available evidence, and resource constraints to identify the best set of matchers to operate upon the query image. The DCE outputs geo-location hypotheses (candidate lists) and backs out supporting evidence from the modules for vetting by the user, who can then use this evidence to reject, refine, or revise a hypothesis or constraint. This procedure continues until the user is satisfied that the system has suitably geo-located the query image or video given the available evidence. The fusers operate upon reference data to generate fused reference data. The matchers extract useful features from the fused reference data and index it in the knowledge base.

The DCE is the central part of the system. The DCE manages communications between different components, selects different matchers and user tools (implemented as web services) and combines their output to form a belief model about the geo-location of the query image. The fusion workflow uses multiple fusers to operate upon reference data to generate fused reference data (FRD). Different matchers extract relevant features from the FRD and index them in the knowledge base.

In Search mode, the user is presented with a web-based user interface that allows uploading an image or video. While uploading the image or video, the user may specify a Region of Interest, and if the query file is a video, a frame in the video. The user is presented with a drawing canvas with a view of the image or video frame, and several drawing tools. The user may then provide information about the images, by providing appropriate tags about the image as a whole, by highlighting and tagging specific elements in the image, or by marking the skyline, coastline, or roads in the image. The user may also draw an overhead view of the scene using simple drawing tools and link them to the semantic tags. To help the user provide accurate tags about the scene or scene elements, the user is presented with virtual SME tools, for example, a Geology SME. Lastly, the user may request that the system search for the location of a specific region in the image, by drawing a polygon around that region in the image. The user then submits the search request. Based on the information provided by the user, the system launches several matchers that exploit the user-provided cues as well as image features to find matches in the reference data and estimate camera geo-location. The system fuses the output of each matcher and generates a candidate list of geo-spatial regions likely to contain the camera location. The system presents the ranked list (based on likelihood) to the user who can then verify the camera location by matching the image features with the available satellite imagery and other GIS layers. The results are returned to the user as a list of candidate regions in a KML file, which are displayed by the user interface using the Google Earth Plug-in. The user may then navigate through the candidate list, and see the regions on the Google Earth map. The user may then also view a summary of the rationale for returning each of the candidate regions. WALDO uses the concept of the "User in the Loop" for verification and validation of the matcher results and decision support output (candidate lists). This involves the user reviewing intermediate results, and ruling out unlikely candidate.

16.2 Approach

The research and development for WALDO was informed by the analysis of different regions of the world, available reference data, the challenges presented in each world region, and the cues identified from research queries. In all world regions, the GIS-based region matcher (Tag2GIS) exploits user-defined tags, such as road or hilly area, and searches indexed GIS databases to identify the regions that contain those tags. The GIS-based pinpoint matcher (TagConfig2GIS) uses a combination of tags (if they exist) and their relative positions to estimate the geo-location. The digital elevation model (DEM)-based region matcher (Im2DEM) exploits user-defined or automatically extracted skyline and ridgeline features, and search indexed DEM-feature set to identify the regions that contains similar features. The ground-image-based matcher (Tag2ImageTag) exploits user-defined and/or automatically extracted tags, such as bridge, to identify geo-tagged images that contain those tags. Another ground-image-based matcher (Visual Place Recognition) attempts to find images in the reference data viewing the same scene as the query. The ProjectLive2D matcher allows the user to find point correspondences between the query and reference data to further refine the camera model. These technologies and several other user tools are detailed in the following subsections.

16.2.1 Tag2GIS—Match Semantic Tags to GIS Data

The Tag2GIS matcher fuses data from various datasets and makes it searchable for semantic tags, for example, developed, grass, crop, barren, wetland, shrub, grass, playa, limestone, etc. The matcher uses a search engine that indexes spatial data over geo-spatial tiles. The tiling provides base unit for search to make aggregation at indexing time feasible and scalable. The system uses Universal Transverse Mercator (UTM) as the tiling scheme to minimize area distortion.

16.2.1.1 Data Indexed by GIS Matcher

We have ingested a combination of land cover and land use data, GIS data, and elevation data into the GIS matcher. The processed data was a fusion of information obtained from free sources on the Internet, extracted from satellite images, and from OpenStreetMap (OSM) [1] data. The list of indexed data is provided below:

- Land Cover/Land Use data: GeoCover, VisNav, Landsat-8, WorldView-2
- GIS data: OpenStreetMap data, GeoNames, Geology data, Administrative boundaries, Google Street View-based road network
- Elevation data: Hilly, Plain, Mountain tops
- OSM-based extracted data: Road intersection cardinality, Divided Highways

We also analyzed the DEM data to determine hilly, plain, and mountain regions. Such analysis can greatly reduce the search region for many queries.

16.2.1.2 OpenStreetMap (OSM) Processing

The OSM processing uses gdal (version 1.10 and higher) [2] to transfer the original XML data into shapefile, which is a popular geo-spatial vector data format for geographic information system (GIS) software. The indexing is straightforward and can be done using simple scripts. We enhanced the processing of key-value pairs to extract more data types such as viaducts, subways, roundabouts, bus stops, etc. Our system also supports non-Latin alphabet queries such as Chinese. For Latin-based alphabets (e.g., Spanish) in addition to indexing original phrases we index ASCII phrases that consist of letters stripped of diacritics (e.g., Cerro Peñón is also indexed as Cerro Penon). The indexing and search systems support *soundex* queries that return results that do not exactly match the query but sound like the query.

16.2.1.3 Extraction of Additional Information Based on OSM Data

Intersection Cardinality: We performed extraction of intersection cardinality (e.g., 3-way intersection, 4-way intersection, etc.) based on OSM information about roads. Many roads do not have exactly the same endpoint in OSM and therefore the vector approach yielded inferior results. To deal with this quality issue, we rasterized road network data (using the gdal_rasterize utility) and extracted intersection cardinality based on raster data. The raster approach yielded much better results creating only some problems in the area of freeway exits.

Divided Highways: We also performed identification of divided highways. We assumed that a divided highway consists of two one-way roads that are in close proximity to one another. We used dilation and erosion combined with minimum area threshold to identify regions in the rasterized image that correspond divided highways.

16.2.1.4 GeoNames Data

We indexed the GeoNames [3] geographical database, which covers all countries and contains over eight million place-names that are available for a free download. Each feature includes a geo-spatial coordinate, a type of feature, and the name, including variants of the name and spellings. There are 667 different object types in the GeoNames database.

16.2.1.5 Building Heights

We have developed a prototype system for accurate reconstruction of high-resolution digital elevation models from stereo satellite imagery. This DEM reconstruction system is built on top of the Ames Stereo Pipeline (ASP), NASA's state-of-the-art open source stereogrammetry software for 3D terrain model reconstruction from planetary imagery. Furthermore, to improve the disparity estimation in areas with high-rise building (HRBs), we have developed a robust algorithm to reliably detect and match HRBs from stereo satellite imagery. The HRB detection and matching algorithm further refines the disparity map from ASP and leads to significant improvement in DEM reconstruction in HRB areas. Once accurate DEM data is extracted, we estimate the heights of buildings in urban regions (by assuming a certain floor height) which can then be used as tags for search queries (Fig. 16.2).

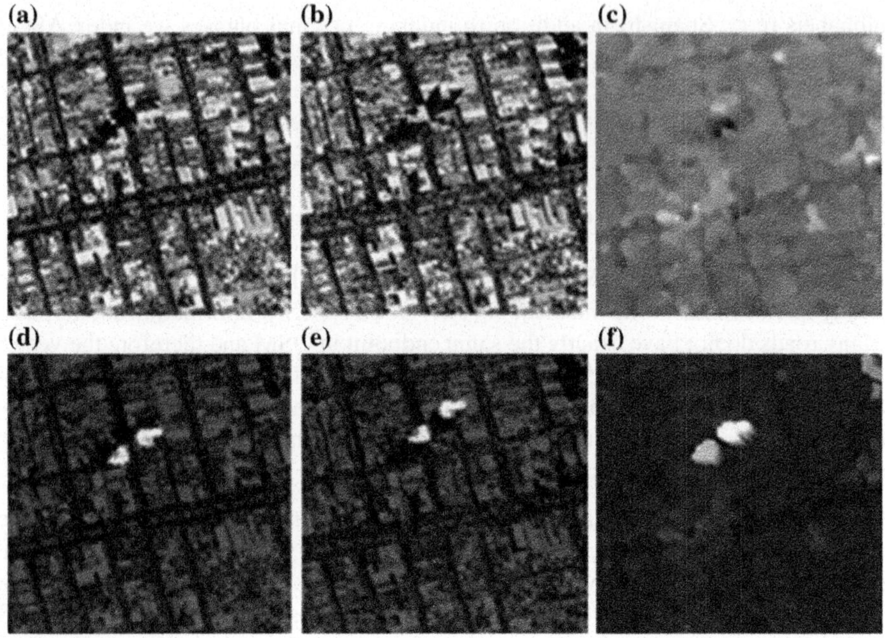

Fig. 16.2 HRB detection and matching and DEM reconstruction from a satellite stereo tile pair. **a**, **b** Show the "*left*" and "*right*" tiles, respectively. **c** Shows the raw ASP DEM reconstruction without HRB detection and matching. **d**, **e** Show the detected and matched HRB from the stereo tile pair and **f** shows the improved DEM reconstruction after integrating HRB detection and matching results in ASP

16.2.1.6 Landsat 8 Land Use Classification

We also downloaded Landsat 8 images for all world, which are freely available [4]. As Landsat 8 was launched in 2013, the coverage is new. We performed Dark Object Subtraction [5] as a rudimentary atmospheric compensation, and mosaiced images. We performed land use classification extracting commercial, high-density residential, and low-density residential areas, as well as high-rise downtown regions. We also extracted other land cover classes such as water, forest, and agriculture.

16.2.1.7 WorldView-2 Land Cover Classification

We selected WorldView-2 satellite coverage to build a land cover model at 2 m resolution. First, we manually selected strips for mosaicing based on cloud cover, recency, and seasonality. The data was orthorectified and atmospherically compensated, and then land cover classification was done using machine learning models built based on manually provided training examples. In addition to traditional land cover classes

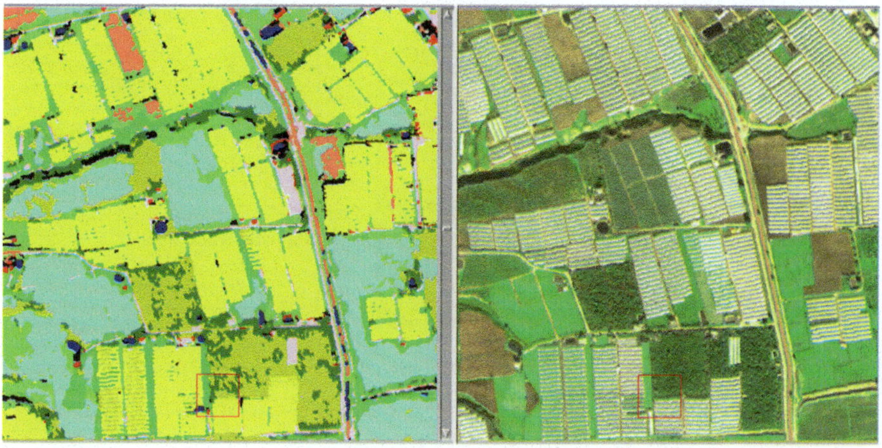

Fig. 16.3 WorldView-2 classification of greenhouses. *Yellow* color regions denote detected greenhouse structures

such as water, forest, shrubs or barren, we extracted orchards and vineyards, and colors of the roofs (Fig. 16.3).

16.2.1.8 Geology Data and Subject Matter Expert Tool

We obtained geology data for world regions, and processed them to represent widely used rock types. We have developed a Geology Subject Matter Expert (SME) knowledge tool to guide the user (using open-source images and relevant questions) to define types of rocks in the query image based on color, vegetation, and terrain.

16.2.2 TagConfig2GIS—Match Configuration of Semantic Tags to GIS Data

Tag2GIS matcher only tests for presence of one or more semantic tags in a given world region, which narrows the search space but may still yield many regions that satisfy the presence constraints. To improve search and to pinpoint the geo-location, we developed a matcher that exploits the topological, order, and metric relationships of different objects in the scene.

In previous semantic tag configuration algorithms [6], the GIS data is generally treated as points, horizontal (viewing directions) and vertical (distance to camera) consistencies are considered, and matching locations are obtained by dynamic programming. We further exploited the geometric and topological relationships between objects present in the query image. In our approach for the TagConfig2GIS matcher,

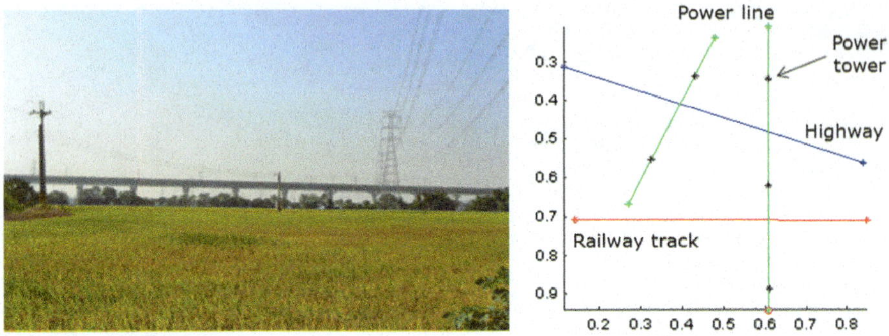

Fig. 16.4 Query image with four classes and the OHS

objects are divided into three groups: points (e.g., water towers, temples), lines (e.g., highways, power lines), and regions (e.g., parks, mountains). The configuration of the objects in the query image is matched to GIS data configuration via a shape context matching approach. GIS data is first extracted from OpenStreetMap and other sources such as DEM. Line-type objects are treated differently from the previous approach to preserve connectivity between nodes. In this way, we can explicitly extract the intersections between different line-type objects (whether or not they belong to the same class) and other geometric features such as tangents and curvatures. Unlike point-type objects which are processed using the raw data obtained from OSM and other GIS data sources, region-type objects are densely sampled. The camera candidates are obtained by the same region-filtering approach (i.e., only regions where all objects are visible and within user-specified distances are considered as camera candidates).

The user then provides an overhead sketch (OHS) of the query image to represent the objects in terms of points, lines, and polygons. The OHS is matched to the GIS data within a sliding window centered at each camera candidate by an algorithm adapted from shape context matching [7]. The heading direction of the camera candidate is estimated from the visible GIS objects as well as orientations of these objects (e.g., tangent direction of line objects) (Fig. 16.4).

The original shape context matching algorithm works on object boundaries and does not discriminate objects in different classes. It is not directly applicable to the semantic configuration matching problem. Therefore, we made the following modifications:

(1) Use all objects to compute the histograms, but only match correspondences for a subset of the data (called key points). Currently, we consider line-type objects, the associated intersections, and point-type objects as key points. In the future the key point set will be expanded to include other features, such as the boundaries of region-type objects, intersections between lines and regions, etc. In the above example, four classes are provided in the OHS, and only the red line and blue dots are used as key points for matching.

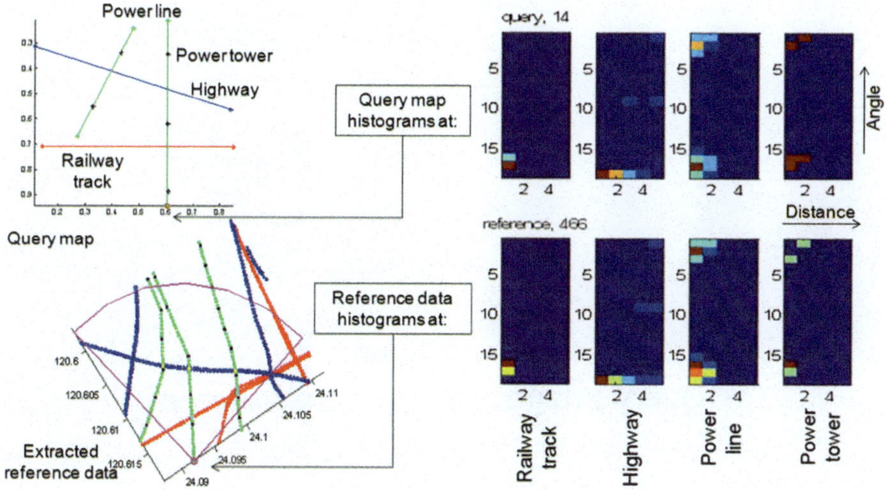

Fig. 16.5 An example of class-dependent histogram matching of a query image with four classes, where four shape context histograms are computed for every key point. *Top row* histograms for a point in query image OHS. *Bottom row* histograms of the matching point in the GIS data

(2) Compute separate shape context histograms for different classes. For each key point in query OHS and GIS data, a histogram is computed for each of the present classes. In the above query example, four histograms are computed per point, as illustrated in Fig. 16.5.

(3) Match only query points and GIS points belonging to the same class, and combine the total matching cost. The line intersections are considered as different classes from the line-type objects. For histogram matching, the "contain semantic" (no penalty as long as the GIS data contains the query points at least) is employed, based on the observation that there are usually many more outliers in GIS data than in the query OHS.

The matching score of each camera candidate is computed by combining the shape context cost, affine cost, deformation, bending energy, and ratio of outliers. After all candidates are processed, they are re-ranked according to the corresponding matching scores.

16.2.3 Tag2ImageTag—Match Semantic Tags with Ground Images

In many world regions, there is a significant amount (order of millions) of Flickr reference data available. One way to match query image with the geo-tagged ground imagery is by using semantic tags. The semantic tags in the reference imagery can be

Fig. 16.6 Query example which was found using Flickr image tag "bridge"

extracted via automated semantic scene labeling methods, detection and extraction of text from images, and using existing Flickr user tags. Tagged imagery is a very useful reference especially for tags that involve discernible landmarks, for example, bridges, water towers, etc. This reference data, when combined with visual appearance of images and GIS information, can significantly reduce the geo-location search space. In addition to landmarks, another use case is when the query or reference data has legible text present. Figure 16.6 show an examples of a query image which was found using Flickr user image tags.

16.2.4 Visual Place Recognition—Match Image Features with Ground Images

The Visual Place Recognition (VPR) module of the WALDO system seeks to recognize the place depicted in a query image using a database of ground images annotated with geo-location information. Collections of geo-tagged web imagery, such as Flickr or Panoramio, open up the possibility of image-based place recognition. Given the query image of a particular scene, the objective is to find one or more images in the geo-tagged database depicting the same place. Place recognition is an extremely challenging task as the query image and images available in the database might show the same place imaged at a different scale, from a different viewpoint or under different illumination conditions.

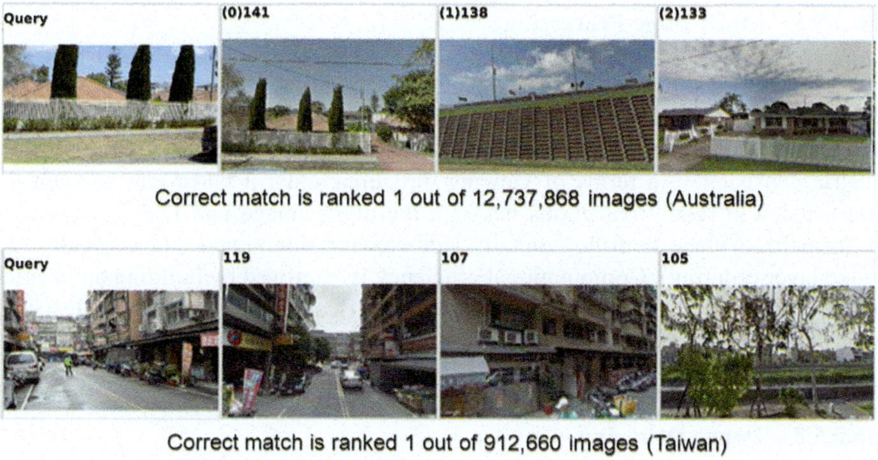

Fig. 16.7 Visual Place Recognition (VPR) results

In order to improve the performance of visual place recognition, we have also experimented with an alternative approach to the bag of visual words approach [8]. The new approach is based on optimized transform coding. We improved transform coding (TC) and use the inverted multi-index [9]. The TC achieves greater speed, simplicity, and generality as compared to the state-of-the-art Optimized Product Quantization (OPQ) [10] with limited accuracy. We have improved the original TC by estimating underlying probabilistic density functions of PCA coefficients, then applying the efficient Lloyd-Max algorithm enabled by the estimated density functions, and optimizing the bit allocation problem enabled by the previous two steps, which would otherwise be computationally prohibitive. Details of our algorithm can be found in [11] (Fig. 16.7).

16.2.5 Im2DEM—Match Skyline Features with 3D Terrain Data

Skyline and ridgeline features are key features that can be identified in many world regions. Our Im2DEM matcher [12] attempts to match these features with pre-indexed 3D terrain features for rough estimation of camera position using the approach of [13]. The method is in principle very similar to Visual Place Recognition (VPR), except that while VPR represents each image using a bag of visual word along with an inverted index and TF-IDF to find matching images, the Im2DEM approach represents each skyline using a bag of curve (contour word) and an inverted index and TF-IDF to find the location, heading, and roll of the camera.

16.2.5.1 Multi-ridge Processing

We investigated a new efficient algorithm for Im2DEM matching to fully exploit multiple mountain ridges labeled from the query image. The approach is based on the contour word framework described above. However, instead of considering geometric consistency in terms of viewing directions only, it takes into account the consistency in relative positions between the query image and DEM database in a normalized plane both horizontally (for viewing directions) and vertically (for ridge layer ordering). Computational efficiency is improved by building an inverted index table for multiple ridge contour words and discarding most intermediate results obtained from the previous approach.

16.2.5.2 Image Dehazing

We have experimented with dehazing algorithms to enhance images where the skyline is not clear in the query image. We used the dehazing algorithm proposed by [14]. Figure 16.8 presents some preliminary image enhancement results, showing that the skyline is clearly visible in the image on the right.

16.2.5.3 Fine Im2DEM

The Im2DEM matcher provides a rough estimate of the geo-location and heading of the camera based on the similarity of the histogram of curve features. The Fine Im2DEM matcher attempts to refine the results of the coarse matcher using additional geometric constraints. The matcher operates on the output of the coarse Im2DEM matcher, which is a list of N (around 1500) camera candidates with the latitude/longitude coordinates, together with the coarse estimation of the camera heading direction (quantized to 3° bins), roll angle (integer from −6 to 6), and

Fig. 16.8 *Left*—original image. *Right*—dehazed image

horizontal field of view (nonuniform 14 samples from 10° to 70°). The refinement process takes the list of coarsely estimated camera candidates as input. The output of the refinement step is a short list of camera candidates (usually no greater than 20) with estimated parameters (latitude, longitude, altitude, heading, roll, tilt, and FOV). The refinement method clusters camera candidates with similar views and considers them jointly to identify stable features and to establish correspondences. A simplified pinhole camera model with intrinsic focal length, position (latitude/longitude/altitude), and orientation parameters (heading, tilt, and roll) parameters is assumed. The coarsely estimated camera candidates are first re-ranked by a 2D Iterative Closest Point (ICP)-like fine alignment step. This is similar to the geometric verification process in [13], but we employ both skyline and ridges. Moreover, the alignment errors are normalized to alleviate the effects from different vertical fields of view. The camera candidates are then re-ranked according to the alignment errors and then grouped into several clusters. A camera model is estimated for each cluster separately.

The complete querying process is illustrated by the examples shown in Fig. 16.9. The skylines and ridges are first extracted from the query image (Fig. 16.9a, b). The coarse estimation step takes the query features and outputs a ranked list of estimated camera candidates in the quantized space (Fig. 16.9c, d). The results from the coarse estimation step are then refined to obtain a small list of more accurately estimated camera parameters (Fig. 16.9e, f).

16.2.6 ProjectLive2D—Refine Camera Model Using Point Correspondences

Once the system has narrowed down the candidate regions to a select few, based on the inputs from the above matchers, the final step is to verify and refine the estimate based on user inputs. The relationship between ground and aerial features is very complex, and therefore it is difficult to automate this process. Using the ProjectLive2D matcher, the user provides point correspondences between the query image and aerial imagery, which are then used to compute the camera parameters. Figures 16.10 and 16.11 show snapshots of the user interface that allows the user to provide correspondences between the image and reference data (including satellite imagery in Fig. 16.10 and Google Street View data in Fig. 16.11). The tool allows the user to estimate extrinsic/intrinsic camera parameters by enforcing constraints on where the camera may be (e.g., not in the water, etc.).

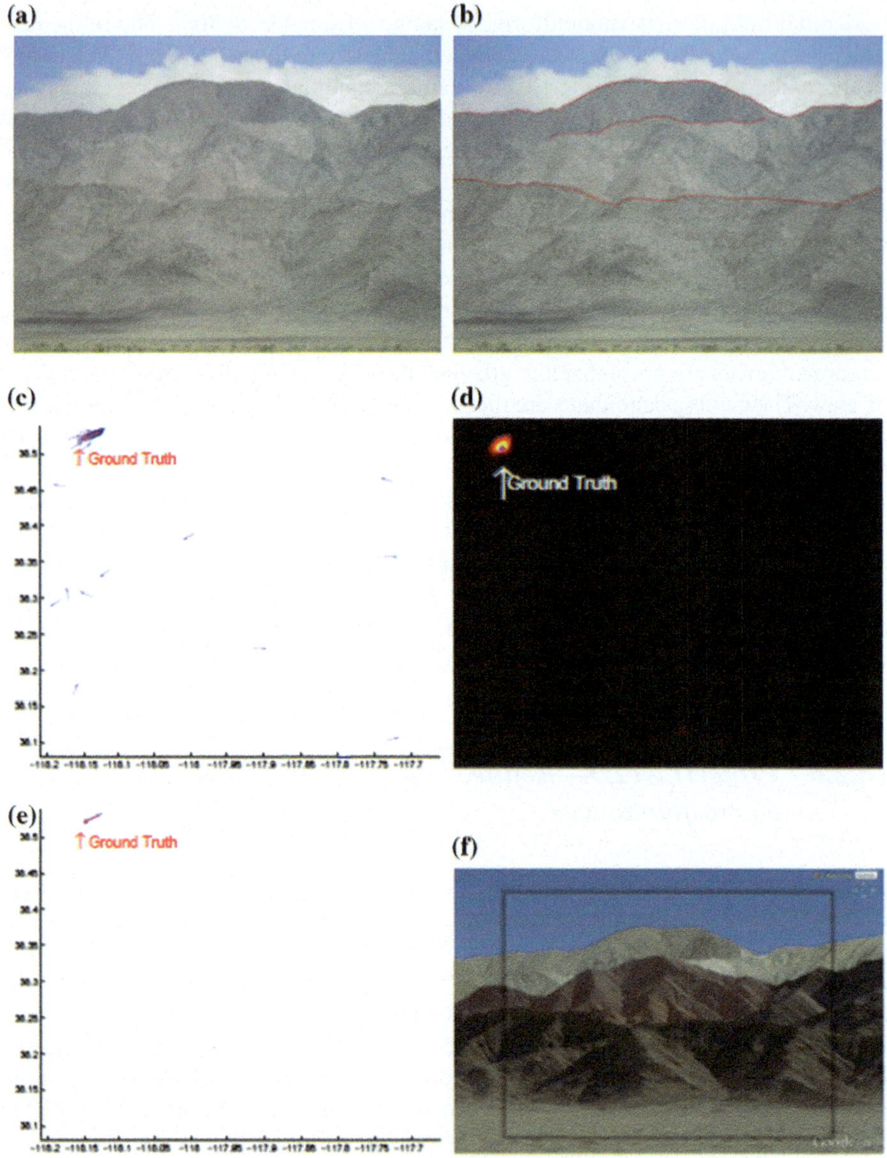

Fig. 16.9 An example of the system input and output: **a** query image, **b** extracted skyline and ridges, **c** coarse estimates of the camera locations and heading (*blue arrows*) and the ground-truth location and headings (*magenta arrow*), **d** probability heat map generated from the coarse estimates, **e** final output: a small list of refined camera estimates (there is only one in the displayed region), and **f** a synthetic view rendered using the refined camera parameters

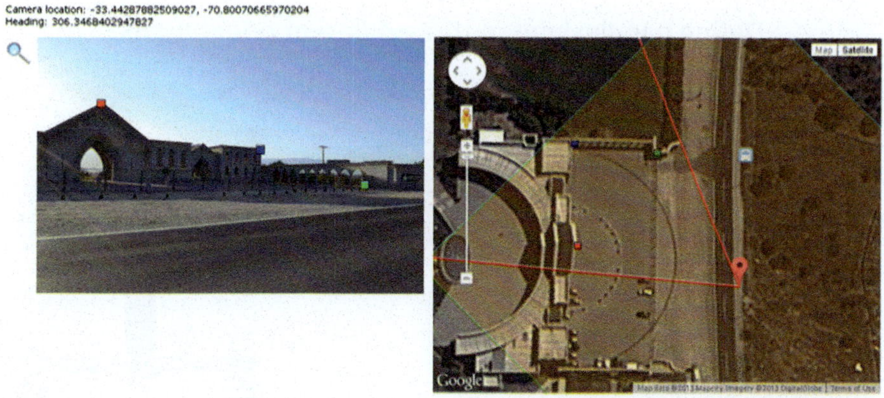

Fig. 16.10 Camera model refinement—point correspondences between image and satellite imagery

Fig. 16.11 Camera model refinement—point correspondences between image and Google Street View imagery

16.2.7 Image Calibration and Mensuration

Determining the calibration of a camera is a fundamental first step in many vision problems. We have developed a web-based user-guided system for accurately estimating camera calibration given a single image. In addition, the system supports estimating lengths, heights, and areas of objects in the image. The motivation is that uncertainty in scale and calibration is a source of geo-localization error, and this system makes it easy to estimate this information. Such estimates can be used to improve the effect of other matchers such as Im2DEM and TagConfig2GIS.

From a user standpoint, interacting with our system is simple. We ask the user to identify and annotate sets of lines in the image that correspond to parallel lines in the world. The first two sets are parallel to the ground. The third set corresponds to vertical lines. Each set of lines must be orthogonal to the others. The user is guided through the calibration process step by step, with visual feedback provided along the way. Figure 16.12 shows some screenshots of the wizard interface.

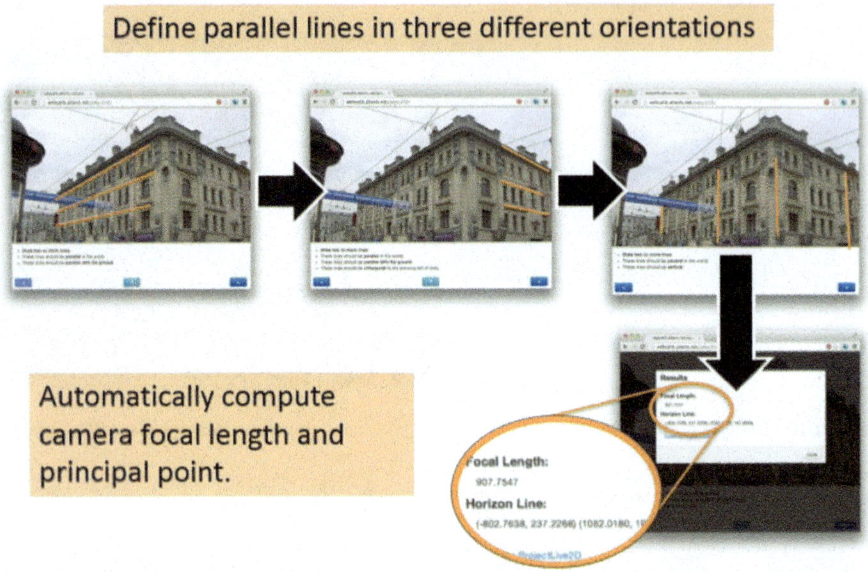

Fig. 16.12 Screen captures of the WebCalibrator "wizard interface" in use. The user is shown an image and guided through the process of calibration and mensuration

We evaluated the effectiveness of our user-in-the-loop methods on two benchmark datasets. The results demonstrate that our system enables novice users to accurately calibrate and measure objects in images. To evaluate the performance of our user-in-the-loop calibration interface, we make use of the Eurasian Cities Dataset (ECD) [15], a publicly available dataset consisting of 103 outdoor images of European and Asian cities. For evaluation of our measurement interface, the lack of an existing mensuration dataset necessitated the construction of our own. We built a still life mensuration dataset called the Lab Mensuration Dataset, which contains objects of known dimensions. We compare the performance of our semiautomatic web-based calibration tool to fully automatic methods using horizon error as in [16]; we compute the maximum difference between the computed horizon and ground-truth horizon in the image, normalized by image height. Our tool, operated by an experienced user, is 1.93 % better than the current state of the art [16].

16.2.8 Video Calibration

Videos with camera motion contain a lot of useful information (feature tracks, etc.) which can be exploited to determine camera parameters. We used the structure-from-motion technique of [17] to extract geometry from video queries. The extracted geometry can be used to compute focal length information and the relative orientations

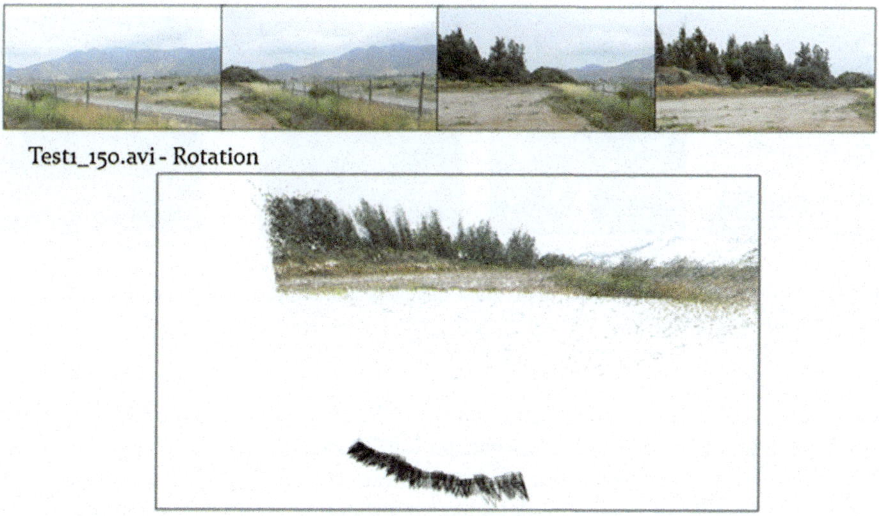

Fig. 16.13 Reconstructed camera geometry for a rotating camera

of different videos frames. Figure 16.13 shows an example of geometry extraction from a query video.

In order to evaluate the effectiveness of the algorithm, we chose a test set consisting of 120 videos (with known ground truth) and compared the estimated focal length with the ground-truth values. Over the entire set, the median focal length error was less than 7.5 %.

16.2.9 Decision Support System

The decision support system of the WALDO system fuses multiple predictions from different matchers for each input image into a geo-distribution. In addition, it exploits the estimated geo-distributions to estimate a list of candidate regions for a given query image. The system does this by converting the output of each matcher into a geo-spatial probability distribution and then fusing each of these geo-distributions to form a combined probability distribution. The basic model for combining the output of various matchers and user inputs into a density over locations can be motivated through a Naive Bayes (NB) approach; under NB, the log probabilities produced by these experiments are added, exponentiated and normalized producing a distribution over locations. The Naïve Bayesian combination rule is given as:

$$
\begin{aligned}
p(C, F_1, \ldots, F_n) &\propto p(C) \cdot p(F_1|C) \cdot p(F_2|C) \ldots p(F_n|C) \\
&\propto p(C) \prod_{i=1}^{n} p(F_i|C)
\end{aligned}
$$

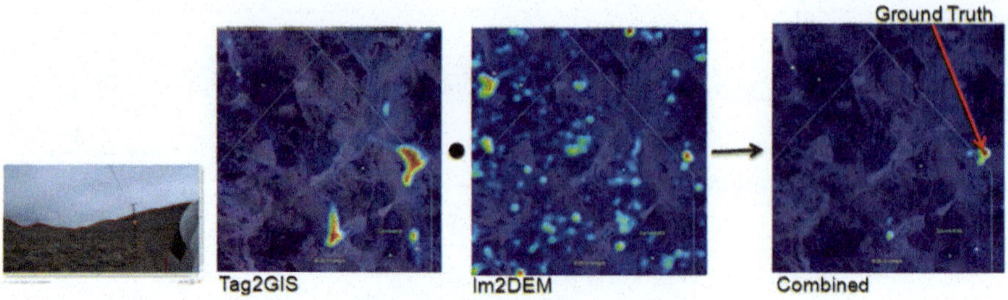

Fig. 16.14 Geo-distributions from individual matchers are combined to estimate the final geo-distribution for thresholding

where $F_1, \ldots F_n$ are feature variables and C is the class variable. The naïve Bayesian approach treats every feature (geo-distribution) as independent and equally reliable. This assumption is very strong and can result in severe overconfidence and inaccurate predictions. Furthermore, the model makes it difficult to assess which features matter most. We implemented the decision support engine using logarithmic opinion pooling of the mixture of experts [18]. The combination rule of logarithmic opinion pooling is given as:

$$\begin{aligned} p(C, F_1, \ldots, F_n) &\propto p(C) \cdot p(F_1|C)^{\lambda_1} \cdot p(F_2|C)^{\lambda_2} \ldots p(F_n|C)^{\lambda_n} \\ &\propto p(C) \prod_{i=1}^{n} p(F_i|C)^{\lambda_i} \end{aligned}$$

Under this model, the feature variables (geo-distributions) are considered as possibly correlated experts, whose opinions are pooled or aggregated. The weight λ_i corresponds to the relative reliability of expert i, and is learned from success rates over time (Fig. 16.14).

Once the final distribution is estimated, the system must compute a candidate list from the distribution. This is done by simply binarizing the distribution using a single threshold on probability mass. The threshold is computed as follows: The system first finds the probability value, p_0, corresponding to a fixed percentage of total probability mass (60 % in our experiments). Then, it finds the point, p_1, corresponding to a significant drop-off in probability value, i.e., the sorted probability values start to become flat after dropping from maximum. The threshold is determined as maximum of p_0 and p_1. The candidate regions are the connected components of the resulting binary map.

16.3 Conclusion

In conclusion, we have made a large number of technical advancements in the field of image geo-localization under this effort. The resulting WALDO system is robust and has been shown to perform well in challenging test cases. The system was also tested independently by novice users using unseen queries. Even with limited training and experience with the system, the novice users were able to geo-locate multiple queries over the course of two weeks of testing. There are still many challenges and possible avenues to improve the system. Some key challenges include:

- Lack of reliable GIS data: There is a need for development of robust knowledge discovery tools to automatically create a rich and diverse GIS dataset.
- Lack of ground images: Ground image coverage from sources such as Google Street View is limited to a few countries. There is global satellite image coverage, and therefore, more focus should be placed on advanced satellite image analysis tools.
- Scalability: As the system is expanded for global operation, scalability will be a big challenge. Distributed computing solutions can alleviate some of these issues.
- Lack of subject matter knowledge: Many query images from certain parts of the world contain very geo-informative scene elements such as tree/rock/vegetation type, but a novice user cannot be expected to be aware of this information. Focus should be placed on developing Virtual Subject Matter Expert (vSME) tools to help the user provide inputs in a more meaningful manner.

References

1. http://www.openstreetmap.org/
2. http://www.gdal.org/
3. http://www.geonames.org/
4. http://landsat.usgs.gov/landsat8.php
5. Campbell J (1993) Evaluation of the dark-object subtraction technique for adjustment of multispectral remote-sensing data. In: Proceedings SPIE, vol 1819
6. Park M, Chen Y, Shafique K (2013) Tag configuration matcher for geo-tagging. In: ACM SIGSPATIAL
7. Belongie S, Malik J, Puzicha J (2002) Shape matching and object recognition using shape contexts. In: IEEE PAMI, vol 24
8. Philbin J, Chum O, Isard M, Sivic J, Zisserman A (2007) Object retrieval with large vocabularies and fast spatial matching. In: CVPR
9. Babenko A, Lempitsky V (2012) The inverted multi-index. In: IEEE conference on computer vision and pattern recognition (CVPR)
10. Ge T, He K, Ke Q, Sun J (2013) Optimized product quantization for approximate nearest neighbor search. In: IEEE conference on computer vision and pattern recognition (CVPR)
11. Park M, Gunda K, Gupta H, Shafique K (2014) Optimized transform coding for approximate KNN search. In: BMVC
12. Qian G, Chen Y, Gupta H, Gunda K, Shafique K (2015) Camera geolocation from mountain image. In: Fusion

13. Baatz G, Saurer O, Koser K, Pollefeys M (2012) Large scale visual geo-localization of images in mountainous terrain. In: ECCV
14. He K, Sun J, Tang X (2009) Single image haze removal using dark channel prior. In: CVPR
15. Barinova O, Lempitsky V, Tretiak E, Kohli P (2010) Geometric image parsing in man-made environments. In: European conference on computer vision
16. Xu Y, Oh S, Hoogs A (2013) A minimum error vanishing point detection approach for uncalibrated monocular images of man-made environments. In: IEEE conference on computer vision and pattern recognition
17. Agarwal S, Snavely N, Simon I, Seitz S, Szeliski R (2009) Building Rome in a day. In: ICCV
18. Hinton G (1999) Products of experts. In: Proceedings of the ninth International Conference on Artificial Neural Networks (ICANN99)

Chapter 17
Photo Recall: Using the Internet to Label Your Photos

Neeraj Kumar and Steven Seitz

Abstract We describe a system for searching your personal photos using an extremely wide range of text queries, including dates and holidays (*Halloween*), named and categorical places (*Empire State Building* or *park*), events and occasions (*Radiohead concert* or *wedding*), activities (*skiing*), object categories (*whales*), attributes (*outdoors*), and object instances (*Mona Lisa*), and any combination of these—all with **no** manual labeling required. We accomplish this by correlating information in your photos—the timestamps, GPS locations, and image pixels—to information mined from the Internet. This includes matching dates to holidays listed on Wikipedia, GPS coordinates to places listed on Wikimapia, places and dates to find named events using Google, visual categories using classifiers either pre-trained on ImageNet or trained on-the-fly using results from Google Image Search, and object instances using interest point-based matching, again using results from Google Images. We tie all of these disparate sources of information together in a unified way, allowing for fast and accurate searches using whatever information you remember about a photo. We represent all information in our system in a layered graph which prevents duplication of effort and data storage, while simultaneously allowing for fast searches, generating meaningful descriptions of search results, and even suggesting query completions to the user as she types, via auto-complete. We quantitatively evaluate several aspects of our system and show excellent performance in all respects. Please watch a video demonstrating our system in action on a large range of queries at http://youtu.be/Se3bemzhAiY.

N. Kumar (✉) · S. Seitz
Department of Computer Science and Engineering, University of Washington, Seattle, WA, USA
e-mail: me@neerajkumar.org

S. Seitz
e-mail: seitz@cs.washington.edu

A.R. Zamir et al. (eds.), *Large-Scale Visual Geo-Localization*, Advances in Computer Vision and Pattern Recognition,
DOI 10.1007/978-3-319-25781-5_17

17.1 Introduction

We have all had the frustrating experience of trying—unsuccessfully—to find photos of a particular event or experience. With typical personal photo collections numbering in tens of thousands, it is like finding a needle in a haystack. Current tools like Facebook, Picasa, and iPhoto only provide rudimentary search capabilities, and that only after a tedious manual labeling process. In contrast, you can type in almost about anything you want on Google Image Search (for instance), and it will retrieve relevant photos. Why should searching your personal photos be any different? The challenge for personal photos, unlike Internet photos, is the lack of descriptive text for use in indexing; people generally label very few of their photos.

The key insight in this work is that a surprisingly broad range of personal photo search queries are enabled by **correlating information in your photos to information mined from the Internet**. For starters, we can find your photos from *Christmas* (Fig. 17.1a) by using lists of holidays and dates, or of *Hawaii* by analyzing GPS (aka *geo-tags*) and matching to online mapping databases. We introduce an extremely powerful version of location search that enables queries ranging from exact place names [*FAO Schwartz*, *Grand Canyon* (Fig. 17.1b)] to rough recollections [*park*, *skiing* (Fig. 17.1c)]. These capabilities alone are very powerful and go beyond what is possible in leading photo tools like Facebook and iPhoto (which only just added location search in their latest version).

More interestingly, there is a broad class of important queries that are not expressed in terms of time or location, but which can be answered *using* photo time and location information, in conjunction with online data sources. For example, suppose you want to find the photos you took of the *Radiohead* concert (Fig. 17.1d). This query does not specify a location or a date; so to answer it, we find all your photos that are taken near performance venues (e.g., stadiums, concert halls, arenas, major parks), search Google for events that occurred at those places on the dates when you took your

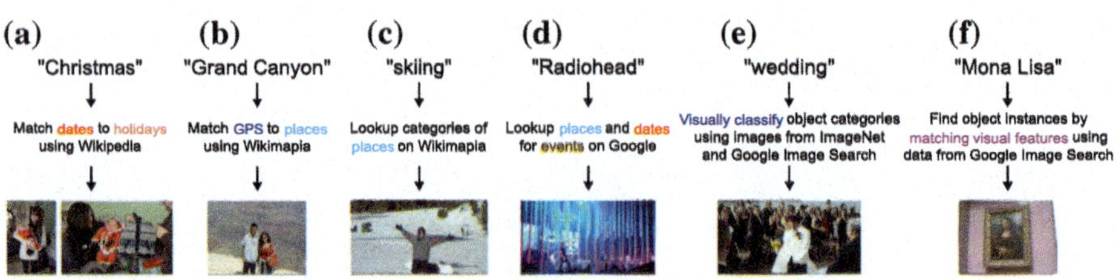

Fig. 17.1 Our system allows users to search their personal photos using queries like **a** "Christmas" or other **holidays**, **b** "Grand Canyon" or other **places**, **c** "skiing" or other **activities**, **d** "Radiohead" or other **events**, **e** "wedding" or other **visual categories**, **f** "Mona Lisa" or other **object instances**, and arbitrary combinations of these—all with **no manual labeling.** We associate images with labels by correlating information in user photos to information online using a variety of techniques, ranging from computer vision, to GPS and map databases, to on-the-fly internet search and machine learning. Users search via natural language text queries—not limited to an arbitrary fixed set—based on whatever information the user remembers about a photo

photos (using a query like, "Key Arena, Seattle, April 9, 2012"), parse the results page (on which many mentions of "Radiohead" occur), and associate the resulting text to the corresponding photos in your collection. This enables searching for a wide range of events you have seen, like *Knicks game*, *Cirque du Soleil*, *Obama's inauguration*, etc.. All of this is transparent to the user: they simply issue the query *Knicks game* and we figure out how to answer it.

An even broader range of queries is enabled by analyzing the pixels in photos and correlating them to other photos on the Internet. For example, when you type in *Mona Lisa* (Fig. 17.1f), we do a Google Image search for "Mona Lisa," download the resulting images, match them to your photos using interest points, and return the results—all in a few seconds. Whereas this is an example of a specific instance, we also support category-level queries using both pretrained and on-the-fly trained classifiers. For example, to find your *wedding* photos, we do a Google Image search for "wedding," download the results, train a visual classifier, and run the classifier on your photos—again, in just seconds. As Table 17.1 summarizes, we support an extremely wide range of queries:

Furthermore, these types of queries can be combined, e.g., *wedding in New York*, *college graduation in Atlanta*, *dogs on Halloween*, *concerts in parks*, *Charles River on St. Patrick's day*, to provide even richer queries and more specific results. The use of complementary types of queries also greatly improves both the robustness and flexibility of our system; the former because uncertain estimates of one modality can be compensated for by others, and the latter because there are often many possible ways to arrive at the same image, allowing the user to search by whatever pieces of information she remembers about the image.

Fundamentally, the use of Internet data enables an enormous shift in user experience, where *the user chooses the search terms* rather than them being limited to a predefined set of options (as is the norm for virtually all prior work in CBIR and object recognition), or requiring manual labeling. More specifically, ours is the first published work to include the following new capabilities, without requiring any labeling:

Table 17.1 Query Types supported by our system

Type of query	Example queries	**Type of query**	Example queries
Dates/holidays	August 2012, Thanksgiving	**Activities**	Skiing, cricket, paintball
Named places	Sea World, FAO Schwartz	**Places by type**	Zoo, hotel, beach
Named events	Radiohead concert, Knicks game	**Events by type**	Wedding, birthday, graduation
Object categories	Whales, green dress, vase	**Attributes**	Portrait, blurry, close-up
Instances	Mona Lisa, Eiffel tower		

1. Named event personal photo search (e.g., *Knicks game*).
2. Object instance search in its full generality, i.e., matching your photos to arbitrary named objects on the Internet.
3. Far more extensive location-based query support (including named and categorical places at all levels of granularity) than any work, by leveraging Wikimapia.

We represent all information in our system as a hierarchical knowledge graph, with layers corresponding to language, semantics, sensors, image, and grouping constructs. The graph provides a unified representation of all data and lets us perform inference operations via propagations through the graph, including search, auto-complete, and query-dependent description of matched images. Queries are performed by propagating downward in the graph (from language down to images), and concise descriptions of returned results are generated by propagating upward from photos to language tags. This architecture enables incorporating different types of information (time, place, text, image) within the same framework, while providing flexibility and efficient searches.

Finally, we quantitatively evaluate the key aspects of our system by testing search performance on manually labeled images. Specifically, we test our coverage of places (both named and categorical), our ability to find named events, and our computer vision-based visual category classifiers. For a qualitative look at many more examples of search results, please see our video demo, shared at http://youtu.be/Se3bemzhAiY.

17.2 Related Work

Our work is inspired by Google Image Search and other Internet search engines aimed at producing relevant content for *any* user-specified query. We seek to provide similar functionality in the domain of personal photos. But while Internet image search engines exploit co-occurrences of images and text on web pages, most personal photo collections have scarce textual information to use as a ranking signal; hence the latter domain is much more challenging.

Thus, consumer photo organization tools like Picasa and Facebook provide only rudimentary search capabilities, almost completely based on manual labeling. Since most users label few if any of their photos, search is largely ineffective. In their latest release, iPhoto introduced the ability to search for place names by "matching terms such as *Seattle* or *Milan*, to a mapping database."[1] While no other technical details have been published, the feature seems to provide similar capabilities to our system with regard to matching place names, but not place categories. In June 2013, Google enabled a new auto-labeling feature that leverages deep learning to classify users' photos using 2,000 pretrained visual classifiers (again, few technical details are provided).[2] Similarly, in 2014, Yahoo's flickr photo service started offering content-

[1] quoted from http://support.apple.com/kb/PH2381.

[2] http://googleresearch.blogspot.com/2013/06/improving-photo-search-step-across.html.

based image search based on deep learning,[3] but no technical details have been published. However, none of these systems support searches for named events (e.g., *Burning Man*), on-the-fly training of arbitrary visual classifiers (e.g., *green dress*), or matching object instances (e.g., *The Last Supper*). While some researchers have explored the use of manual tags, applications, we focus here on purely automated approaches.

In the research community, there is a large body of work on content-based image retrieval (CBIR). See Datta et al. [8] for a survey of this field. While much of this literature involves novel browsing interfaces or visual search techniques (e.g., query-by-example, similar images), we focus specifically on related work aimed at *text-based search* in particular, which requires indexing based on semantic content, and discuss the most relevant techniques. Naaman et al. [21] first proposed using a database of named geographic locations to automatically label geo-tagged photos, although their capabilities were limited to cities, states, and parks, and they did not support search. More recent work has looked at combining personal tags with community tags, using GPS-tagged photos on flickr to find nearby tags [13, 28]. Many authors have used date and time information to organize photo collections and group them into events, e.g., [14]. Image content has also been used to improve event clustering [6]. There is also much work specific to finding faces in photos [15, 35]. For assigning labels to images based on visual content, existing work assumes a labeled dataset is accessible beforehand to learn image annotation models using sophisticated optimization methods [17, 33]. In contrast, we use standard off-the-shelf vision algorithms applied to Google Images results, allowing for more generality.

We strongly leverage the recent progress in the computer vision community on object recognition. Modern techniques, mostly based on low-level features and discriminative classifiers [18, 22, 26, 32], perform increasingly well on computer vision benchmarks such as the Pascal Challenge [11] and ImageNet [9]. They have not been evaluated, however, in the context of unconstrained image search, where the user can type in *any* query term, although some have tried learning object classifiers for relatively "clean" images using Google Image search—but at far slower (non-interactive) speeds [12], or improving classifiers with incremental learning from web search results [16]. Others have looked at using text from the web to learn better classifiers [29]. Even with rising classifier accuracies, however, the problem is that classifiers tend to perform well for many queries but fail catastrophically for others; hence, it is not obvious whether these classifiers will perform reliably in enough cases to give a positive user experience. We present the first system that uses visual classifiers for unconstrained personal photo search, and show it to be remarkably effective, especially when combined with time and place cues.

Our layered graph data representation, and more generally, our focus on tying all kinds of cues together are partially inspired by Vannevar Bush's idea of the *memex* [4]—a hypothetical device for assigning tags to individual articles (or other forms of information), allowing for elaborate cross-referencing via a network of connections between these different articles. As will be made clear in the following

[3]http://code.flickr.net/2014/10/20/introducing-flickr-park-or-bird/.

sections, our system takes exactly this same approach for organizing photo collections, except that whereas Bush envisaged the labels coming from solely the user and her colleagues, the Internet today offers a vast repository of labels, assuming one can find the right way to index them. In the vision community, others have also referenced the idea of a "visual memex," most prominently Malisiewicz and Efros [19], who reason about object relationships within single images by looking at what kinds of connections exist between them across large datasets. Our work is different both in application and in scope—we focus on personal photos for the purpose of search and organization (much more akin to Bush's original vision) rather than labeling objects within single images (for the most part; see Sect. 17.3.5 for how we also identify object instances in our system).

Finally, our technique of training classifiers on-the-fly adapts a line of recent work on using Google Image Search results to train classifiers for people, scenes, and objects directly at test time [1, 5, 24]; we apply similar techniques but in a different domain: searching personal photo collections. Others have also started exploiting Google Images for training classifiers directly, e.g. [10].

17.3 Data Sources

Our personal photo search system takes text queries as input and returns a ranked list of matching images; this requires associating text labels with images. In most existing systems, users must assign these labels directly to the photos, a manual and time-consuming process. To avoid this tedium, we find existing sources of text labels and associate these to the right photos. One of the keys to enabling the wide variety of queries we support is that we make much more extensive use of *all* data stored in images. In the vision and graphics communities, images are generally thought of as simply arrays of pixels; however, photos taken with modern cameras (such as mobile phones) also store two other critical pieces of information: a timestamp denoting *when* the photo was taken, and a GPS coordinate denoting *where* the photo was taken. (In some cases, the altitude and orientation of the phone are also stored; however, as these are currently much less reliable, we ignore them in this work.) While GPS coordinates are not present in all existing photographs, the explosive growth of mobile phones means that we can safely assume that most future images will have reliable GPS and timestamp data. We therefore take advantage of all three types of "sensor readings"—time, location, and pixels.

As much as possible, we associate labels with images in an initial indexing step on the user's photos, i.e., once, when new photos are added to the system. This means that at query time, searches are near-instantaneous. This is true for holidays (Sect. 17.3.1), places (Sect. 17.3.2), events (Sect. 17.3.3), and visual categories (Sect. 17.3.4) via ImageNet. It also allows for query suggestions to the user in the form of auto-complete (see Sect. 17.4.4). Nevertheless, other powerful sources of data are unamenable to this kind of pre-indexing, as they depend on the query itself. This includes the general visual category classification and specific object instance matching, both done via Google Image Search (Sect. 17.3.5). Due to the speed of modern computer vision techniques, this can be done in only a few seconds, making it practical for our prototype system. We anticipate that the time required for these methods in a commercially fielded system will be reduced down to a second or less. The details of how all of this data are stored and searched over are described in Sect. 17.4; for now, it is safe to assume that we simply associate images with text labels along with an optional weight for the label. Figure 17.2 visually summarizes all of our data sources and indexing methods.

17.3.1 Holidays

Time is one of the most important qualities about a photo. When you think of a particular event or memory from your life, you probably remember *when* it happened: yesterday, last year, during Halloween, etc.. How do we get descriptions such as these from a particular photo, which only records a timestamp, like `2011-08-13 13:05:20`? We map dates to holiday names using Wikipedia, the online crowd-sourced encyclopedia. We use the article, "Public holidays in the United States"[4] to do our holiday matching. (Similar lists exist for other countries; here we chose to focus on the United States.) As shown in Fig. 17.2a, tables on this page show the dates corresponding to various holidays. Using the datestamp stored in each of the user's photos, we see if they occur on or around any of these dates (within ± 2 days), and if so, associate them with the name of the holiday. This allows for searches like *Saint Patrick's Day* (Fig. 17.2a), *Christmas* (Fig. 17.3a), and *Independence Day* (Fig. 17.3b).

17.3.2 Places

Another primary way of describing photos is by *where* they happened. As with time, we have to work our way up from the raw sensor data recorded in the image metadata, in this case the GPS coordinates of a photo (e.g., 51° 30′ 2.2″ N, 0° 7′ 28.6″ W). While these coordinates precisely define the location (point on the globe) where a photo was taken, *places* are what people care about. For example, no one remembers the above location as where they took a particular photo, but rather that it was at the Big Ben clock tower in the Westminster area of London, England. Notice that places can be defined at many levels of granularity, from specific buildings to neighborhoods and cities, all the way up to countries. We get labels at all of these granularities using Wikimapia,[5] another online crowd-sourced database, but one which focuses exclusively on geographic information. It currently has data on 20 million places and

[4] http://en.wikipedia.org/wiki/List_of_US_holidays.

[5] http://wikimapia.org.

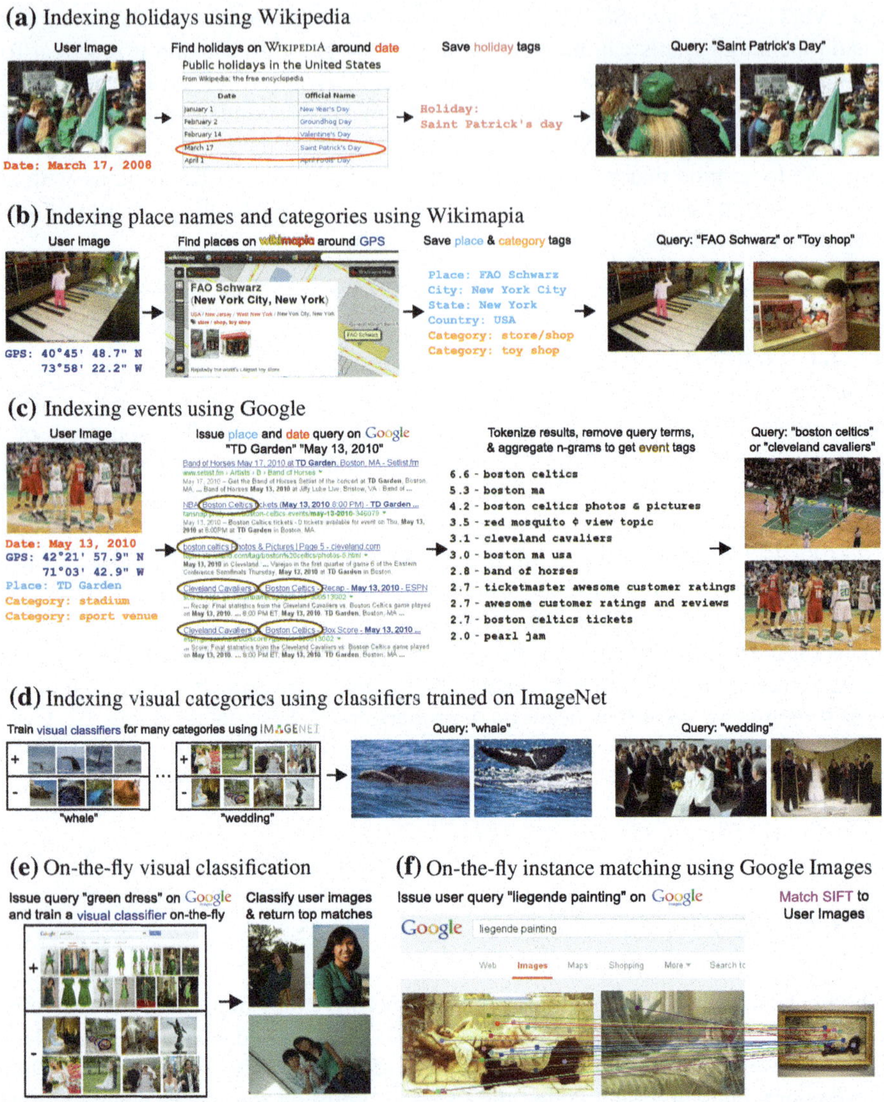

Fig. 17.2 Our system associates images with labels by matching different types of image data to various online sources, either in an initial indexing step (**a–d**), or on-the-fly when the user issues a query (**e–f**). **a** We match the datestamps stored in photos to a list of holidays from Wikipedia to allow searches by holidays. **b** We use GPS coordinates from photo metadata to get place names and categories from Wikimapia. **c** We issue searches on Google for dates and place name pairs to find what event took place there. We parse the results and accumulate frequent n-grams to get event tags. **d** We pretrain thousands of binary visual classifiers using categories from ImageNet to allow for querying by image content. **e** For categories not covered in ImageNet, we issue queries on Google Images and train binary visual classifiers on-the-fly. **f** For finding object instances rather than categories, we match SIFT descriptors on-the-fly from Google Image search results. Despite several sources of noise in the data and matching processes, we are able to return accurate results

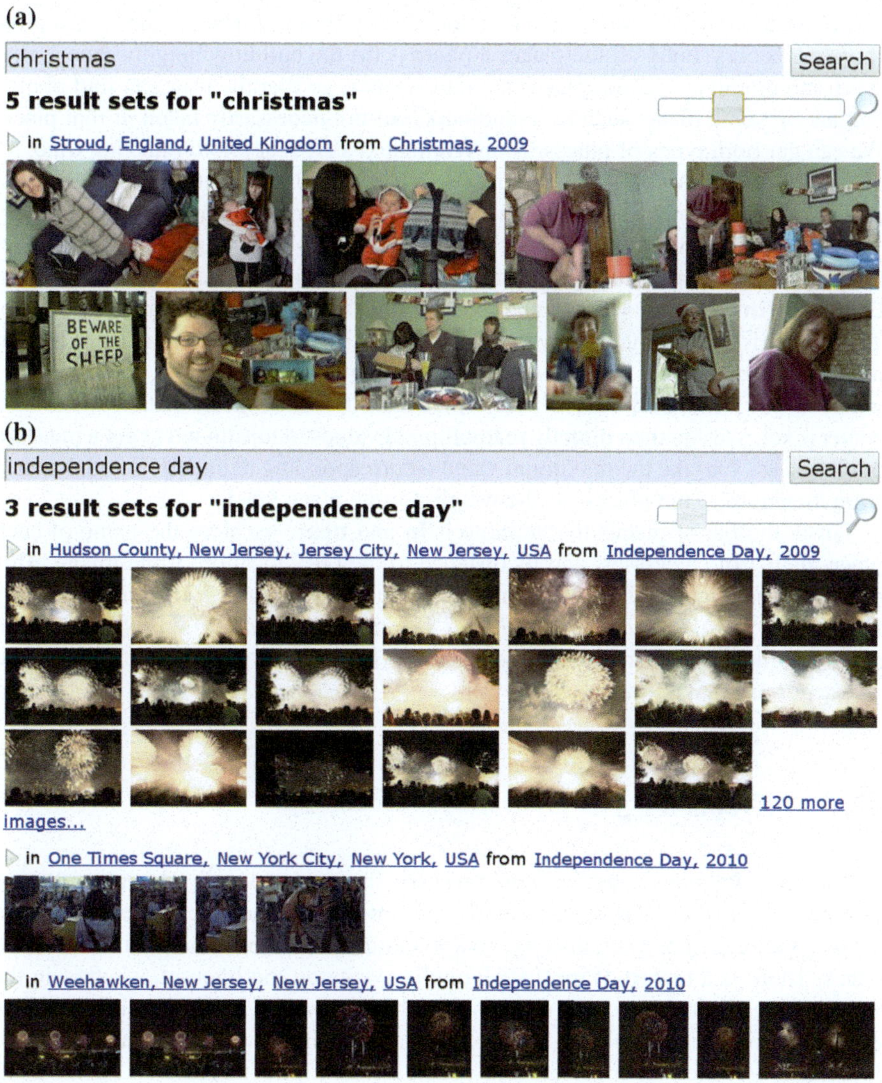

Fig. 17.3 Search results for time-related queries: **a** "Christmas" results. **b** "Independence Day" results. Note that our system groups results by year and also displays the location of each set of results

is growing rapidly. Each place includes the place name, a text description, a list of type categories describing the use of the place (such as *park*, *hotel*, *mountain*, etc.), links to related websites, and other miscellaneous information. Places are delineated by polygons, which users can create and edit on the Wikimapia website. All of this information, including the polygonal extents, are available through a public API.

For each photo, we would like to associate two types of labels along with corresponding scores. First, all the places a photo is **in**: the building, neighborhood, city, state, and country. Second, places a photo is **near**, so that the user can find photos that are *of* some place, such as a landmark, but not necessarily taken *at* that place. We can get both types of places from Wikimapia's API, at photo indexing time, by querying for places near each photo's GPS location. Unfortunately, the Wikimapia API does not currently distinguish between *in* and *near*, so we must do it ourselves.

We create a binary image $P_i(l, r)$ around each photo's GPS location l and project the polygonal extent of each place i from Wikimapia within r meters of the location. For testing **in**, we set $r = 50\,\text{m}$, and then simply check if any pixel in P_i is nonzero. This allows for slight uncertainty in GPS locations (often necessary, particularly with mobile phone images). For **near**, we set $r = 1$ km, and then divide P_i by $D(l, r)$, where $D(l, r)$ is a real-valued image that measures Euclidean distance from the center pixel. We can then directly read off pixel values in this image to get a measure of closeness. We take the maximum value—corresponding to the minimum distance from the photo to the place—as the weight for the place label.

Given the list of places that a photo is **in** and **near**, we store the name of each place, its list of categories, and its distance to the GPS location of the image. This allows the user to search using all manners of queries, such as *FAO Schwarz* or *toy shop* (Fig. 17.2b), *Grand Canyon* (Fig. 17.4a), *railway station* (Fig. 17.4b), and even activities, like *skiing* (Fig. 17.4c) and *paintball* (Fig. 17.4d), which are often included as categories on Wikimapia.

17.3.3 Events

Putting time and place together yields *events*—the natural way in which people tend to group many of their photos. Several of our most cherished memories come from shared public events like concerts, sports, cultural activities, and business conferences. There is a record of most such public events on the Internet, whether in the form of large domain-specific databases such as www.last.fm for music and www.espn.com for sports, global aggregators of events like www.ticketmaster.com, or individual websites for specific events, such as www.london2012.com for the 2012 Olympics (held in London). Fortunately, people do not have to know what the relevant sites are for a particular event, because all of these websites—and billions more—are indexed by search engines such as Google and Bing. While searching for the name of an event will obviously bring up relevant websites (with information such as the date and venue of the event), *the reverse is also true*: searching for the date and venue on Google returns results containing the name of the event and often other relevant information. This additional information includes, for example, the names of performing artists at concerts, operas, and dances; and the names of participating teams at sporting events. Figure 17.2c shows an example from a basketball game: a search for *"TD Garden" "May 13, 2010"* (the name of the venue and the date of the event) brings up a page of results, many of which contain the terms *Boston Celtics*

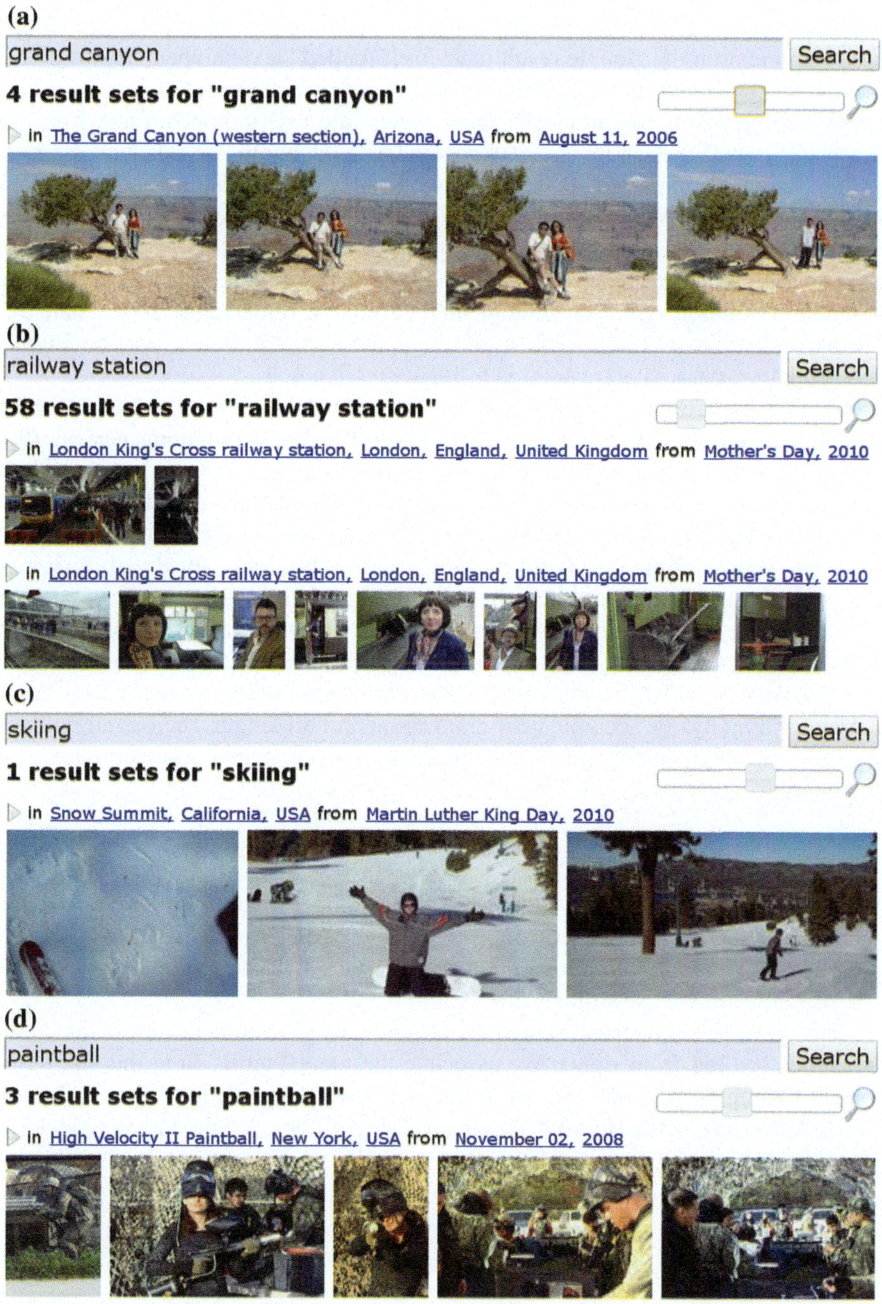

Fig. 17.4 Search results for place- and activity-related queries: **a** "Grand Canyon" uses place names from Wikimapia. **b** "Railway Station" shows that we can also find categories of places, in addition to their proper names. **c** "Skiing" and **d** "Paintball" highlight our ability to find photos based on the activities being performed. These activities are often stored as categories on Wikimapia

and *Cleveland Cavaliers*, the two teams playing that night. Note that we are talking here about text on the Google result page itself (called "text snippets")—we do not need to follow any of the result links to their respective webpages.

We exploit this to automatically label events in a user's photos. First, from the timestamp stored in each image, we know the date the photo was taken on. Second, as described in the previous section, we have already mapped photos' GPS locations to place names via Wikimapia. Third, we also know the categories of each place returned from Wikimapia. We thus select those places that are likely to be venues for events—stadiums, theatres, parks, etc.—and find all photos taken at those places. Our system then submits queries on Google in the form "< venue name >" "< date >". We parse the text on the Google result page and accumulate the most commonly occurring terms across all results, as follows.

We first tokenize the text for each result using the Natural Language Toolkit (NLTK) [2]. We lower-case each token and drop tokens containing periods (this gets rid of website names, like *espn.com*) and query terms. We then split this single stream of tokens into separate "phrases" using a predefined list of delimiters, things like "|" and "–", which are frequently used in webpage titles to separate the page title from the website name. We build n-grams from each phrase, for $n = 1, 2, 3, 4, 5$. For example, the third search result shown in Fig. 17.2c, "*boston celtics Photos & Pictures | Page 5 - cleveland.com*", gets tokenized as two phrases: [{*boston, celtics, photos, &, pictures*}, {*page, 5*}]. This prevents us from creating n-grams such as *pictures page 5*, which would almost never correspond to a term that a user might search for. Finally, we accumulate n-gram counts (of all lengths) across all returned results into a single histogram. We weight each instance of an n-gram term using the following weight:

$$\text{weight}_{\text{term}} = n \cdot (1 - \frac{\text{rank}}{2R}), \tag{17.1}$$

where n is the length of the n-gram (thus encouraging longer n-grams), rank is the rank of the Google result the term appeared in (i.e., earlier search results are probably more relevant), and R is the total number of results (dividing by $2R$ means that the last search result we use still has a weight of 0.5).

This gives us a ranked and scored list of candidate terms for each event, as shown in Fig. 17.2c (2nd from right). We associate the top 10 terms from this list to the image. Examples of events we can index in this way include the basketball game shown in this figure, the *Burning Man* festival (Fig. 17.5c), and the *Italian Grand Prix* (Fig. 17.5d).

17.3.4 Visual Categories

Many other photos you would like to be able to find—from your sister's wedding, to portraits of your daughter, to the exotic flowers you saw in Brazil—correspond to visually distinctive categories. For example, wedding photos tend to have formal

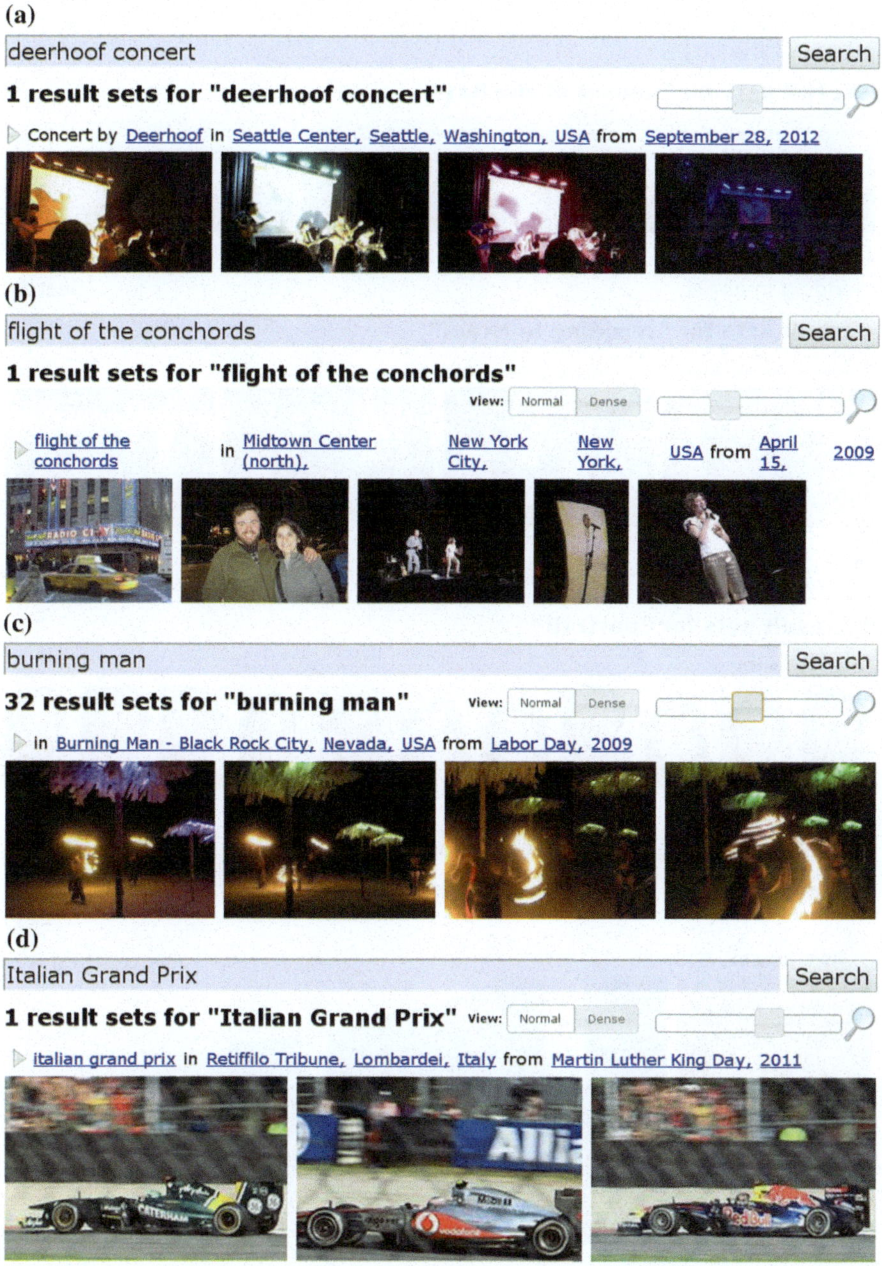

Fig. 17.5 Search results for event-related queries. Our system can handle all sorts of events, including concerts like (**a**) "Deerhoof concert," comedy shows like (**b**) "Flight of the Conchords," named festivals like (**c**) "Burning Man," and sports events like (**d**) "Italian Grand Prix"

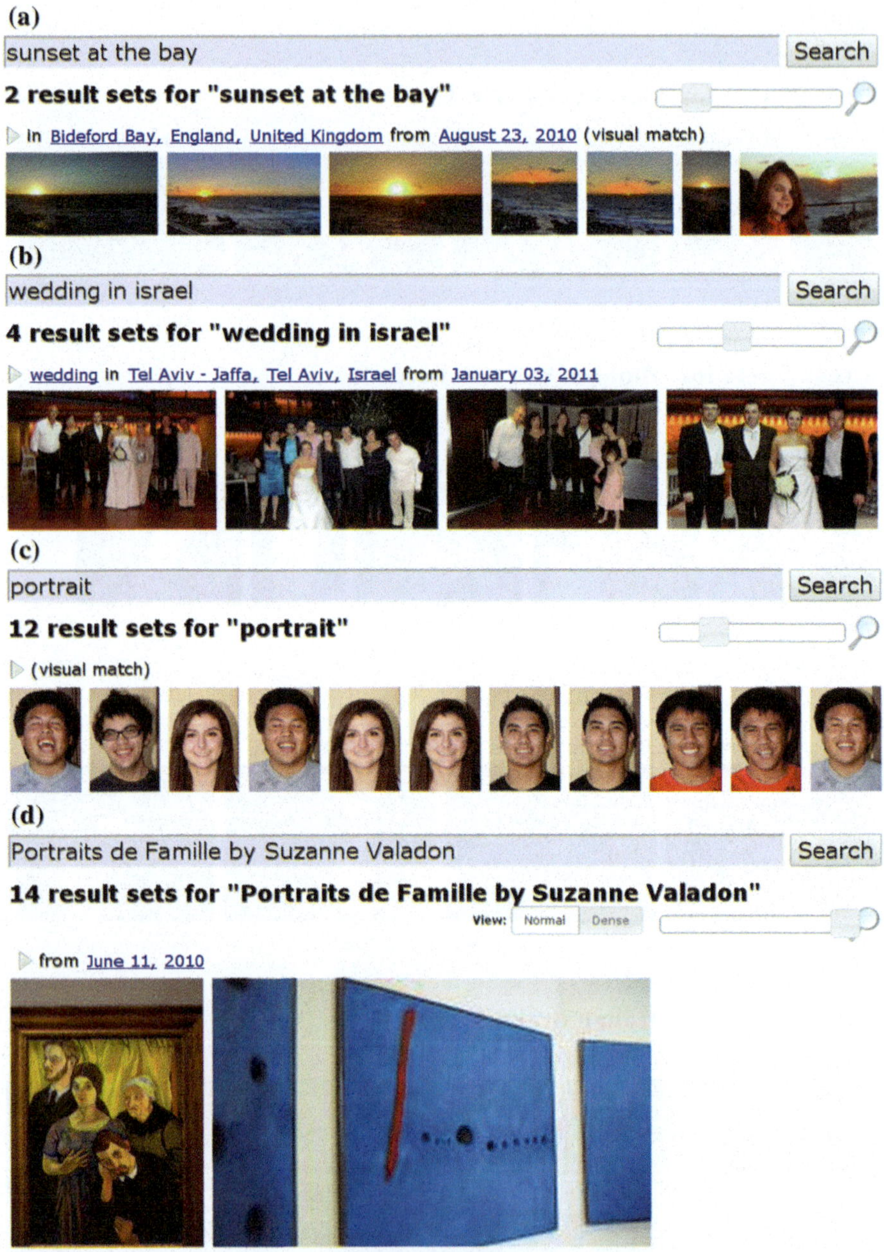

Fig. 17.6 Search results for visual categories and object instances, trained on-the-fly using results from Google Image Search: **a** "Sunset at the Bay" and **b** "wedding in Israel" highlight the power of combining visual categories with our place data. **c** "Portrait" shows that we can also find photos by type or attribute (based on visual features alone). **d** An object instance match for the painting, "Portraits de Famille" by Suzanne Valadon, showing how we can generalize feature-point based methods to any recognizable named object on the Internet

dresses, veils, flowers, and churches. Portraits contain faces, and are usually shot in one of a few traditional views: close-up, three-quarters, profiles, or whole-body. Flower photos contain bright colors, regular geometric structures, and often exhibit narrow depth-of-field. Many of these characteristics are amenable to classification by modern computer vision recognition techniques. In the computer vision literature, this problem is variously referred to as recognition, classification, or categorization of objects, scenes, or attributes; here we use the more generic term "visual categories" to refer to all of them, as they all tend to use the same techniques. In particular, we follow the standard supervised learning pipeline typical in computer vision systems today: features extracted from labeled images are used as positive and negative examples to train binary classifiers. This approach requires a source of labeled examples: images of the category we wish to learn as positive training data, and images of other categories as negative data.

Our first data source is ImageNet [9], a large dataset that aims to collect thousands of curated images for all nouns from the Wordnet [20] project. ImageNet currently has 14 million images for 21, 841 different categories, the latter of which are organized into a hierarchy. Some of these categories are quite useful for our task, such as *wedding* or *whale* (Fig. 17.2d). However, ImageNet is still missing many interesting ones, from random omissions (like *fireworks*, or *graduation ceremony*), to instances of particular classes (e.g., *2007 Toyota Camry* or *Mona Lisa*) or modern cultural artifacts (such as *Superman* or *iPhone*), and of course, it does not cover adjectives (e.g., *blurry* or *shiny*). Conversely, ImageNet also contains many categories which are irrelevant for personal photo search.

Fortunately, there exists a much larger set of images covering all the types of queries users might want to do today or in the future: the Internet, as indexed by Google Image Search. By taking advantage of the rich structure of HTML webpages, links between pages, and text surrounding images, Google Image Search can return relevant image results for an extraordinary range of queries. While not perfect, the top ranked images for most queries tend to be quite relevant. We thus use these results as a source of labeled data, albeit a noisy one. When a user performs a search on *our* system, we issue the same query on Google Image Search, immediately. We then download the top results, and run the entire classifier training and evaluation pipeline, described next.

Implementation Details: We pretrained classifiers for nearly a fourth of ImageNet—4,766 categories (those that seemed relevant to our system). For each category, also referred to as a *synset*, we trained a binary classifier by using images of the synset as positive examples and images of other synsets (excluding ancestors or descendents) as negatives. Our low-level features are histograms of color, gradient magnitude, and gradient orientation, along with a holistic representation of the image computed using the gist [23] descriptor. For our classifier, we use linear support vector machines (SVMs) [7] trained via stochastic gradient descent [3], as implemented by Scikit-Learn [25]. We normalize the outputs of these classifiers using the Extremal Value Theory-based $w\text{-}score$ technique [30], which requires no labeling of calibrated outputs (as is needed by the more commonly used Platt scaling [27]). The w-scores give us the probabilities of each image belonging to the given ImageNet synset; we store the names of synsets that match with probabilities above 80 %.

We follow much the same procedure for on-the-fly training using Google Image Search results, with some small changes for speed. We only use thumbnails of the first 64 images from Google as positive examples, and a random set of 30,000 images from ImageNet as negatives. By using thumbnails, we not only reduce the size of the inputs (both for faster downloads and faster training), but also far less latency, as the thumbnails are downloaded directly from Google as opposed to arbitrary websites, which might take longer to respond. We also process results in a parallelized streaming manner—interleaving downloads and feature extraction across several concurrent threads. Consequently, the entire process takes under 10 s on a single machine in our prototype system, which is fast enough to be usable. (In a commercially deployed version of our system, we can easily expect an order of magnitude improvement in speed, making it almost indistinguishable from a pretrained classifier.) In addition, we cache these classifiers, so that future searches of the same query return instantly. We associate each highly classified user image with the search query and the probability returned from the classifiers. This has the side-benefit of also matching future queries that are very similar to the current one, e.g., *sunset at the bay* (Fig. 17.6a) would also match existing classifiers for both *sunset* and *bay* (in addition to Wikimapia place names and categories, of course). Through Google, we can cover an extremely wide range of queries, including very specific visual categories like green dress (Fig. 17.2e), combinations of places and categories, like *wedding in Israel* (Fig. 17.6b), cultural icons like *Santa Claus*; photos exhibiting particular attributes, like face portraits (Fig. 17.6c), or some objects of interest, like flowers. Obviously, our performance on visual queries is far from perfect—in particular, by tuning for speed, we have sacrificed some accuracy, but we found that the user experience was hurt much more by very long waits than by lower accuracy. In fact, the 10 s it currently takes already feels very long to wait and it is extremely unlikely users would wait much longer.

17.3.5 Object Instances

The binary classification approach described in the previous section is well suited for finding categories of images, but not for specific instances of objects, such as the *Mona Lisa* painting. This is because our classifiers use features that discard most spatial information in order to generalize across different instances of the same category. But if you are searching for a specific instance, we do not even need to train a classifier; instead, we can simply match distinctive local features from a labeled image (again, we use Google Image Search results) to those in your images and return the images that have the most consistent of such feature matches.

We follow the standard matching pipeline first proposed by Sivic and Zisserman [32]: features are quantized into visual words and stored in an inverted index for fast ranking of search results (see their paper for details). We construct a vocabulary of size $k = 10{,}000$ using mini-batch k-means [31]. We use the Scikit-Learn implementation [25] with its default parameters. Since we are matching features against those returned from the thumbnails of Google Image Search results, we downsample user images to thumbnail size before extracting features (both to increase speed and to reduce ambiguities due to large-scale differences). At query time, we issue the user query on Google Images and download thumbnails of the top 10 results. We extract SIFT features on these, project them using the learned vocabulary into a list of visual words, and use the inverted index to accumulate scores for each user image. This process ranks the user images with the most features in common with the Google results highest (weighted by distinctiveness of feature). Examples of queries we can support using this technique include Figs. 17.1f and 17.2f.

17.4 Layered Knowledge Graph

The previous section described several methods for obtaining labels for images; how should we store all of this data? A naive implementation might simply store a mapping from images directly to a list of {label, weight} pairs, and perhaps a reverse mapping to allow for fast searches. But this approach has several drawbacks—most notably, that it would be difficult to interpret search results due to lack of context for *why* a particular result was returned for a given query.

Instead, we store *all* data—not just labels and weights—in a hierarchical knowledge graph (see Fig. 17.7). The graph consists of nodes, each denoting a single conceptual piece of knowledge, connected via weighted edges, denoting the strength of the relation between two concepts. There are different types of nodes corresponding

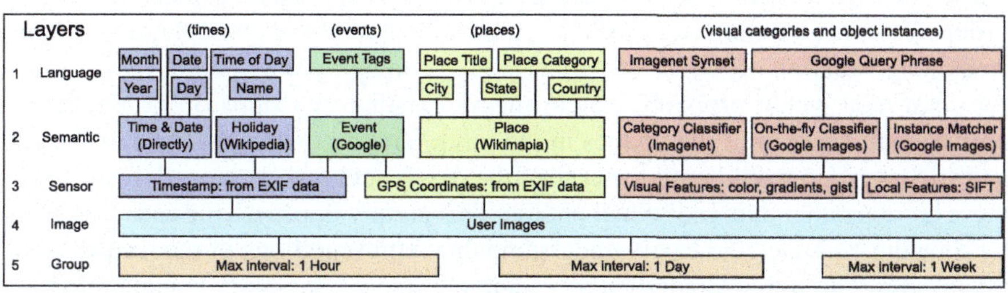

Fig. 17.7 We represent all information in our system as nodes in a layered graph. Each colored box contains many nodes—individual bits of information—of a particular type (denoted by the name in the *box*). *Lines* between *boxes* indicate weighted connections between nodes of the two layers. Images are connected to their sensor values—timestamp and GPS, and low-level visual features. These are mapped into semantic concepts (i.e., the things that people care about) through the use of Internet data sources, shown in parentheses. Finally, semantic nodes are exposed to search queries through the language layer, which contains text tags. By unifying all sources of information in this graph, we can easily incorporate new types of data to support novel types of queries, and perform fast and accurate search using any combination of terms

to different types of data, and each group of nodes of one type are assigned to a specific layer in the hierarchy. For example, a **place node** stores all the information about a given place from Wikimapia, which is connected above to **language nodes** denoting the place title, category, city, etc., and below to **GPS coordinate nodes** that are close to the given place. In this way, even though there might be hundreds of photos near a particular landmark, we only need to store a single copy of the information about that landmark. In total, the size of the graph is linear in the number of *unique* entities in the system—images, places, text terms, etc.. There are a number of operations we perform using this graph, including **search**: given a user query, find matching images (Sect. 17.4.2); **result labeling**: assign a description to each result (Sect. 17.4.3); and **auto-complete**: offer suggestions as the user types (Sect. 17.4.4). But first, we describe a simple addition to our graph that greatly improves the user experience: image groups.

17.4.1 Image Groups

In Fig. 17.7, note that there is an additional layer below the image layer: groups. These groups are automatically created from images based on timestamps. By looking at the time intervals between successive photos, we build up a hierarchy of image groups. For example, images taken within minutes of each other, within hours, within days, and within weeks. These might correspond to meaningful groups such as, the cake-cutting at your son's birthday party, the day you spent exploring Rome, your spring break trip to the Canary Islands, or your sabbatical in England. Each group is connected to all of its contained images with a weight of $1/N$, where N is the cardinality of the group. This ensures that large groups don't dominate search results. It also helps in picking the appropriate group-interval to return for a query, by favoring groups in which the majority of images match the query. For example, if you took a week long vacation of which you spent 2 days photographing a coral reef, a search for *coral reef* would probably match images existing in groups of all lengths, but only a small fraction of the images in the week-length groups would be of the reef, and so that group would rank lower than the day-length groups. Search results show these groups rather than individual photos. This greatly improves the user experience by showing compact and easily understandable representations of the results; rather than wading through a hundred nearly identical shots of the party you attended, you can see a small sampling of those images. (A toggle button lets you see the rest of the group's images.)

Groups also function as a form of smoothing. Consider the case of missing GPS tags. Since the *places* and *events* parts of our system rely on GPS coordinates to index into Wikimapia, images without GPS would normally be much less accessible. However, if they were taken within close temporal proximity of a photo that does have GPS (e.g., if you took one or two photos with your mobile phone, which recorded GPS, but then switched to your high-quality DSLR, which did not), they would be assigned to the same group. And since search results are displayed as groups rather than images, even place-based queries would show the non-GPS'ed images as well. Another instance where this smoothing is critical is with the visual classifiers. As recognition is still an extremely challenging problem (and an active area of research), many classifiers are not very reliable. They might correctly label only a fraction of, say, *wedding* photos. By returning groups instead of images, users will be able to see photos of not only the highest classified (i.e., most prototypical) wedding photos—like the bride and groom at the altar—but also other, less wedding-like photos.

17.4.2 Search

Search on the graph consists of assigning scores to each node in the graph, starting at the top (layer 1: language) and propagating them down to the bottom (layer 5: groups), and then returning the top-scoring groups in sorted order. To understand this process, let us consider an example with two images: I_1 taken at a wedding in New Jersey, and I_2 at a wedding in Israel, at the "Hotel Tal" in Tel Aviv. Figure 17.8a shows a subset of nodes in this graph, along with their edge weights (precomputed using the methods described in Sect. 17.3). Given the user query, "*wedding in Israel*," we would like to return I_2 as the best match. To do the search, we first tokenize the query and drop stop words like "in" to get two terms: "wedding" and "Israel." We match each of these to all nodes in the language layer of our graph, via string similarity. This gives a score between 0 and 1 for each language node (most will be 0). In this simplified example, let's say we match two nodes exactly: "wedding" (visual category name) and "Israel" (country name). Thus we assign these nodes scores of 1, and all other language nodes a score of 0. Propagation to the next layer is accomplished by simply multiplying scores with the edge weights, summing up scores at each target node. Formally: we represent the edge weights between layers i and j as matrix E^i_j. Given scores s_i at layer i, we compute the scores at layer j as $s_j = E^i_j s_i$. Notice that due to our restricted set of connections (only between adjacent layers), we don't have to worry about complex ordering constraints or loopy behavior, as might happen in a fully connected graph. In practice, E^i_j tends to be quite sparse, as most nodes only connect to a few other nodes. This makes propagation extremely fast.

We repeat this process for each layer, until every node in the graph has an assigned score. We call a complete assignment of scores a *flow*, in this case, the *search flow*, F_{search}. Scores for each node in this flow are shown in the 3rd column of Fig. 17.8b and the flow is shown visually in Fig. 17.8c. Notice that the final scores for the two images are 0.8 and 1.3, respectively, which means that we would display both images in the results, but with I_2 first, as it has the higher score. This is exactly what we wanted. (In our actual system, we flow all the way down to the group layer, and return those in the appropriate order.)

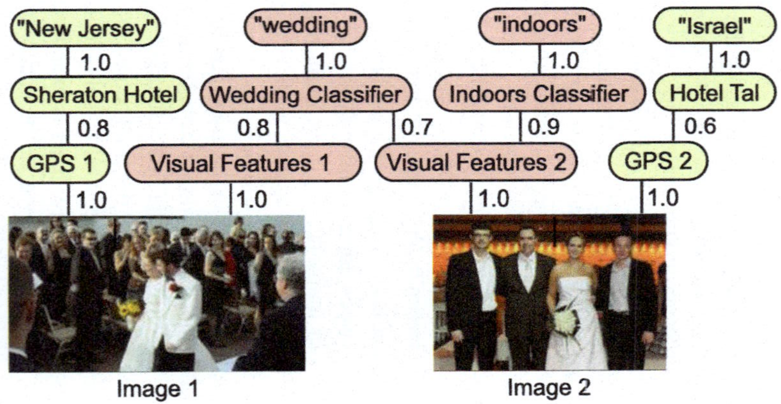

(a)

Layer	Node	$F_{search}(\downarrow)$	$F_{I_2}(\uparrow)$	$F_{desc_{I_2}}$
1	"New Jersey"	0.0	0.0	0.0
1	"wedding"	1.0	0.7	**10.7**
1	"indoors"	0.0	0.9	0.9
1	"Israel"	1.0	0.6	**10.6**
2	Sheraton Hotel	0.0	0.0	0.0
2	Wedding Classifier	1.0	0.7	10.7
2	Indoors Classifier	0.0	0.9	0.9
2	Hotel Tal	1.0	0.6	10.6
3	GPS 1	0.0	0.0	0.0
3	Visual Features 1	0.8	0.0	8.0
3	Visual Features 2	0.7	1.0	8.0
3	GPS 2	0.6	1.0	7.0
4	Image 1	0.8	0.0	8.0
4	Image 2	**1.3**	1.0	14.0

(b)

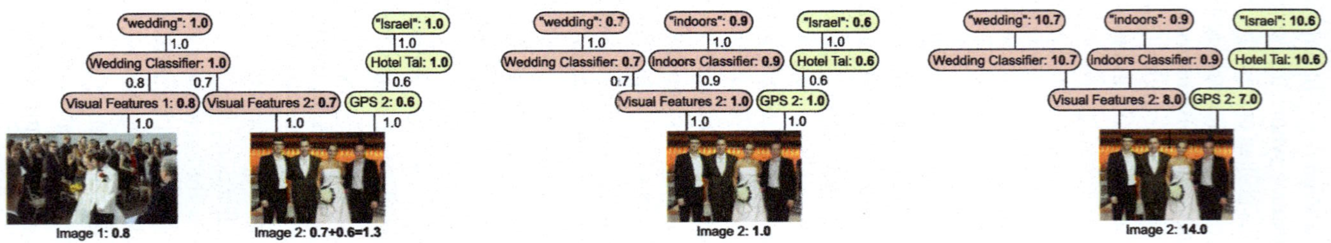

(c) (d) (e)

◄ **Fig. 17.8** Given the graph shown in **a** with specified edge weights, we examine what happens when the user does a search for *wedding in Israel*. First, the search flow F_{search} [(**c**) and 3rd column of (**b**)] is computed by assigning scores at the language layer (*top*) based on string similarity, and then propagating scores *down*. Since Image 2 gets the higher score, it will be displayed first in the results. Then, to generate the description for this image, we first compute its Image flow F_{I_2} by propagating *up* the layers (**d** and 4th column of (**b**)), and then add this to $\lambda F_{\text{search}}$ ($\lambda = 10$) to obtain the final description flow $F_{\text{desc}_{I_2}}$ [(**e**) and 5th column of (**b**)]. We pick the highest ranked language nodes of each type (*what, place, city, country*, etc.) to fill out the description template. In this case, that results in a description of "*Wedding at Hotel Tal, Israel.*" **a** Simplified graph with precomputed edge weights. **b** Computed flows for *wedding in Israel*. **c** Search flow. **d** Image 2 flow. **e** Description flow

17.4.3 Result Labeling

Simply showing the resulting image groups without any context would be confusing to a user, especially if the results are not obviously correct—she might wonder, "why did I get this result?" Therefore, we label each returned image group with a short description. This description always includes *where* and *when* the image group is from, and optionally additional context related to the user's search terms, such as *what* it is of. These descriptions are akin to the snippets shown in search engines that highlight elements from the results matching the user's query. Like Google, we generate query-specific descriptions, and thus can't simply use a single description per image/group. Concretely, we create these descriptions using a template system. Given a template such as:

what at **place**, **city**, **country** on **date**, **year**

the goal is to fill in each bolded component with a label from our graph. Since our language nodes are already separated into different types (see Fig. 17.7), this reduces down to simply choosing one node within each type of language node. We want to pick labels that are relevant to the images in the group, i.e., language nodes connected to the image nodes via nonzero edges. We also want to bias toward the terms used in the query, when applicable, so that it is clear why the images matched. Notice that we already know which labels these might be—they are the ones with nonzero scores in the search flow. Formally, we can write this down as a *description flow for Image I*,

$$F_{\text{desc}_I} = F_I + \lambda \text{F}_{\text{search}}, \tag{17.2}$$

where F_I is the *image flow*, described next, and λ is a weight determining how much to favor the query terms in the generated description. F_I is the flow created by applying a score of 1 to the image node (or a group, although in our simplified example, we will use an image), and propagating scores *up* through the graph until we get to the language nodes. This flow describes how relevant each label is for the given image. By changing λ, we can tune how much we want to adhere to the query terms used. Finally, generating the description is simply a matter of picking the highest scoring node in the description flow for each component in the template.

Let us now continue the example started in the previous section, to see how we generate the description for image I_2. (The process is identical for all results). We first compute image flow F_{I_2} by assigning a score of 1 to the node for I_2 and 0 for all other nodes at that layer. We then propagate scores *up* the graph by multiplying with the inverse edge matrices, until we have the complete result flow (shown in the 4th column of Fig. 17.8b, and visually in Fig. 17.8d). This creates nonzero scores at all language labels that apply to the image—the date, year, names of any places the photo is in or near, the categories of the place, any visual classifiers it ranked highly for, etc.(Notice that this is independent of the query, and can thus be precomputed.) Now we add the search flow we computed in the previous section, weighted by an appropriate λ to get the description flow, shown in the last column of Fig. 17.8b and visually in Fig. 17.8e. We pick the highest scoring node for each template component to get the final description, which yields, "wedding at Hotel Tal, Tel Aviv, Israel on January 3, 2011." (Note that not all nodes needed to generate this description are shown in the figure.)

17.4.4 Auto-Complete

To help users decide what to search for, we offer auto-complete functionality. Similar to many search websites today, our system will show you the most likely options to complete your currently typed-in search term. We want completions to match what you have typed so far, ranked by how likely you are to be searching for that completion. In a fielded system, completion likelihoods could be measured empirically. However, for our prototype system, we approximated this by using a measure of label "reach"—how many images a given label connects to, i.e., for each label, we assign a score of 1 to only that label's node and propagate scores down to get a complete flow, and then we count the percentage of image nodes that have a nonzero score in this flow. This measures how many images would be returned if we searched using only this label. Intuitively, we expect terms with greater reach to be more likely completions for what a user has typed.

17.5 Evaluation

To see how well our proposed method works, we quantitatively evaluated the major components of our search system. In the following sections, we evaluate our coverage of places, both by name and by category/activity (Sect. 17.5.1); our ability to find specific named events (Sect. 17.5.2); and our computer vision-based capabilities for finding visual object classes (Sect. 17.5.3). For more examples of search results, please see our video demo, shared at http://youtu.be/Se3bemzhAiY.

17.5.1 Places Evaluation

As described in Sect. 17.3.2, we use the online crowd-sourced website Wikimapia for locating places around the GPS coordinates of photographs. We wanted to measure what kind of coverage it offers, both with regard to searching for places by name (e.g., *Stuyvesant High School*) and by category or activity type (e.g., *toy store* or *paintball*). To accomplish this, we gathered geo-tagged images from flickr, manually labeled them, and then compared these ground-truth annotations with search results generated using our system.

To obtain a wide sampling of places, we used a subset of place categories from the Scene UNderstanding (SUN) database [34]. This database contains a fairly exhaustive list of 908 scene categories, such as *amphitheatre*, *beach*, and *sawmill*, along with images and some object segmentations. Unfortunately, these images do not contain GPS information, and so we cannot use the database directly. Instead, we use a subset of these category names as search queries to get geo-tagged images from flickr. In all, we downloaded 68,925 images of 406 categories. For each image, we looked through the photographer-provided title, tags, and description to find the proper name of the place. We skipped images that did not have this information, images that were not of the specified category, images of places that no longer exist, and images taken from too far away. In total, we labeled 1185 images of 32 categories. We then queried Wikimapia for all places around the GPS location for each labeled photo, as described in Sect. 17.3.2, and checked whether the annotated title of the place matched any of those found on Wikimapia, and also whether the scene category matched. These correspond to a user running searches by specific place name and by category type, respectively. We use the same soft string-matching function used in our search system, to allow for small variations in naming. For computing category matches, we also use NLTK's [2] wordnet `path_similarity()` function to ensure that if a category was defined as, e.g., "church" in the SUN dataset, and is called a "cathedral" in the Wikimapia categories, we count that as a match (since they will be close to each other in the wordnet [20] hierarchy).

These results were aggregated over different groups of categories and are summarized in Table 17.2. In particular, we find that 75.4 % of all places were successfully found by our system when searching by name and 69.8 % when searching by category. In the table, we also show averages for various subsets of categories. We do slightly worse on man-made places than on natural categories when searching by category. This is because natural places typically have standard category names, whereas man-made structures can have a variety of different categories, with a particular one missing in many cases. We emphasize that our overall performance on this task is already quite impressive; as the list of places from flickr is extremely diverse, it spans dozens of countries and includes several obscure places. As Wikimapia continues to expand, we expect that recall rates will also increase.

The primary failure modes for place matching, apart from the places not being on Wikimapia, are due to incorrect GPS. To handle small offsets of the GPS tags from the true location, we use fairly liberal search bounds when querying Wikimapia.

Table 17.2 Quantitative evaluation results

Places	Titles matched (%)	Cats matched (%)
Indoors	72.2	64.0
Man-made outdoors	77.5	75.9
Natural outdoors	74.9	60.8
Overall	75.4	69.8
Events	Tags matched	Images found
Overall	17.3 %	30.2 %
Visual classes	Queries with a top-5 match	Queries with a top-10 match
Overall	42.4 %	49.6 %

On top, place matching accuracy by title and by category, for various subsets of place types. In the middle, events: percentage of tags matched, and percentage of images found using at least one tag (many events have multiple tags). At the bottom, percentage of visual category queries which resulted in at least one relevant result within the top 5 and 10 matches. Each component of our system is already quite accurate individually; in queries combining multiple cues, we expect performance to increase still further

This works quite well, especially in our search setting, where recall is more important than precision. There were a few instances, however, of the GPS being drastically wrong—dozens of miles away. There is not much we can do in such situations.

17.5.2 Events Evaluation

One of our most novel contributions is a general method for labeling photos of specific public events using a combination of place information from Wikimapia and simple NLP applied to the results of queries on Google (see Sect. 17.3.3). We evaluate this capability using a methodology similar to that for our places evaluation described in the previous section. However, since there is no established academic dataset of events (or even a listing of event types), we manually created a list of 226 event search queries—such as *charity walk*, *opera*, and *NFL football*—and downloaded a total of 37,918 geo-tagged flickr images from these queries. To get the ground-truth terms related to the specific event in each photo, we manually labeled the name of the event and/or key search terms (e.g., *New York Knicks* and *Boston Celtics* for a basketball game) for a subset of these images. We then download Wikimapia places around the GPS location of each image, and ran our event processing pipeline as described in Sect. 17.3.3 to get tags corresponding to the events. Finally, we evaluated whether the ground-truth labels matched the top n-grams. We compute what fraction of ground-truth tags were successfully matched, and also what fraction of images were matched with at least one of the ground-truth tags. These measures distinguish between the situation where a venue is not found (in which case no event tags would be found), and that when the Google results (or our parsing of them) are insufficient to get the right tags (in which case we might find some tags, but perhaps not all).

The results are summarized in Table 17.2. We successfully matched 17.3 % of all tags, and 30.2 % of all labeled images had at least one tag match. The biggest problem was generally that the venue of the event was often not present on Wikimapia. We also found that older events (more than 2–3 years ago) tend to be less well covered on Google. Still, given the large variety of event types we tested, our performance on matching event tags is quite reasonable, especially given its generality. Also, note that our system does not rely on any fixed datasets—future events can be found without changing a single line of code. As webpages are created to describe new events, Google will index these pages and thus our system will be able to find these new events.

17.5.3 Visual Categories Evaluation

For evaluating visual classifiers, we downloaded the personal photo collections of 55 users from Google's Picasa Web Albums website who chose to make their photos publicly available under a Creative Commons license, encompassing a total of 258,040 photos (individual collections ranged from 587 to 25,158 photos). We labeled subsets of these images with the visual categories found in them using Amazon Mechanical Turk and used these for a quantitative evaluation. In total, we showed 50,950 images to 337 unique workers, collecting 20 labels per set. We filtered the raw labels to remove misspellings, etc., and sorted these "normalized" labels by total number of occurrences. To a first approximation, the most frequently used tags serve as a rough estimate of the visual content of these images (since dates, places, and events are often not identifiable from the photo alone).

To run our quantitative evaluation, we selected the photo collections of the 5 users with the most GPS-labeled photos. As not all collections contained ground-truth examples of all tags, we chose the 100 most-used tags (globally) that also existed in each user's collection, dropping tags that were used fewer than 10 times in all. We then queried our search engine with these tags, and measured performance using recall @ k—the fraction of the top k returned result sets that were correct (had a ground-truth annotation for that query). We feel that this is a fair metric for evaluating a system like ours, as a user is likely to be reasonably satisfied if she sees a relevant result on the first "page" of results. As shown in Table 17.2, 42.4 % of queries returned at least one relevant result in the top 5, and 49.6 % in the top 10. This is quite remarkable because visual recognition is extremely challenging, even in "closed-world" datasets like ImageNet, where every image belongs to exactly one of a fixed set of classes. In contrast, our system is "open-world," allowing a nearly infinite number of possibilities. Also, the best performing published methods are highly tuned and often slow. In contrast, we download images, extract features, train a classifier, and run it on a user's collection in under 10 s—a very stringent operating scenario—and so we have favored speed over accuracy in our implementation (limited number of training examples, thumbnail-sized images, using a relatively weak linear classifier). Furthermore, the top Google results are not always good training examples, either due to ambiguous query terms, or because the images might not be very representative of typical personal photos, which makes our search accuracy even more impressive. We would like to emphasize that actual users of our system would frequently combine visual category searches with times and places, both of which are usually far more reliable (see above).

17.6 Conclusion

In this work, we described a system for searching personal photo collections in a flexible and intuitive way. By taking advantage of all types of sensor data associated with an image—timestamp, GPS coordinates, and visual features—we gather and generate a large set of semantic information that describes the image in the ways that people care about. This includes the *time* of the photo, specified as dates, holidays, or times of day; its *place*, in terms of names, categories, and common activities; *events* that this photo is a part of, such as concerts; *visual categories* exhibited in this photo, such as weddings, fireworks, or whales; and *object instances*, such as the Mona Lisa. We label images by taking advantage of large online data sources, including Wikipedia for mapping dates to holidays, Wikimapia for mapping GPS locations to place information, Google for finding events, ImageNet and Google Image Search for training our visual classifiers and doing instance-level matching.

We believe that using the Internet to label personal photos is transformative: the user now gets to decide how to search, and does not need to spend time tediously labeling photos. Additionally, by allowing combinations of multiple query terms, we make it easy to find photos using whatever aspects of the photo she remembers. Handling people and faces is an obvious future work. We are also interested in creating new compelling ways of browsing through image collections using some of the techniques discussed here.

Acknowledgments This work was supported by funding from National Science Foundation grant IIS-1250793, Google, Adobe, Microsoft, Pixar, and the UW Animation Research Labs.

References

1. Arandjelović R, Zisserman A (2012) Multiple queries for large scale specific object retrieval. In: BMVC
2. Bird S, Klein E, Loper E (2009) Natural language processing with Python. O'Reilly Media
3. Bottou L (2010) Large-scale machine learning with stochastic gradient descent. In: COMP-STAT'2010. Springer
4. Bush V et al (1945) As we may think. The Atlantic monthly 176(1):101–108
5. Chatfield K, Zisserman A (2012) Visor: towards on-the-fly large-scale object category retrieval. In: ACCV

6. Cooper M, Foote J, Girgensohn A, Wilcox L (2005) Temporal event clustering for digital photo collections. ACM Trans Multimedia Comput Commun Appl 1(3):269–288. doi:10.1145/1083314.1083317
7. Cortes C, Vapnik V (1995) Support-vector networks. Mach Learn 20(3)
8. Datta R, Joshi D, Li J, Wang JZ (2008) Image retrieval: ideas, influences, and trends of the new age. ACM Comput Surv 40(2), 5:1–5:60. doi:10.1145/1348246.1348248
9. Deng J, Dong W, Socher R, Li LJ, Li K, Fei-Fei L (2009) ImageNet: a large-scale hierarchical image database. In: CVPR
10. Divvala S, Farhadi A, Guestrin C (2014) Learning everything about anything: Webly-supervised visual concept learning
11. Everingham M, Van Gool L, Williams CKI, Winn J, Zisserman A (2010) The pascal visual object classes (VOC) challenge. IJCV 88(2):303–338
12. Fergus R, Fei-Fei L, Perona P, Zisserman A (2005) Learning object categories from google's image search. In: International conference on computer vision, vol 2, pp 1816–1823
13. Joshi D, Luo J, Yu J, Lei P, Gallagher A (2011) Using geotags to derive rich tag-clouds for image annotation. In: Social media modeling and computing, pp 239–256. Springer
14. Kirk D, Sellen A, Rother C, Wood K (2006) Understanding photowork. In: SIGCHI, pp 761–770. doi:10.1145/1124772.1124885
15. Kumar N, Belhumeur PN, Nayar SK (2008) Facetracer: a search engine for large collections of images with faces. In: European conference on computer vision (ECCV)
16. Li LJ, Fei-Fei L (2010) Optimol: automatic online picture collection via incremental model learning. Int J Comput Vis 88(2):147–168
17. Li X, Chen L, Zhang L, Lin F, Ma WY (2006) Image annotation by large-scale content-based image retrieval. In: ACM international conference on Multimedia, pp 607–610
18. Lowe D (2004) Distinctive image features from scale-invariant keypoints. Int J Comput Vis
19. Malisiewicz T, Efros A (2009) Beyond categories: the visual memex model for reasoning about object relationships. In: Advances in neural information processing systems, pp 1222–1230 (2009)
20. Miller G et al (1995) Wordnet: a lexical database for english. Commun ACM 38(11):39–41
21. Naaman M, Song YJ, Paepcke A, Molina HG (2004) Automatic organization for digital photographs with geographic coordinates. In: ACM/IEEE joint conference on digital libraries
22. Nister D, Stewenius H (2006) Scalable recognition with a vocabulary tree. In: IEEE conference computer vision and pattern recognition (CVPR), pp 2161–2168
23. Oliva A, Torralba A (2001) Modeling the shape of the scene: a holistic representation of the spatial envelope. IJCV 42:145–175. http://dx.doi.org/10.1023/A:1011139631724
24. Parkhi OM, Vedaldi A, Zisserman A (2012) On-the-fly specific person retrieval. In: International workshop on image analysis for multimedia interactive services
25. Pedregosa F, Varoquaux G, Gramfort A, Michel V, Thirion B, Grisel O, Blondel M, Prettenhofer P, Weiss R, Dubourg V, Vanderplas J, Passos A, Cournapeau D, Brucher M, Perrot M, Duchesnay E (2011) Scikit-learn: machine learning in python. J Mach Learn Res 12:2825–2830
26. Philbin J, Chum O, Isard M, Sivic J, Zisserman A (2007) Object retrieval with large vocabularies and fast spatial matching. Comput Vis Pattern Recog
27. Platt J et al (1999) Probabilistic outputs for support vector machines and comparisons to regularized likelihood methods. Advances Large Margin Class 10(3):61–74
28. Quack T, Leibe B, Van Gool L (2008) World-scale mining of objects and events from community photo collections. In: International conference on content-based image and video retrieval, pp 47–56. ACM
29. Rohrbach M, Stark M, Szarvas G, Gurevych I, Schiele B (2010) What helps where–and why? semantic relatedness for knowledge transfer. In: Computer vision and pattern recognition (CVPR), pp 910–917
30. Scheirer W, Kumar N, Belhumeur PN, Boult TE (2012) Multi-attribute spaces: calibration for attribute fusion and similarity search. In: CVPR
31. Sculley D (2010) Web-scale k-means clustering. In: International conference on world wide web, pp 1177–1178. ACM

32. Sivic J, Zisserman A (2003) Video google: a text retrieval approach to object matching. In: ICCV
33. Weston J, Bengio S, Usunier N (2011) Wsabie: scaling up to large vocabulary image annotation. In: International joint conference on artificial intelligence, pp 2764–2770
34. Xiao J, Hays J, Ehinger KA, Oliva A, Torralba A (2010) SUN database: large-scale scene recognition from abbey to zoo. In: CVPR
35. Zhang L, Hu Y, Li M, Ma W, Zhang H (2004) Efficient propagation for face annotation in family albums. In: ACM international conference on multimedia, pp 716–723

Index

A.R. Zamir et al. (eds.), *Large-Scale Visual Geo-Localization*,
Advances in Computer Vision and Pattern Recognition,
DOI 10.1007/978-3-319-25781-5

V

W

The manufacturer's authorised representative in the EU is Springer Nature Customer Service Centre GmbH, Europaplatz 3, 69115 Heidelberg, Germany. If you have any concerns regarding our products, please contact ProductSafety@springernature.com

Printed and bound by CPI Group (UK) Ltd, Croydon, CR0 4YY
01/10/2026
02225860-0008